SYNTHETIC PEPTIDES

Advances in Molecular Biology

Series Editor
Allan Jacobson, University of Massachusetts Medical Center

Editorial Advisory Board
Joan Brugge, ARIAD Pharmaceuticals, Inc.
Henry Erlich, Roche Molecular Systems
Stanley Fields, University of Washington
Larry Gold, NeXstar Pharmaceuticals, Inc.

The Yeast Two-Hybrid System, edited by Paul L. Bartel and Stanley Fields

A Comparative Methods Approach to the Study of Oocytes and Embryos, edited by Joel D. Richter

Synthetic Peptides: A User's Guide, Second Edition, edited by Gregory A. Grant

SYNTHETIC PEPTIDES

A User's Guide

Second Edition

Edited by
Gregory A. Grant

OXFORD
UNIVERSITY PRESS
2002

OXFORD

UNIVERSITY PRESS

Oxford   New York
Auckland Bangkok Buenos Aires Cape Town Chennai
Dar es Salaam Delhi Hong Kong Istanbul Karachi Kolkata
Kuala Lumpur Madrid Melbourne Mexico City Mumbai Nairobi
São Paulo Shanghai Singapore Taipei Tokyo Toronto

and an associated company in Berlin

Library of Congress Cataloging-in-Publication Data
Synthetic peptides: a user's guide / edited by Gregory A. Grant.—2nd ed.
    p.   cm. — (Advances in molecular biology)
  Includes bibliographical references and index.
  ISBN 0-19-513261-0
  1. Peptides—Synthesis. 2. Protein engineering. I. Grant, Gregory A., 1949– II. Series.
QP552.P4 S963 2002
572'.6545—dc21   2001036643

9 8 7 6 5 4 3 2 1
Printed in the United States of America
on acid free paper

For Marie

Contents

Contributors

George Barany
Departments of Chemistry and
  Laboratory Medicine and
  Pathology
University of Minnesota
Minneapolis, Minnesota 55455

Gregg B. Fields
Department of Chemistry and
  Biochemistry and the Center for
  Molecular Biology and
  Biotechnology
Florida Atlantic University
Boca Raton, Florida 33431

Gregory A. Grant
Department of Molecular Biology
  and Pharmacology
Washington University School of
  Medicine
St. Louis, Missouri 63110

Victor J. Hruby
Department of Chemistry
University of Arizona
Tucson, Arizona 85721

Janelle L. Lauer-Fields
Department of Chemistry and
  Biochemistry and the Center for
  Molecular Biology and
  Biotechnology
Florida Atlantic University
Boca Raton, Florida 33431

Rong-qiang Liu
Departments of Chemistry and
  Laboratory Medicine and
  Pathology
University of Minnesota
Minneapolis, Minnesota 55455

Terry O. Matsunaga
Department of Pharmacology and
    Toxicology
School of Pharmacy
University of Arizona
Tucson, Arizona 85721

Michael L. Moore
Smith-Kline Beecham
    Pharmaceuticals
King of Prussia, Pennsylvania 19406

SYNTHETIC PEPTIDES

1

Synthetic Peptides

Beginning the Twenty-first Century

Gregory A. Grant

This second edition of *Synthetic Peptides* is being published at the beginning of the twenty-first century and marks nearly 100 years since the beginnings of the chemical synthesis of peptides. From the first decade of the twentieth century up to the present time, the evolution of the development, analysis, and use of synthetic peptides has been steady and remarkable.

It has been about 10 years since the first edition of this book was published. Much remains unchanged, such as the basic principles of peptide structure, the basic chemistry for assembling a peptide chain, and many of the techniques used to evaluate synthetic peptides. However, during that time we have seen a switch from primarily the use of Boc chemistry for routine synthesis to that of Fmoc chemistry. Mass spectrometry has also matured with the development of more user-friendly and affordable instrumentation to the point that it is now the premier analytic method for synthetic peptides. Methods for the production of very long peptides, such as chemoselective ligation, are maturing, although they are still not an everyday thing for most peptide chemists, and better chemistries for producing peptides with "post-translational modifications," such as phosphates, sugars, and specific disulfide bonds, are now well within reach. As a result, you will find many sections of this book largely unchanged, but you will also find many new sections that document the developments of the last ten years with the inclusion of new information and methodologies. However, a good feeling for what the beginning of the twenty-first century offers can best be appreciated by considering the developments that have led to this point.

Emil Fischer introduced the concept of peptides and polypeptides and presented protocols for their synthesis in the early 1900s (Fischer, 1902, 1903, 1906). Although others also made contributions in those days, most notably Theodor Curtius, the work of Fischer and his colleagues stands out, and he is generally regarded as the father of peptide chemistry. An excellent account of the history of peptide synthesis which treats the subject much more comprehensively than can be attempted here can be found in a book by Wieland and Bodanszky (1991). Fischer's place in the history of synthetic peptides is eloquently summed up in one sentence from that book which simply states, "To Emil Fischer we owe the systematic attack on a field of natural substances that had previously been avoided by chemists." In those early days, the chemistries developed by Fischer and others led to the production of molecules containing as many as 18 amino acids, such as Leucyl (triglycyl) leucyl (triglycyl) leucyl (octaglycyl) glycine (Fischer, 1907). Nonetheless, the syntheses performed were difficult and limited to simple amino acids, and progress was slow for many years. Then, the discovery of an easily removable protecting group, the carbobenzoxy group, by Bergmann and Zervas in 1932 (Bergmann and Zervas, 1932) provided new impetus by opening the way to the use of polyfunctional amino acids. As a result, the synthesis of naturally occurring small peptides such as carnosine (Sifferd and du Vigneaud, 1935) and glutathione (Harington and Mead, 1935) were soon achieved. Almost 20 years later, the synthesis of an active peptide hormone, the octapeptide oxytocin, by du Vigneaud (du Vigneaud et al., 1953) was acclaimed as a major accomplishment and spurred the advancement of peptide synthesis once again. The synthesis of the 39 residue porcine adrenocorticotropic hormone in 1963 (Schwyzer and Sieber, 1963) by solution-phase segment condensation methods was viewed as no less than sensational at the time; and in 1967, Bodanszky and colleagues succeeded in synthesizing the 27 residue secretin peptide by solution phase stepwise addition methods (Bodanszky and Williams, 1967; Bodanszky et al., 1967). These accomplishments in peptide synthesis were considered to be monumental at the time in that they were exceptionally difficult undertakings that pushed the prevailing technology to its limits.

In 1963, Bruce Merrifield published a landmark paper (Merrifield, 1963) describing the development of solid-phase peptide synthesis. This technique, for which he was awarded the Nobel Prize in

Chemistry in 1984, was responsible, more than anything else, for open-ing the way to the widespread use of synthetic peptides as reagents in chemical and biomedical investigations. Theodor Wieland (Wieland, 1981) once described the advances in peptide chemistry in the follow-ing way: "[T]he synthesis of glutathione opened the door to peptide synthesis a crack, and the synthesis of oxytocin pushed the door wide open." To extend that analogy, the introduction of the solid-phase method blew the door off its hinges. Since that time, many other developments, such as improved synthesis chemistry, automated instrumentation for the unattended production of peptides, and improved purification and analytical methods have contributed to our ability to exploit the potential of synthetic peptides. Not only has there been an explosion in the number of synthetic peptides being produced, but reports of the synthesis of larger and larger pep-tides (Gutte and Merrifield, 1969, 1971; Clark-Lewis et al., 1986; Nutt et al., 1988; Schneider and Kent, 1988, Muir et al., 1997), some exceed-ing 100 residues in length, are becoming more commonplace.

Now, and for the last 10 years or so, synthetic peptides are available not only to those actively involved in developing the field, but to virtually every investigator in any field who perceives a need for them and wishes to pursue their use.

This book is intended primarily for those investigators who wish to utilize synthetic peptides in their research but who themselves are not already intimately involved in the field. This group would encompass researchers who either simply want access to synthetic peptides as tools, or who wish to become actively involved in the production of the peptides, themselves. As such, it strives to provide practical back-ground information, answer basic questions, and address common problems relating to all aspects of synthetic peptides. At the same time, however, it is intended to be current, comprehensive, and sophis-ticated in its treatment of the subject and should be of general interest to researchers and educators at all levels.

The process of obtaining a synthetic peptide for use in biochemical or biomedical research involves several discrete steps which can be represented diagrammatically as shown in figure 1-1. The process starts with the design of the peptide and follows through with the chemical synthesis, evaluation, and purification of the product, and finishes with application or the actual use of the peptide in an experi-mental situation. As indicated in the diagram, the process can be viewed as pivoting on the evaluation step which actually is part of a more comprehensive evaluation–purification cycle. It is at this point that the product is characterized and its disposition determined. That disposition can be either (1) use of the peptide as is, (2) additional purification and subsequent evaluation, eventually leading to experimental usage, or (3) changes in either the design or synthesis

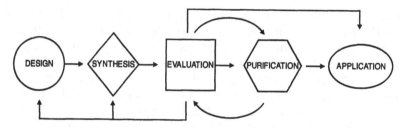

Figure 1-1. Diagram of the steps involved in obtaining a synthetic peptide.

protocols, or both, followed by reiteration of the latter steps of the process.

Each subsequent chapter of this book deals with one major aspect of this process, that is, design, synthesis, evaluation/purification, and application.

The book begins in chapter 2 with a discussion of the fundamental aspects of peptide structure and uses that as a foundation to build into both basic and more advanced considerations in the design of synthetic peptides. The design process is a constantly evolving one which is something of an art in itself. In many cases, there are no hard-and-fast rules and function usually dictates design. Perhaps the most commonly asked question from general users of synthetic peptides is how to choose a sequence for the production of antibodies. This chapter contains an expanded section on this aspect that draws from what has been learned from the last 10 years. In addition, a section on combinatorial peptide libraries has been added. This is an area that presents great promise for screening very large numbers of peptides for de novo discovery as a complement to rational design and for optimizing structure–activity relationships.

Chapter 3 deals with the synthesis chemistry itself. It presents a logical and straightforward explanation of the solid-phase method of peptide synthesis and discusses the overall state of the chemistry today. This chapter places particular emphasis on the Boc and Fmoc protection strategies as the two most useful approaches available and provides specific recommendations for undertaking routine synthesis. It is immediately evident that the chemistry is at the same time very sophisticated and complicated, yet relatively straightforward in the overall approach and amenable to routine procedures. Yet, one must always be mindful that, although routine synthesis using a uniform set of reagents is commonly performed successfully, a vast array of problems can occur. At the same time, there are a large number of alternatives that can be utilized to prevent or circumvent these problems. This chapter therefore presents both recommended "recipes" for routine synthesis and more advanced chemical strategies and considerations for the maturing practitioner. This chapter also contains a useful section on the automation of solid-phase synthesis. It gives an informative overview of the range of instruments available today and illustrates the choice of capabilities, from single peptide synthesis to as many as 96 at one time and from microgram to gram quantities.

Chapter 4 deals with the evaluation of the finished peptide. It presents a discussion of the latest methods available for the routine characterization of the synthetic product and illustrates their utilization through examples of the evaluation of actual peptides. The message that should be very clear from this chapter is that although the chemistry can in many cases be preprogrammed on modern peptide synthesizers, and these machines do most of the work that used to be done laboriously on the benchtop, the process is by no means foolproof or trivial. Many problems from a large variety of sources can be manifest even in a "routine" synthesis and the final product can never be taken for granted. Every peptide produced must be rigorously evaluated with the ultimate goal being the proof that the peptide obtained is the one intended. In addition, problematic residues and peptide solubility are very important considerations in the evaluation and use of the peptides, which are very often ignored or overlooked. This chapter also contains a more in-depth description of mass spectrometry as an evaluation tool and highlights its use in determining the sequence of a peptide.

Chapter 5 presents an excellent survey of the use and application of synthetic peptides. It not only illustrates the diversity in the use and applications of synthetic peptides in modern research but, perhaps, serves as a basis from which new applications and ideas may develop. These include the antigenic and immunogenic use of synthetic peptides, the use of peptides as enzyme inhibitors, structure/function studies involving synthetic peptides, peptide-based vaccines, antisense peptides, and the use of peptides for affinity labeling of receptors or "acceptors" and their use in structure/function studies of receptors. Undoubtedly, some uses for synthetic peptides have not been included, but the areas that are discussed are among the more exciting and successful applications being investigated today. Furthermore, many of the concepts and approaches discussed can be easily adapted to other systems and areas of exploration. Also, since the use very often dictates the design, many aspects of peptide design are discussed in this chapter and, as such, it complements chapter 2 very nicely.

Peptide synthesis has come a long way from the beginning of the twentieth century to the present day. It has spanned the twentieth century and has enhanced the passing of that period. As we start the new millennium, it is tempting to speculate on what may lie ahead. However, history has told us that such attempts invariably fall far short of the eventual reality. After all, even Merrifield admitted in a 1986 article (Merrifield, 1986) that he did not foresee the impact of his technique: "From the accumulated data presented, I conclude that the solid phase synthesis of peptides up to 50 or somewhat more residues can be readily achieved in good yield and purity; this is a far better situation than I could have expected when this technique was first proposed" (p. 345).

Undoubtedly, in the future great things will be accomplished in peptide synthesis and in the use of the peptides themselves. It is, perhaps, sufficient to have had the privilege of taking part in a small bit of that process.

References

Bergman, M., and Zervas, L. (1932). Über ein allgemeines Verfahren der Peptid-Synthese. Ber. Dtsch. Chem. Ges. 65:1192–1201.

Bodanszky, M., and Ondetti, M. A. (1966). *Peptide Synthesis*, New York, John Wiley and Sons.

Bodanszky, M., and Williams, N. J. (1967). Synthesis of secretin I. The protected tetradecapeptide corresponding to sequence 14-27. J. Am. Chem. Soc. 89:685–689.

Bodanszky, M., Ondetti, M. A., Levine, S. D., and Williams, N. J. (1967). Synthesis of secretin II. The stepwise approach. J. Am. Chem. Soc. 89:6753–6757.

Clark-Lewis, I., Aebersold, R., Ziltener, H., Schrader, J. W., Hood, L. E., and Kent, S. B. H. (1986). Automated chemical synthesis of a protein growth factor for hemopoietic cells, Interleukin-3. Science 231:134–139.

Fischer, E. (1902). Ueber einige Derivate des Glykocolls Alanins und Leucins. Ber. Dtsch. Chem. Ges. 35:1095–1106.

Fischer, E. (1903). Synthese von Derivaten der Polypeptide. Ber. Dtsch. Chem. Ges. 36:2094–2106.

Fischer, E. (1906). Untersuchungen über Aminosäuren, Polypeptide, und Proteïne. Ber. Dtsch. Chem. Ges. 39:530–610.

Fischer, E. (1907). Synthese von Polypeptiden XVII. Ber. Dtsch. Chem. Ges. 40:1754–1767.

Gutte, B., and Merrifield, R. B. (1969). The total synthesis of an enzyme with ribonuclease A activity. J. Am. Chem. Soc. 91:501.

Gutte, B., and Merrifield, R. B. (1971). The synthesis of ribonuclease A. J. Biol. Chem. 246:1922.

Harington, C. R., and Mead, T. H. (1935). Synthesis of glutathione. Biochem. J. 29:1602–1611.

Merrifield, R. B. (1963). Solid phase peptide synthesis I. The synthesis of a tetrapeptide. J. Am. Chem. Soc. 85:2149–2154.

Merrifield, R. B. (1986). Solid phase synthesis. Science 232:341–347.

Muir, T. W., Dawson, P. E., and Kent, S. B. H. (1997). Protein synthesis by chemical ligation of unprotected peptides in aqueous solution. Meth. Enzymol. 289:266–298.

Nutt, R. F., Brady, S. F., Darke, P. L., Ciccarone, T. M., Colton, C. D., Nutt, E. M., Rodkey, J. A., Bennett, C. D., Waxman, L. H., Sigal, I. S., Anderson, P. S., and Veber, D. F. (1988). Chemical synthesis and enzymatic activity of a 99 residue peptide with a sequence proposed for the human immunodeficiency virus protease. Proc. Natl. Acad. Sci. U.S.A. 85:7129–7133.

Schneider, J., and Kent, S. B. H. (1988). Enzymatic activity of a synthetic 99 residue protein corresponding to the putative HIV-1 protease. Cell 54:363–368.

Schwyzer, R., and Sieber, P. (1963). Total synthesis of adrenocorticotrophic hormone. Nature 199:172–174.

Sifferd, R. H., and du Vigneaud, V. (1935). A new synthesis of carnosine, with some observations on the splitting of the benzyl group from carbobenzoxy derivatives and from benzylthio ethers. J. Biol. Chem. 108:753–761.

Vigneaud, V. du, Ressler, C., Swan, J. M., Roberts, C. W., Katsoyannis, P. G., and Gordon, S. (1953). The synthesis of an octapeptide amide with the hormonal activity of oxytocin. J. Am. Chem. Soc. 75:4879–4880.

Wieland, T. (1981). From glycylglycine to ribonuclease: 100 years of peptide chemistry. In *Perspectives in Peptide Chemistry*, A. Eberle, R. Geiger, and T. Wieland, eds., Basel, Karger, pp. 1–13.

Wieland, T., and Bodanszky, M. (1991). *The World of Peptides: A Brief History of Peptide Chemistry*, Berlin, Springer-Verlag.

2

Peptide Design Considerations

Michael L. Moore
Gregory A. Grant

Peptides have become an increasingly important class of molecules in biochemistry, medicinal chemistry, and physiology. Many naturally occurring, physiologically relevant peptides function as hormones, neurotransmitters, cytokines, and growth factors. Peptide analogs that possess agonist or antagonist activity are useful as tools to study the biochemistry, physiology, and pharmacology of these peptides, to characterize their receptor(s), and to study their biosynthesis, metabolism, and degradation. Radiolabeled analogs and analogs bearing affinity labels have been used for receptor characterization and isolation. Peptide substrates of proteases, kinases, phosphatases, and aminoacyl or glycosyl transferases are used to study enzyme kinetics, mechanism of action, and biochemical and physiological roles and to aid in the isolation of enzymes and in the design of inhibitors. Peptides are also used as synthetic antigens for the preparation of polyclonal or monoclonal antibodies targeted to specific sequences. Epitope mapping with synthetic peptides can be used to identify specific antigenic peptides for the preparation of synthetic vaccines, to determine protein sequence regions that are important for biological action, and to design small peptide mimetics of protein structure or function.

A number of peptide hormones or analogs thereof, including arginine vasopressin, oxytocin, luteinizing hormone releasing hormone

(LHRH), adrenocorticotropic hormone (ACTH), and calcitonin, have already found use as therapeutic agents, and many more are being investigated actively. Peptide-based inhibitors of proteolytic enzymes, such as angiotensin converting enzyme (ACE) and human immuno-deficiency virus (HIV) protease, have widespread clinical use, and inhibitors of renin and elastase are also being investigated for therapeutic use. Finally, peptides designed to block the interaction of protein molecules by mimicking the combining site of one of the proteins, such as the fibrinogen receptor antagonists, show great therapeutic potential as well.

With the development of solid-phase peptide synthesis by Bruce Merrifield (1963) and the optimization of supports, protecting groups, and coupling and deprotection chemistries by a large number of researchers, it has become possible to obtain useful amounts of peptides on a more or less routine basis. With the increasing ease of synthesizing peptides, it has become all the more important to understand the underlying principles of peptide structure and physical chemistry that govern solubility, aggregation, proteolytic resistance or susceptibility, secondary structure stabilization or mimicry, and interaction with nonpolar environments like lipid membranes.

The design of any specific peptide depends primarily on the use for which it is intended as well as on synthetic considerations. A number of aspects of the peptide sequence and structure can be manipulated to affect solubility, proteolytic resistance, or stability, and the pre-dilection to adopt specific secondary structures to produce peptides with specific properties. It is this topic that will be addressed in this chapter.

The distinction between what constitutes a peptide and what con-stitutes a protein becomes increasingly fuzzy as peptides increase in length. An operational definition in the context of this chapter might be that a peptide is any sequence that the researcher can conveniently synthesize chemically. Although such proteins as ribonuclease A (124 amino acids) (Hirschmann et al., 1969; Yajima and Fujii, 1981), acyl carrier protein (74 amino acids) (Hancock et al., 1972), and the HIV protease (99 amino acids) (Nutt et al., 1988; Wlodawer et al., 1989) have been chemically synthesized, most peptide synthesis generally involves peptides of 30 amino acids or less. The underlying principles applied to peptide design will be dependent in some degree on peptide length, especially those relating to peptide secondary structure. For example, shorter peptides do not tend to exhibit preferred solution conformations and possess a large amount of segmental flexibility. If secondary structure is important in a small peptide's design, it must be approached by chemical modifications designed to decrease conforma-tional flexibility. As peptides increase in length, they have a greater tendency to exhibit elements of secondary structure, with a consequent decrease in segmental flexibility. In such cases, secondary structure can often be induced by optimizing the peptide sequence.

Protein and Amino Acid Chemistry

The physical and chemical properties of proteins and peptides are determined by the nature of the constituent amino acid side chains and by the polyamide peptide backbone itself. Twenty protein amino acids are coded for by DNA, which are translationally incorporated into proteins. Amino acids can be modified in the protein post-translationally to yield new amino acids. Also, many peptides are synthesized enzymatically rather than ribosomally, especially in lower eukaryotes, and those peptides often contain highly unusual amino acids.

Protein Amino Acids

The structures of the 20 primary protein amino acids (those coded for by DNA) are given in table 2-1 along with their three-letter abbreviations, one-letter codes, and a general grouping by physical properties. Amino acids generally can be divided into hydrophobic and hydrophilic residues. The hydrophobic residues include those with aliphatic side chains, such as alanine, valine, isoleucine, leucine, and methionine, and those with aromatic side chains, such as phenylalanine, tyrosine, and tryptophan. The hydrophilic residues include amino acids with (1) neutral, polar side chains, such as serine, threonine, asparagine, and glutamine; (2) those with acidic side chains, such as aspartic acid and glutamic acid; and (3) those with basic side chains such as histidine, lysine, and arginine. It can be appreciated that these categories are not entirely exclusive. Alanine, with its small aliphatic side chain, and glycine can be found in hydrophilic regions of peptides and proteins. Conversely, the long alkyl chains of lysine and arginine can give those residues an overall hydrophobic character with just the terminal charged group being hydrophilic.

Two amino acids, cysteine and proline, have special properties that set them apart. Cysteine contains a thiol moiety that can be oxidatively coupled to another cysteine thiol to form a disulfide linkage. Disulfides are the principal entities by which peptide chains are covalently linked together to stabilize secondary or tertiary structure or to hold two different peptide chains together. Although the disulfide form is the most stable form under normal aerobic conditions, free thiols are also present in some proteins, where they often serve as ligands for metal chelation, as nucleophiles in proteolytic enzymes, such as papain, or as carboxyl activators in acyl transferases. The secondary amino acid proline has specific conformational effects on the peptide or protein

Table 2-1 Amino acid structures and properties

	$H_2N-\overset{R}{\underset{H}{C}}-CO_2H$	
H	CH_3	CH_3 CH_3
Glycine Gly G	Alanine Ala A	Valine Val V
Neutral, hydrophobic, aliphatic		

CH_3 H CH_3	CH_3 CH_3	S CH_3
Isoleucine Ile I	Leucine Leu L	Methionine Met M
Neutral, hydrophobic, aliphatic		

	OH	
Phenylalanine Phe F	Tyrosine Tyr Y	Tryptophan Trp W
Neutral, hydrophobic, aromatic		

(*continued*)

Table 2-1 (*Continued*)

OH	CH₃ H OH	CONH₂
Serine Ser S	Threonine Thr T	Asparagine Asn N
Neutral, hydrophilic		

CONH₂	CO₂H	CO₂H
Glutamine Gln Q	Aspartic Acid Asp D	Glutamic Acid Glu E
Neutral, hydrophilic	Acidic, hydrophilic	

	NH₂	H₂N NH / NH
Histidine His H	Lysine Lys K	Arginine Arg R
Basic, hydrophilic		

SH	HN CO₂H
Cysteine Cys C	Proline Pro P
Thiol-containing	Imino acid

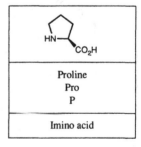

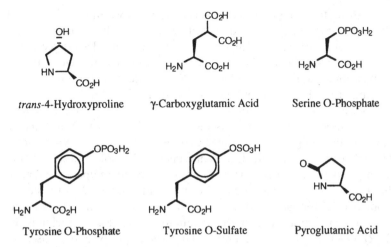

Figure 2-1. Some amino acids formed by post-translational modification.

backbone both because of its cyclic structure and because of the alkylation of the amino group. It often plays an important role in stabilizing or influencing the secondary structure of proteins, as will be discussed later.

Some amino acids can be modified enzymatically after incorporation into certain proteins to give rise to new amino acids, some of which are shown in figure 2-1. In collagen, for example, proline can be hydroxylated to yield trans-4-hydroxyproline (Hyp). Glutamic acid is carboxylated by a vitamin K-dependent carboxylase to yield γ-carboxyglutamic acid (Gla) in a number of proteins involved in blood coagulation, where the malonyl moiety of the Gla residue is thought to be important in providing a bidentate ligand for calcium ions. The hydroxyl functions of tyrosine, serine, and threonine can be reversibly phosphorylated by kinases and phosphatases, a process thought to be an important regulator of biological activity in the target proteins. The hydroxyl of tyrosine may also be sulfated in peptide hormones such as gastrin and cholecystokinin.

One important nonenzymatic transformation of a protein amino acid occurs with glutamine, which is chemically unstable at the amino terminus of a peptide or protein in aqueous solution. Glutamine will spontaneously cyclize to form pyroglutamic acid (pyrrolidone carboxylic acid, also shown in figure 2-1). This transformation typically occurs when an internal glutamine residue is exposed by proteolysis of an X-Gln bond (where X represents any amino acid) or during protein sequencing.

All amino acids (except for glycine) are chiral molecules with an asymmetric center at the α-carbon. The protein amino acids all have the L absolute configuration at this center, as shown in the structure at the top of table 2-1. The L does not refer to the direction of optical rotation but, rather, to having the same stereochemical arrangement as L(-)-glyceraldehyde. In the more unambiguous Cahn–Ingold–Prelog convention, L-amino acids have the S absolute configuration (except cysteine, for which this configuration is defined as R).

Two amino acids, threonine and isoleucine, have a second asymmetric center at the β-carbon, as shown in table 2-1. Threonine has the R absolute configuration at the β-carbon; that is, it is 2-(S)-amino-3-(R)hydroxybutanoic acid. Isoleucine has the S absolute configuration at the β-carbon; that is, it is 2-(S)-amino-3-(S)-methylpentanoic acid. Because there are two asymmetric centers in the molecule, four stereoisomers exist for each of these amino acids. The D-amino acid has the opposite configuration at both asymmetric centers; D-isoleucine is the 2-(R)-3-(R) isomer, for example. The trivial prefix allo is used to denote inversion at only one asymmetric center. Thus D-allo-isoleucine is the 2-(R)-3-(S) isomer. D-allo-Isoleucine occasionally forms as an artifact during hydrolysis of isoleucine-containing peptides or proteins. Because it is a diastereomer of isoleucine rather than an enantiomer, it will appear as a separate peak in amino acid analyses.

Nonprotein Amino Acids

Literally hundreds of naturally occurring amino acids exist that are not found in proteins. Some of them, like γ-aminobutyric acid, have important functions as neurotransmitters. Others, like ornithine, appear as intermediates in metabolic pathways or, like dihydroxyphenylalanine, are precursors to amino acid-derived products, catecholamines in this case.

By far the largest number of nonprotein amino acids are found in prokaryotes and lower eukaryotes, especially in algae, sponges, yeasts, and fungi, although peptides with unusual amino acids have been isolated from chordates, such as tunicates, as well. These amino acids are incorporated into peptides by enzymatic synthesis rather than ribosomally. More than 700 of these nonprotein amino acids are known, and the structural variations are immense (Hunt, 1985). Those amino acids and the peptides containing them are usually the products of secondary metabolism, and their function in the producing organism is often obscure. Those unusual amino acids can confer unusual biological activities on peptides that contain them. Such peptides have been the basis either of biochemical tool molecules or of antibiotics, immunomodulators or antineoplastic agents that have therapeutic utility.

Nonprotein amino acids can be roughly categorized into a few basic structural types, representative members of which are depicted in

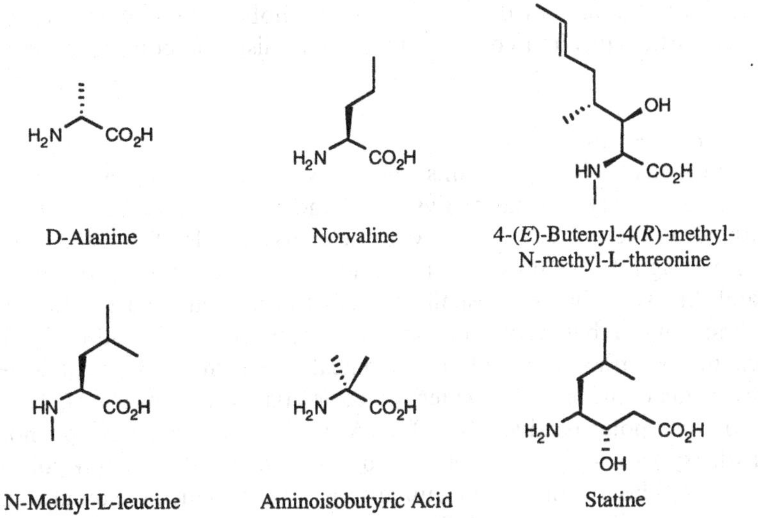

Figure 2-2. Some nonprotein amino acids.

figure 2-2. Some nonprotein amino acids are simply the enantiomeric D-amino acid analog of a protein L-amino acid, for example, D-alanine and D-glutamic acid, which are important constituents of the proteoglycan bacterial cell wall. Others have a normal α-amino acid structure but with a novel side chain. The side chain can be a simple alkyl group, such as in norvaline, or it can be quite unusual, such as in 4-(E)-butenyl-4(R)-methyl-N-methyl-L-threonine (MeBmt), a critical constituent of the immunosuppressive peptide cyclosporin A (Rüegger et al., 1976). Some nonprotein amino acids deviate from the normal α-amino acid structure and are methylated on their amine function, such as the N-methylleucine (MeLeu) of the tunicate-derivated antineoplastic peptide didemnin B (Rinehart et al., 1981), or on their α-carbon, such as the aminoisobutyric acid (α-methylalanine, Aib) of the ionophoretic antibiotic peptide alamethicin (Payne et al., 1970; Pandey et al., 1977). In addition, there are large numbers of amino acids in which the amino group is not on the α-carbon but at some other position in the molecule. Sometimes several of these features are combined in one amino acid, such as statine [3-(S)hydroxy-4-(S)-amino-6-methylheptanoic acid, Sta], the critical residue in the fungal protease inhibitor pepstatin (Morishima et al., 1970).

Side-Chain Interactions

Side chains interact with each other, with the amide peptide backbone, with bulk solvent, and through noncovalent interactions, such as hydrogen bonds, salt bridges, and hydrophobic interactions. Cysteine also participates in a covalent interaction, disulfide bond formation.

Hydrogen Bonds and Salt Bridges

In proteins, polar side chains tend to be extensively solvated. Acidic (Asp and Glu) and basic (Lys, Arg, and His) residues generally are found on the protein surface with the charged ends of the side chains projecting into the bulk solvent, although the alkyl portion of the Lys and Arg side chains is usually buried. Internal charged residues are almost invariably involved in salt bridges, where acidic and basic side chains are either directly bonded ionically to each other or connected by a single intermediary water molecule (Baker and Hubbard, 1984). Nonionic polar residues (Ser, Thr, Asn, Gln, and Tyr at the phenolic hydroxyl) are also extensively hydrogen-bonded, either to bulk solvent or to backbone, other side-chain groups, or to specifically bound water molecules (Thanki et al., 1990). In helices, the side chains of Ser, Thr, and Asn often make specific hydrogen bonds to the carbonyl oxygen of the third or fourth residue earlier in the sequence, which may help to stabilize helical segments (Gray and Matthews, 1984). In shorter peptides, side-chain hydration occurs mostly through the bulk solvent, although the ability to form low-energy intramolecular hydrogen-bonded structures or salt bridges may be an important factor in the association of peptides with macromolecular targets or receptors.

Consideration of solvation and desolvation effects is a potentially critical, but often overlooked, aspect of peptide design and structure–activity relationships. One case will serve to illustrate the importance of solvation in interpreting structure–activity studies. Bartlett and coworkers prepared a series of thermolysin inhibitors that were based on the peptide substrate sequence Cbz-Gly-Leu-Leu-OH but contained a phosphoryl moiety in place of the Gly carbonyl function (Morgan et al., 1991). The phosphoryl moiety was designed to mimic the tetrahedral transition state for amide bond hydrolysis and was suggested by a naturally occurring glycopeptide inhibitor of thermo-lysin, phosphoramidon, which contains a similar phosphoryl group. The inhibitors that were prepared are shown in table 2-2. Compound 1, which contains a phosphoramidate linkage (PO_2NH), was found to be a potent inhibitor. X-ray crystal structure analysis of the enzyme-inhibitor complex showed a hydrogen bond between the phosphor-amidate NH and a backbone carbonyl in the enzyme. Compound 2 contained a phosphonate linkage (PO_2O). It was unable to form the same hydrogen bond as compound 1 because it lacks the corresponding hydrogen. It was found to be three orders of magnitude less potent than compound 1, but X-ray crystal analysis showed that it bound to the enzyme in an identical fashion (except for the absence of the

Table 2-2 Phosphorus-containing
thermolysin inhibitors

Compound	X	Ki(nM)
1	NH	9.1
2	O	9000
3	CH$_2$	10.6

hydrogen bond). Compound 3 contained a phosphinate linkage
(PO$_2$CH$_2$). Like compound 2, it could not form that same hydrogen
bond. Unlike compound 2, however, it was essentially equipotent with
the phosphoramidate 1. The explanation comes from consideration of
solvation and desolvation effects. Both the phosphoramidate (1) and
the phosphonate (2) will be solvated in aqueous solution and require
desolvation to bind to the enzyme active site. The phosphoramidate
(1) can recover some of the energy required for desolvation by forming
a hydrogen bond with the enzyme, but the phosphonate (2) cannot.
The phosphinate (3) is somewhat less polar due to the presence of
a methylene group instead of an amine or oxygen moiety. It is
correspondingly less solvated in aqueous solution, and, therefore, it
requires less desolvation when it binds to the enzyme. The net change
in free energy in going from solution to enzyme-bound states is
roughly comparable for compound 1 (with the PO$_2$X group going
from solvated to solvated states) and compound 3 (going from
unsolvated to unsolvated states) and is more favorable than for
compound 2 (going from solvated to unsolvated states).

Hydrophobic Interactions
Just as hydrophilic residues tend to be solvated, hydrophobic side
chains have an equally strong tendency to avoid exposure to the
aqueous environment. This effect is largely entropic, reflecting the
unfavorable free energy of forming a water–hydrocarbon interface,
where the side chain would penetrate the aqueous solvent (Tanford,
1973; Burley and Petsko, 1988). It has long been recognized that the
interiors of soluble proteins are highly hydrophobic (Kauzmann, 1959)

and that proteins fold in such a way as to minimize the exposure of hydrophobic side chains on the protein surface. It has been proposed that protein folding proceeds first through the formation of hydrophobic clusters which direct further folding of the peptide chain into the various low-energy secondary structures, such as helices and β-structure (Rose and Roy, 1980). The importance of hydrophobic interactions to protein stability has been studied by comparing the susceptibility to denaturation of a series of proteins in which site-specific mutagenesis was used to modify single residues in the protein sequence (Kellis et al., 1988). It was found that the absence of even a single methyl group (Ile to Val substitution) destabilized the protein by 1.1 kcal/mol, underscoring the important cumulative effect of hydrophobic interactions on overall protein structure and stability.

Although the sequestering of nonpolar residues away from the aqueous environment is largely an entropy-driven process, specific interactions of hydrophobic side chains occur as well. These are typically induced dipole-induced dipole interactions, known as van der Waals or London interactions (Burley and Petsko, 1988). Although they are much weaker than the salt bridges and hydrogen bonds involving polar residues, they can be important in local secondary structure and protein interactions. For example, a number of DNA-binding proteins that function biologically as dimers share an unusual sequence, featuring a leucine every seventh residue in a 30-residue segment and having a relatively high probability of helical structure. This sequence would create a helix in which the leucine residues occupied every other turn of the same side of the helix. The structure was termed a "leucine zipper" based on the hypothesis that the leucines from such a helix in two monomers could interdigitate like a zipper, holding the monomers together (Landschutz et al., 1988). Although it was subsequently found that the Leu residues do not interdigitate, but rather align themselves parallel to each other in pairs along the helical interface (O'Shea et al., 1989), the term leucine zipper has remained, and it is an important structural motif in hydrophobic protein-protein interaction.

Aromatic residues have an inherent dipole with the electron-rich π-cloud lying parallel to and above and below the plane of the ring and with positively polarized hydrogen atoms in the plane of the ring (Burley and Petsko, 1988). Although aromatic rings in proteins do interact with each other, the arrangement is not typified by the parallel stacking of base pairs in DNA helices. Rather, the edge of one ring interacts with the face of the other in a roughly perpendicular arrangement (Burley and Petsko, 1985), allowing a favorable interaction between the positively polarized hydrogens and the negatively polarized π-cloud.

Peptides, in general, are not long enough to allow the hydrophobic residues to arrange themselves in such a way that they are totally shielded from solvent. This occurrence undoubtedly contributes to the poor aqueous solubility of many peptides compared to proteins. Many peptides require the presence of strong organic co-solvents like dimethylsulfoxide (DMSO), dimethylformamide (DMF), or ethanol to achieve sufficient solubility for biological testing, which may be a limiting factor in some biological test systems. Peptides that have substantial hydrophobic character also tend to aggregate with increasing concentration. Solubility generally increases with peptide length because of the peptide's consequent ability to adopt stable secondary structures and to segregate nonpolar residues.

Disulfide Bonds, Thioesters, and Thioethers
Unlike side-chain interactions of other residues that are noncovalent in nature, cysteine residues form a number of covalent linkages with other amino acid side chains. The most common of these is the disulfide, which involves oxidative coupling of two cysteine thiol groups to form cystine (figure 2-3). Like the amide bond, the sulfur-sulfur bond in a disulfide is not freely rotatable. Rather, it exists in one of two rotamers, with torsional angles in the vicinity of either $+90°$ or $-90°$. The entire disulfide moiety, $CH-S-S-CH_2$, can rotate as a unit by simultaneous rotation about the side-chain angles $\chi 1$ ($NH-C\alpha H-CH_2-S$) and $\chi 2$ ($C\alpha H-CH_2-S-S$). In proteins, disulfide bonds are important in the stabilization of tertiary structure. In disulfide-containing peptides like oxytocin, vasopressin, and somatostatin (see figure 2-4), the disulfide has proportionally greater importance in maintaining a biologically active conformation because the peptides are too small to maintain a stable conformation otherwise. Replacement of the cysteine residues with isosteric alanine residues results in a dramatic loss in biological activity (Walter et al., 1967; Polàcek et al., 1970; Sarantakis et al., 1973). A great deal of conformational flexibility still remains in cyclic disulfide-containing peptides like vasopressin, and it does not exhibit a single preferred conformation (Hagler et al., 1985), but the disulfide does constrain the peptide to folded conformations that otherwise would have a low probability for linear peptides.

Cysteine residues can also form covalent thioester and thioether linkages. The thioester is found as the activated form of acyl groups in acyl transferases and as an acyl enzyme intermediate in thiol proteases such as papain. Thioesters also are found in the reactive binding sites of complement C3b and α2-macroglobulin, where the tetrapeptide sequence Cys-Gly-Glu-Glu contains a thioester involving the Cys-1 thiol and the Glu-4 side-chain carboxylate (figure 2-3) (Sothrup-Jensen et al., 1980; Tack et al., 1980; Howard, 1981). Cysteine forms aliphatic thioether linkages, such as in lanthionine, a

HO2C
NH2
S
S
H2N
CO2H

Cystine

CO2H
NH2
S
H2N
CO2H

meso-Lanthionine

CO2H
H
N
O
H
N
H
N
O
O
ICO--
--HN
S
O

Cys-Gly-Glu-Glu thioester

(HO)H
CO2H
HN
NH2
S
H2N
CO2H

Cys-Trp thioether of phalloidin
(or amanitin)

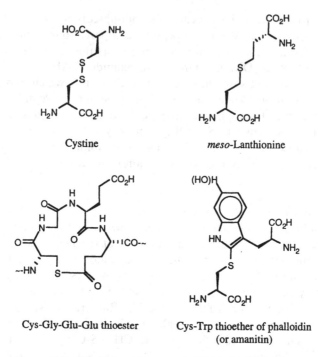

Figure 2-3. Disulfide, thioester, and thioether linkages.

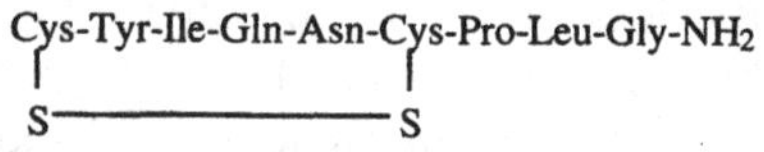

Cys-Tyr-Ile-Gln-Asn-Cys-Pro-Leu-Gly-NH2
S
S

Oxytocin

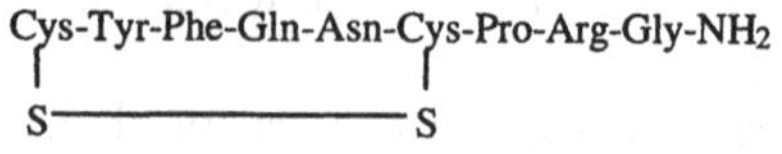

Cys-Tyr-Phe-Gln-Asn-Cys-Pro-Arg-Gly-NH2
S
S

Arginine Vasopressin

Ala-Gly-Cys-Lys-Asn-Phe-Phe-Trp-Lys-Thr-Phe-Thr-Ser-Cys-OH
S
S

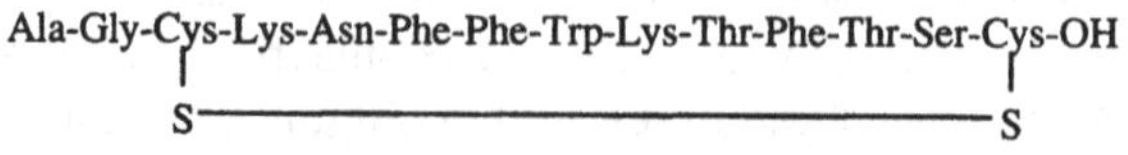

Somatostatin

Figure 2-4. Disulfide-containing peptides.

constituent of the peptide antibiotics subtilin (Alderton and Fevold, 1951) and nisin (Berridge et al., 1952). Aromatic thioether linkages also occur, such as in the Cys-Trp conjugate found in mushroom toxins such as amanitin and phalloidin (Wieland, 1968). A novel thioether Cys-Tyr conjugate recently has been identified in the enzyme galactose oxidase (Ito et al., 1991).

The Amide Bond

The polyamide peptide backbone is also an important contributor to overall protein and peptide structure. A substantial double-bond character is found in the carbon-nitrogen peptide bond due to the resonance structure shown in figure 2-5(a). This structure gives the amide bond several characteristics that are important in peptide and protein structure. The amide bond is flat, with the carbonyl carbon, oxygen, nitrogen, and amide hydrogen all lying in the same plane. No free rotation occurs about the carbon-nitrogen bond because of its partial double-bond character (the barrier to rotation is about 25 kcal/mol). The torsional angle of that bond, ω, is defined by the peptide backbone atoms Cα-C(O)-N-Cα. Because of the partial double-bond character, there are two rotational isomers for the peptide bond: *trans* ($\omega = 180°$) and *cis* ($\omega = 0°$), as shown in figure 2-5(b). The lower-energy isomer is the *trans* peptide bond, which is the isomer generally found for all peptide bonds not involving proline. In the case of amide bonds involving proline, the energy of the *trans* X-Pro bond is somewhat elevated, and both the difference in energy between *cis* and *trans* isomers and the barrier to rotation is lowered. Proline-containing peptides thus will often exhibit *cis-trans* isomerism about the X-Pro bond. This can be detected by nuclear magnetic resonance (NMR) studies because the chemical shifts of some hydrogens, especially those on the proline δ-carbon, are often different. The equilibrium ratio of *cis* to *tans* isomers and the rate of isomerization are highly dependent on the exact peptide sequence. Empirically, it has been found that the *cis* content of X-Pro peptides generally increases when X is a bulky, hydrophobic amino acid (Harrison and Stein, 1990).

It has been appreciated only recently that *cis* X-Pro peptide bonds may be important in proteins as well. The X-ray crystal structure of several proteins has revealed specific *cis* X-Pro peptide bonds (Frömmel and Preissner, 1990). Proline isomerization has also been implicated as a slow step in protein refolding (Kim and Baldwin, 1982).

The resonance forms of the amide bond give it one other character- istic that is extremely important in peptide and protein structure. The

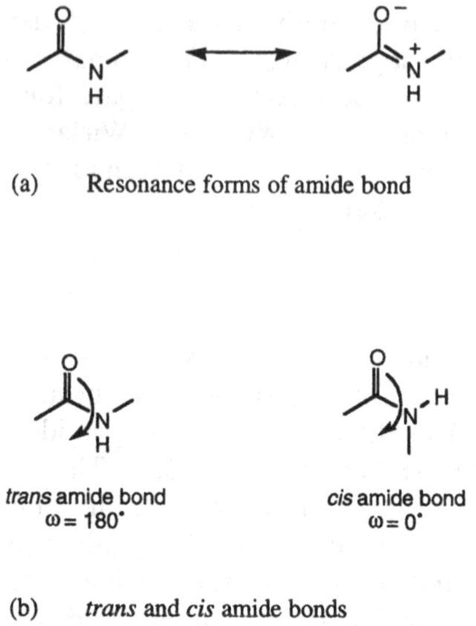

(a)   Resonance forms of amide bond

trans amide bond
$\omega = 180°$

cis amide bond
$\omega = 0°$

(b)   *trans* and *cis* amide bonds

Figure 2-5. Characteristics of amide bonds.

amide bond is quite polar and has a significant dipole moment, which makes the amide carbonyl oxygen a particularly good hydrogen-bond acceptor and the amide NH a particularly good hydrogen-bond donor. Hydrogen bonds involving the peptide backbone are an important stabilizing factor in protein secondary structures. Peptide bonds have a strong tendency to be solvated, by either bulk solvent or specifically bound water molecules, or by internal hydrogen bonds. This is especially true in the hydrophobic interior of soluble protein molecules. Peptide bonds of regular secondary structures, such as helices and β-sheets, are internally solvated by the hydrogen bonds that stabilize those structures. The peptide bonds of the so-called random coil, or irregular structures, also participate in an extensive network of hydrogen bonds involving internal polar side chains, bound water molecules, and backbone interactions.

The polarity of the amide bond also can impart a net dipole to regular structures containing peptide bonds, such as helices. The overall dipole points from the amino terminus (positive partial charge) to the carboxyl terminus (negative partial charge) in a helix, which can be an important factor in protein tertiary structure as well as a contributor to catalytic activity in enzymes. The dipole can be used to stabilize interactions with substrates or to modify the pK_a, or nucleophilicity, of catalytically active residues (Knowles, 1991).

Protein Structure

Protein structure is organized in several hierarchical levels of increasing structural complexity. The most basic level is the primary structure, or amino acid sequence, of the protein. Secondary structure deals with the folding up of short segments of the peptide, or protein chain, into regular structures such as α-helices, β-sheets, and turns. Tertiary structure describes how the secondary structural elements of a single protein chain interact with each other to fold into the native protein structure. Quaternary structure involves the interaction of individual protein subunits to form a multimeric complex.

In the design of peptides, we are concerned mainly with the primary and secondary structures. The primary structure, or sequence, of a protein, is readily obtainable by chemical sequence analysis of the isolated, purified protein or by translation of the corresponding DNA coding sequence. Since it is easier to sequence DNA than to sequence protein, most new protein sequences are now obtained in this way. Computer programs are available, which will be discussed later (see below, Prediction of Protein/Peptide Structure), to help analyze and compare protein primary structures. Protein secondary structure has also become increasingly well understood, to the point that it can be predicted or optimized on the basis of the primary sequence reasonably well by using computer algorithms. In addition, several techniques have been applied to either induce or stabilize secondary structure in peptides, which will be discussed in more detail later (see below, Conformational Design and Constraint).

Protein Secondary Structure

Chains of amino acids can fold into several types of regular structures that are stabilized by intrachain or interchain hydrogen bonds in the amide backbone. Helices and turns are formed from continuous regions of protein sequence and are stabilized by intrachain hydrogen bonds. Sheets are formed from two or more chains, which are separated by intervening sequences (regions of secondary structure as well), and are stabilized by interchain hydrogen bonds. The parts of a protein sequence that do not appear in helices, sheets, or turns are said to be in random coil, which is meant to imply only that no regular, repeating structure exists. It should not be taken to mean that the conformation of the protein is random in those regions, because in any individual protein, the random coil sequences are just as highly ordered and reproducibly formed as other regions of secondary structure.

The conformation of the peptide backbone can be described by three torsional angles: ϕ, which is the angle defined by C(O)-N-Cα-C(O); ψ, which is defined by N-Cα-C(O)-N; and the amide torsional angle ω,

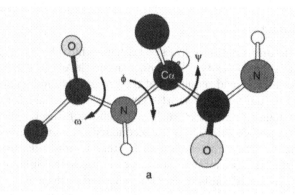

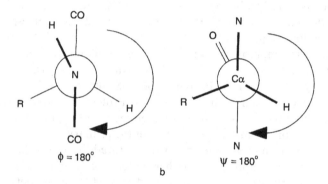

Figure 2-6. Torsional angles of the peptide backbone.
(a) Backbone torsional angles ϕ, ψ, and ω. (b) Newman
projections of torsional angles ϕ and ψ.

which was discussed previously (see above, The Amide Bond). These
angles are shown graphically in figure 2-6(a). The convention in protein chemistry is that these angles are defined with respect to the peptide backbone. The angle ϕ is 180° when the two carbonyl carbons are
trans to each other, and the angle ψ is 180° when the two amide
nitrogens are *trans* to each other as in figure 2-6(b). The angle ψ,
therefore, is +180° compared to the usual chemical definition. A
fully extended peptide chain, however, would have backbone torsional
angles all of 180°, just like a fully extended carbon chain.

Helices
One of the most common types of secondary structure in proteins is
the helix in which amino acid residues are wrapped around a central

Table 2-3 Torsional angles for helices

	ϕ	ψ
α-Helix (right-handed)	$-57°$	$-47°$
α-Helix (left-handed)	$57°$	$47°$
3_{10} Helix	$-60°$	$-30°$
Collagen helix	$-51°$	$153°$
	$-76°$	$127°$
	$-45°$	$148°$
Polyproline	$-78°$	$149°$

Angles taken from IUPAC-IUB Commission on Biochemical Nomenclature (1970).

axis in a regular pattern. Given planar, trans peptide bonds, helices will be uniquely defined by their ϕ and ψ angles. Several types of helices are found in protein structures, the α-helix being the most common. These helices and their characteristic torsional angles are listed in table 2-3. Helices are also characterized by the number of residues per turn and the hydrogen bonding pattern, specifically the number of atoms in the cyclic structure formed by the hydrogen bond. The α-helix is a 3.6_{13} helix, meaning there are 3.6 residues per turn of the helix, and 13 atoms are included in each repetitive hydrogen-bonding structure. Likewise, a 3_{10} helix has three residues per turn and 10 atoms in the cyclic hydrogen-bonded structure. These are the two most relevant helices for protein and peptide structure. There is also a handedness to helical structure, which is defined by the screw sense of the helix. Looking down the helical axis from the amino terminal end, the peptide backbone can be traced from amino terminus to carboxyl terminus in a clockwise sense in a right-handed helix, which is the form that generally occurs.

Helical structures are stabilized by intrachain hydrogen bonds. In a helix, the carbonyl bonds and amide NH bonds lie parallel to the helix's axis with the carbonyls pointing downward, in the direction of the carboxyl terminus of the helix, and the NHs pointing upward, in the direction of the amino terminus. In an α-helix, the carbonyl of any residue i is hydrogen-bonded to the amide NH of the $i + 4$ residue; that is, the residue four amino acids farther down along the peptide chain. In a 3_{10} helix, the ith residue carbonyl is hydrogen-bonded to the NH of the $i + 3$ residue. These relationships are shown graphically in figures 2-7 and 2-8 for six-residue segments of α and 3_{10} helices as viewed from the side, perpendicular to the helical axis. For clarity, only the peptide backbone atoms are shown.

The helix forms a polyamide cylinder. The uniform arrangement of carbonyl and NH groups imparts a strong dipole moment to the helix,

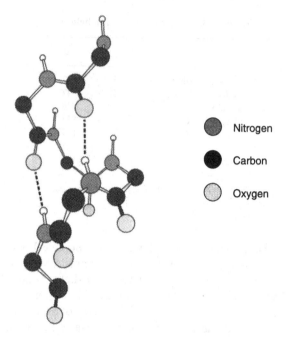

Figure 2-7. Backbone structure of an α-helix. Note that the NH and carbonyl groups are aligned parallel to the axis of the helix. Each hydrogen-bonded segment contains 13 atoms.

Nitrogen

Carbon

Oxygen

with the positive end at the amino terminus and the negative end at the carboxyl terminus. The side chains of the residues are arranged radially outward from the helix, looking down the helix axis (figure 2-9). Because all the backbone amide groups are involved in intrachain hydrogen bonds, the interactions of helices with other peptide chains or small molecules occurs predominantly through side-chain interactions (hydrophobic interactions, salt bridges, and hydrogen-bonding interactions; see above, Side-Chain Interactions) and interactions with the helix dipole itself. Because of this arrangement of intrachain hydrogen bonds, the amide bonds of the helix can be thought of as being internally solvated, resulting in the helix being more readily accommodated in a nonpolar environment, such as a lipid bilayer. The membrane-spanning regions of transmembrane proteins are generally thought to be helical, with the axis of the helix perpendicular to the plane of the membrane (Eisenberg, 1984; Wickner and Lodish, 1985). The membrane pore-forming peptide antibiotics, such as alamethicin, also have a highly helical structure, and a high degree of helix potential exists in the

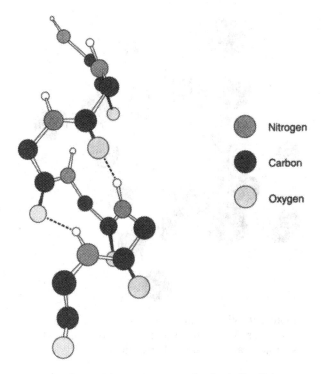

Nitrogen	
Carbon	
Oxygen	

Figure 2-8. Backbone structure of a 3_{10} helix. It is slightly elongated compared to the α-helix, with three residues per turn of the helix instead of 3.6. Each hydrogen-bonded segment contains 10 atoms.

hydrophobic leader sequences of proteins that are transported through membranes.

β-Structure

Although helices are a form of secondary structure resulting from one strand of peptide chain, β-structure requires two strands. The conformation of each peptide chain is generally extended, and interchain hydrogen bonds occur between the carbonyl oxygen and the amide NHs of every other residue in the backbone. Two possible arrangements of adjacent strands occur: parallel, in which each peptide strand runs in the same direction, and antiparallel, in which the peptide strands run in opposite directions. Both can be found in protein structures. These two arrangements are shown diagrammatically in figure 2-10.

The β-structure can be thought of as forming a surface or sheet, although there is a slight twist to it because the peptide backbone is

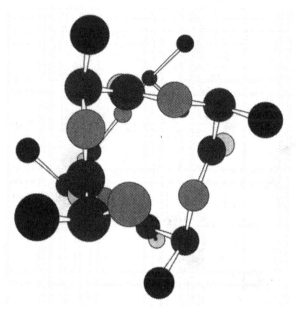

Figure 2-9. View of an α-helix looking down the helical axis. The residue side chains, shown in black, all project radially outward from the helix.

not fully extended (typical angles are given in table 2-4). The side chains extend above and below the rough plane of the sheet, with every other side chain on one surface. Because the interchain hydrogen bonds between two adjacent chains involve only every other residue, there is the possibility of larger structures forming that involve many strands. Such extended β-structures are common features in proteins. The twist of the β-structure also allows sheets to fold into cylindrical structures, called β-barrels, which is another common structural motif.

Turns

Strands of α-helix or β-structure do not extend indefinitely in proteins but rather fold back on themselves. Several regular structures, called turns, are involved in changing the direction of the peptide chain. Turns are classified by the number of residues that are involved in the regular structure; β-turns contain four amino acid residues, while γ-turns contain three residues. Each structure is stabilized by a hydrogen bond extending across the turn, in effect holding the two ends together. The first residue of a turn is usually designated as *i*. In a β-turn, the hydrogen bond is between the carbonyl of the *i* residue and the NH of the *i* + 3 residue, giving the equivalent of a 10-membered ring. There

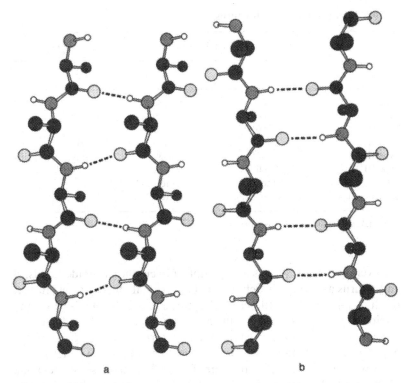

Figure 2-10. Structure of β-sheet. Amide bonds lie in the plane of the sheet, with side chains alternately projecting above and below the plane of the sheet. (a) Parallel sheet, with both strands running in the same direction. (b) Antiparallel sheet, with strands running in opposite directions.

may be an additional hydrogen bond between the NH of the i residue and the carbonyl of the $i+3$ residue, as if they were an incipient antiparallel β-structure. In a γ-turn, the hydrogen bond is between the carbonyl of the i residue and the NH of the $i+2$ residue, giving the equivalent of a seven-membered ring. (Turns are sometimes named for the number of atoms in the ring structure formed by the hydrogen bond. Thus a β-turn would be a C_{10} turn and a γ-turn a C_7 turn.)

Table 2-4 Torsional angles for β-structure

	ϕ	ψ
Parallel β-sheet	−119°	113°
Antiparallel β-sheet	−139°	135°

Angles taken from IUPAC-IUB Commission on Biochemical Nomenclature (1970)

Table 2-5 Torsional angles for turns

	$i+1$		$i+2$	
	ϕ	ψ	ϕ	ψ
Type I	$-60°$	$-30°$	$-90°$	$0°$
Type I$'$	$60°$	$30°$	$90°$	$0°$
Type II	$-60°$	$120°$	$80°$	$0°$
Type II$'$	$60°$	$-120°$	$-80°$	$0°$
Type III	$-60°$	$-30°$	$-60°$	$-30°$
Type III$'$	$60°$	$30°$	$60°$	$30°$
γ-Turn	70 to 85°	-60 to $-70°$		
Inverse γ-turn	-70 to $-85°$	60 to 70°		

Angles taken from Smith and Pease (1980).

Several types of β-turns are possible. Given planar amide bonds, β and γ turns are uniquely defined by the torsional angles of the internal residues $i+1$ and, in the case of β-turns, $i+2$. Table 2-5 lists the torsional angles associated with the various types of β- and γ-turns. Note that there is an inverse form for each turn in which the sign of the internal torsional angles is reversed. Note also that the type III β-turn is equivalent to a 3_{10} helix structure. Figure 2-11 depicts the backbone of types I, I$'$, II, and II$''$ β-turns and figure 2-12 depicts that of γ- and inverse γ-turns.

Turns have several important structural features. The peptide backbone in a turn forms a rough plane that contains the intramolecular hydrogen bond. In a β-turn, the amide bond between the $i+1$ and $i+2$ residues lies perpendicular to this plane. Since it is not part of the hydrogen-bonding structure of the turn, its hydrogen bonding needs must be satisfied elsewhere. In proteins, β-turns are most often found on the protein surface, where the $i+1 - i+2$ peptide bond can be solvated by bulk solvent. When such a turn is located in the (generally hydrophobic) interior of a protein, it is often solvated by a bound water molecule (Smith and Pease, 1980). Besides the amide bond, the side chains of the $i+1$ and $i+2$ residues, which form the "corners" of the turn, are also particularly exposed. In small peptides, such as vasopressin or somatostatin, where a turn has been postulated based on the cyclic nature of the peptide chain, the corner residues are implicated as important elements in the expression of the peptide's biological activity (Rivier et al., 1976; Walter et al., 1977).

Side chains will adopt either a pseudo-axial or pseudo-equatorial arrangement around the cyclic hydrogen-bonded turn structure. One diagnostic criterion for the importance of a turn involves changing the chirality of a corner residue to allow the side chain to be in the more

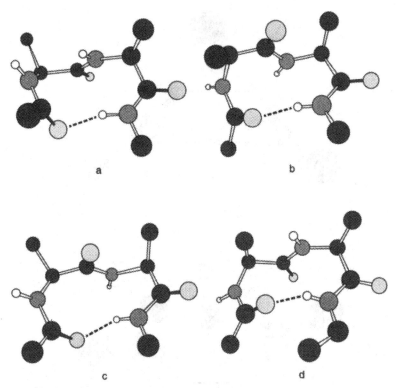

Figure 2-11. (a) Type I β-turn. (b) Type I′ β-turn. (c) Type II β-turn.
(d) Type II′ β-turn. The turn is stabilized by a hydrogen bond from the
carbonyl of the ith residue to the NH of the $i + 3$ residue. The amide
bond between the $i + 1$ and $i + 2$ residues is perpendicular to the plane of
the turn. The side chain of the $i + 1$ residue in a type I′ and II′ turn is in
a pseudo-axial orientation. Substitution of a D-amino acid would give the
more stable pseudo-equatorial orientation. The NH and side chain of the
$i + 2$ residue of the type II′ turn are both on the same side of the ring, an
arrangement in which a proline residue could be accommodated as well.
The NH and side chain of the $i + 2$ residue of a type I′ turn are on
opposite sides of the ring, which would exclude the possibility of a proline
residue at that position.

stable equatorial configuration without otherwise altering the peptide
backbone structure. In a type II′ β-turn, for example, the $i + 1$ side
chain is in a pseudo-axial configuration (see figure 2-11). Replacement
of this residue with a D-amino acid should allow for equatorial placement of the side chain while the peptide backbone remains unchanged,
lowering the overall free energy of that structure and making it more
stable. Replacing the $i + 1$ residue with a glycine, with no side chain,
should have a similar effect.

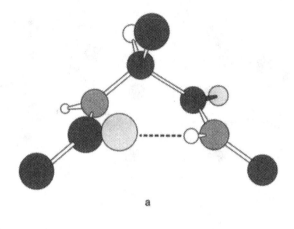

a

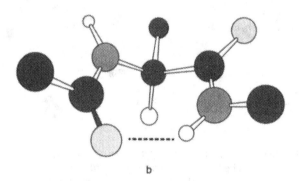

b

Figure 2-12. Structure of a γ-turn (a) and inverse
γ-turn (b). The turns are stabilized by a hydrogen
bond from the carbonyl of the ith residue to the
NH of the $i + 2$ residue. The side chain of the $i + 1$
residue adopts a pseudo-axial orientation in the
γ-turn and a pseudo-equatorial orientation in the
inverse γ-turn.

Amphipathic Structures

Given the regular placement of side chains in helical or β-structures, it
is possible to obtain forms in which the polar side chains are segregated on one face of the structure and hydrophobic side chains are on
the other. This type of arrangement is termed an *amphiphilic*, or
amphipathic, structure.

In an ideal amphipathic helix, the hydrophobic residues are all
found on one side of the helix cylinder, and the hydrophilic residues
are found on the other side of the cylinder. In the linear sequence,

these hydrophilic and hydrophobic residues will be interdispersed, so it can achieve the desired arrangement when folded into a helix. One idealized amphipathic helix, designed by Kaiser and coworkers as a model for apolipoprotein A, has the helical sequence Lys-Leu-Glu-Glu-Leu-Lys-Glu-Lys-Leu-Lys-Glu-Leu-Leu-Glu-Lys-Leu-Lys-Glu-Lys-Leu (Fukushima et al., 1980). The arrangement of side chains can best be shown schematically in a helix wheel projection, which is depicted looking down the helix axis with the side chains arranged radially outward, as shown in figure 2-13(a).

In β-structure, the side chains alternate above and below the plane of the sheet. In an amphipathic strand of β-structure, every other residue in the linear sequence would be hydrophobic and the remaining residues would be hydrophilic. This arrangement would give a structure in which one face of the sheet would be hydrophobic and the other face would be hydrophilic, as in figure 2-13(b).

In proteins, amphipathic structures can be important structural elements, allowing the hydrophobic residues to be sequestered in the interior of the protein while a hydrophilic face is presented to the aqueous environment at the protein surface. This arrangement can be reversed in transmembrane proteins, with amphipathic helices clustered so that the hydrophobic face is exposed to the lipid environment and the hydrophilic faces are sequestered in the interior of the protein, forming a hydrophilic pore, or channel, in the membrane.

The recognition that amphipathic structures are important to the biological activity of smaller peptides is due in large part to the pioneering work of E. T. Kaiser. He recognized potential amphipathic helices in such membrane-active peptides as apolipoprotein A and mellitin, as well as peptide hormones such as calcitonin, corticotropin releasing factor, and β-endorphin, which interact with membrane-bound receptors. Structure–activity studies of synthetic analogs of these peptides have demonstrated the importance of their amphipathic structures by designing sequences optimizing amphipathicity or replacing it with nonhomologous sequences adopting the same amphipathic structure (Kaiser and Kézdy, 1984). Examples of the design of amphipathic peptides will be discussed later (see below, Hydrophobicity Measurements).

Prediction of Protein/Peptide Structure

Protein Sequence Homology
The primary sequence of a peptide, or protein, is its most fundamental structural element. It has long been presumed that all the information content of a protein—folding, translocation, and function—are encoded within the amino acid sequence of the primary translation

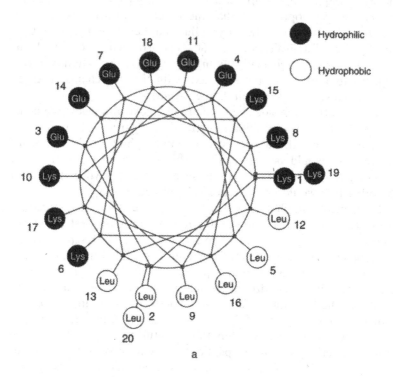

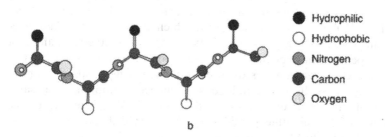

Figure 2-13. (a) Helical wheel projection of a designed amphipathic helix. The residues with hydrophilic side chains are segregated on one face of the helix, while the hydrophobic residues are segregated on the other. (b) Side view of one strand of an amphipathic β-sheet. The hydrophilic side chains project on one side of the sheet, while the hydrophobic side chains project on the other side. The puckering of the backbone, which gives rise to the often-used name β-pleated sheet, is clearly evident in this side view.

product. Many proteins can be denatured reversibly and renatured with complete, recovery of biological activity, lending credence to this concept. The situation in protein biosynthesis is more complex, because nascent peptides can begin to interact with cellular proteins, such as signal recognition particles (which mediate protein translocation across biological membranes), chaperonins, and heat shock proteins (which are involved in protein folding and assembly) even before the full-length translation product has come off the ribosome (Landry and Gierasch, 1991). The elements involved in the recognition of the nascent polypeptide chain by these signal recognition particles and chaperonins still must derive from the primary sequence of the peptide chain.

Although the rules relating protein sequence to folding, translocation, and function are not well understood, the ability to compare protein sequences and determine homology is quite useful. Homology can help to suggest protein class or function in newly identified protein sequences. Identification of a homologous protein for which structural information is known, particularly X-ray crystal structures, can aid in the construction of three-dimensional models for the protein structure. A case in point was the early identification of the HIV-1 protease as an aspartic protease, based on sequence homology to other aspartic proteases, and the construction of a model structure, based on the known X-ray crystal structures of fungal aspartic proteases, such as penicillopepsin and rhizopus pepsin (Pearl and Taylor, 1987).

Our ability to search protein databases for protein similarities is based on the pioneering effort of Margaret Dayhoff at the National Biomedical Research Foundation (NBRF), who began publishing compilations of protein sequences during the 1960s (Dayhoff et al., 1965). Today, several electronic databases are available that are updated on a regular (daily) basis such as the NBRF protein database, the Swiss-Prot database and the Gen-Bank database. These databases, along with software to search for sequence homology, such as BLAST, are available over the internet on an interactive basis. Some particularly useful sites for this purpose are the National Center for Biotechnology Information (http://www.ncbi.nlm.nih.gov/), the ExPASy Molecular Biolgy server of the Swiss Institute of Bioinformatics (http://www.expasy.ch/), and the European Molecular Biology Laboratory Protein and Peptide Group (http://www.mann.embl-heidelberg.de/Default.html). In addition, a comprehensive discussion of computational methods for protein sequence analysis with specific instructions can be found in Michaels and Garian, 1995. With the tremendous advances in nucleotide sequencing, many new protein sequences are available from a translation of the nucleotide sequence rather than from conventional protein sequencing techniques.

Two basic operations exist for sequence homology searching. One is a matching function that allows searching for occurrences of a specific sequence. For example, one could search the protein database for all occurrences of the sequence Arg-Gly-Asp-Ser. The number of mismatches allowed in the sequence can be specified, allowing retrieval of such sequences as Arg-Ala-Asp-Ser. It is also usually possible to specify a list of amino acids that would be considered a match at a given position in the sequence, for example, Arg-Gly-(Asp or Glu)-Ser. Finally, one can also search for patterns like Ser-X-X-Tyr-Pro when one is interested in sequences in which a Ser is found three residues upstream from a Tyr-Pro sequence and where the nature of the intervening amino acids (X) is not important. BLAST is a modern version of a sequence searching algorithm available through the National Center for Biotechnology Information that is widely used today. A detailed description and instructions for its use can be found in Wolfsberg and Madden, 1995.

Although matching is useful for identifying regions of contiguous sequence, many proteins may have homologous regions separated by sequences of varying lengths or may have small gaps or insertions within a region of homology. For this reason, an alignment algorithm is used to compare two entire protein sequences and to determine the best fit of the two sequences. Several variations occur on sequence alignment algorithms: one is based on methodology originally derived by Needleman and Wunsch (1970), and another (FASTA), developed by Pearson and Lipman (Lipman and Pearson, 1985; Pearson and Lipman, 1988), is optimized for speed, something of concern when one wants to check a new sequence against the entire protein database. Modern algorithms have become more and more complex and their inner workings are beyond the scope of this chapter. However, a basic illustration of how they were originally designed can be considered. Most methods relied on the use of a scoring table or matrix that assigns a numerical value to each pair of amino acids. The simplest scoring table would have 1 for identical amino acids and 0 for any other combination. If you were to create a matrix with the sequence of one protein running along the columns and that of the other protein running along the rows, each cell (i, j) would be either 0 or 1 depending on whether residue i from the first protein matched residue j from the second protein. If the two proteins were identical, there would be 1's all along the diagonal of the matrix. If there were more limited regions of homology, they would appear as stretches consisting mainly of 1's along or parallel to the diagonal. Any offset from the diagonal reflects gaps or insertions of amino acids into one of the protein sequences. This often indicates the presence of variable length loops of sequence that connect conserved regions of secondary structure in homologous proteins.

It is readily apparent that the scoring table does not have to be limited to values of 0 and 1. The comparisons can be graded so that a conservative substitution (e.g., valine for isoleucine) would have a high score, and substitution of a very dissimilar amino acid (such as aspartic acid for isoleucine) would have a very low score. A commonly used scoring table (Dayhoff et al., 1978) takes into account conservative substitutions, the frequency of appearance of amino acids, the frequency of single base mutations to change the genetic three-base code from one specific amino acid to another, and the conservation of rare amino acids (like tryptophan) or structurally significant amino acids (like cysteine). Scoring values range from a high of 17 (conservation of the relatively uncommon tryptophan in a sequence) to 2 (conservation of the commonly occurring and not generally structurally significant alanine) to lows of −6 to −8 (substitution of the structurally significant cysteine). This sort of table is ideal for studying the evolutionary relatedness of various proteins. If one were interested in convergent evolution, where a peptide or protein might have independently evolved to a similar structure and function, one could construct a scoring matrix which emphasized chemically conservative substitutions. The regions of sequence homology are not the only regions of interest or importance in protein function. For example, if there are species differences in the activity of homologous proteins, the structural basis for these differences may lie in the regions of non-homologous amino acid sequence. In addition, loops, such as those connecting conserved regions of secondary structure, have a high probability of lying on the protein surface and may serve as important sites for post-translational modification or as antigenic epitopes (see below, Epitope Mapping and Production of Protein-Specific Antibodies from Peptides).

Secondary Structure Prediction
Protein sequence alignment alone cannot suggest any of the structure of a protein unless a homologous protein has been studied by X-ray crystallography, nuclear magnetic resonance, circular dichroism or any of the other physical methods of protein structure determination. Structural information of this type is not available for most protein sequences. However, algorithms have been developed that allow prediction of secondary structures based on the primary sequence of the protein. Secondary structure prediction algorithms are generally available as part of the protein sequence analysis software packages such as that found at the ExPASy site mentioned above. They are based on statistical analyses of the frequency of residues or sequences being found in particular secondary structures in a dataset of known protein structures. Prediction of overall secondary structure is

generally good, but specific details, such as exact helix boundaries, may be less reliable. A comprehensive discussion of secondary structure prediction methods with specific instructions can be found in Krystek et al., 1995b.

Perhaps the most widely used secondary structure prediction scheme is that of Chou and Fasman (1978). It is based on a statistical analysis of the frequencies of individual amino acid residues being found in specific secondary structures (α-helix, β-structure or β-turn) compared to the overall frequency in which they appear in protein sequences. This allows each amino acid to be assigned a numerical probability of being found in a given secondary structure and to be classified as a strong helix-former, helix-breaker, weak β-former, β-breaker, and so on. The Chou–Fasman probabilites for β-turns are listed in table 2-6. Since an α-helix or β-structure is a regular repeating one, each position in it is essentially equivalent. To a first approximation, then, the probability of a given amino acid appearing in a helix or a β-structure is not dependent on the position it would occupy in that structure. Each amino acid, therefore, has a single probability value for α-helix or β-structure. Since the conformational requirements for each position in a β-turn are different, however, each amino acid will have a different probability of appearing in each of the four positions of a β-turn.

Secondary structure is predicted by applying a series of empirically determined rules to the primary sequence. Helices are predicted by searching for nucleation sites—clusters of residues with a high helix forming potential. The helix propagates out in both directions until a cluster of helix-breaking residues is encountered. There are other rules for locating helix boundaries, such as the general tendency for negatively charged amino acids (Asp and Glu) at the amino terminal end (where they help stabilize the helix dipole) and positively charged amino acids (His, Lys and Arg) at the carboxyl terminal end. β-sheet structure is predicted in a similar fashion. Turns are predicted for a sequence of amino acids i to $i + 3$ by considering the sum of their position-dependent probability values. In practice, algorithms supplied with protein database software generally evaluate helix potential, β-sheet potential, and β-turn potential (also, sometimes helix and sheet boundary conditions) separately and plot each against residue number. Helix and sheet potential are averaged over a user-defined window of residues (typically four) and normalized. Average probabilities greater than one indicate a propensity for that sequence to adopt that type of secondary structure. The graphical output allows regions with a high probability of adopting helix, β-structure, or β-turns to be readily identified, and plotting of residues often found at helix or sheet boundaries helps to define the extent of secondary structures.

Table 2-6 Hydrophobic and β-turn indices of amino acids

Amino acid	Symbol	Hydrophobicity value[a]	β-Turn propensity[b]
Arginine	(Arg, R)	−4.5	0.95
Lysine	(Lys, K)	−3.9	1.01
Aspartic acid	(Asp, D)	−3.5	1.46
Glutamic acid	(Glu, E)	−3.5	0.74
Asparagine	(Asn, N)	−3.5	1.56
Glutamine	(Gln, Q)	−3.5	0.98
Histidine	(His, H)	−3.2	0.95
Proline	(Pro, P)	−1.6	1.52
Tyrosine	(Tyr, Y)	−1.3	1.14
Tryptophan	(Trp, W)	−0.9	0.96
Serine	(Ser, S)	−0.8	1.43
Threonine	(Thr, T)	−0.7	0.96
Glycine	(Gly, G)	−0.4	1.56
Alanine	(Ala, A)	1.8	0.66
Methionine	(Met, M)	1.9	0.60
Cysteine	(Cys, C)	2.5	1.19
Phenylalanine	(Phe, F)	2.8	0.60
Leucine	(Leu, L)	3.8	0.59
Valine	(Val, V)	4.2	0.50
Isoleucine	(Ile, I)	4.5	0.47

[a] Kyte and Doolittle (1982).
[b] Chou and Fasman (1978).

It often happens that there are ambiguities in the assignment of secondary structure to regions that may have high probabilities of adopting either helix or β-structure. There are additional rules by which these ambiguities can be resolved, including considering which type of secondary structure has the highest numerical probability for that sequence, the length of the sequence (α-helices, being more compact structures, typically are longer than regions of β-structure), and so forth. Garnier et al. (1978) developed a method, again based on a statistical analysis using a known set of protein structures, in which position-dependent probability values are used to generate a numerical value which specifies helix, β-structure, β-turn, or random coil for each residue of the protein. An algorithm based on this method is also often included in protein sequence analysis software packages.

It is not currently possible to predict the secondary structure of a protein from the sequence alone with absolute certainty. These predictive methods generally have an overall accuracy of about 70% when applied to proteins of known structure that are not in the "knowledge set" used to generate the predictive rules, that is, about 70% of the secondary structure present is predicted in an overall sense. Helical and turn structures are reasonably well predicted, probably in part because they require only one strand of peptide, while β-structures, which require the presence of at least one complementary strand, are not as well predicted. It must be remembered that these methods rely on statistical analyses and empirically determined rules. Predictions of overall helical content or β-sheet content may be reasonably accurate, but details such as boundaries are generally much less precise. There are many sequences that are ambiguous as to secondary structure preferences and, indeed, many peptide sequences are able to adopt different conformations or structures depending on their environment, solvent, and so on. Also, while random coil or irregular structure can be inferred from the absence of any propensity to form a regular secondary structure, there is no conformational or structural information available for such sequences to aid in model building. Finally, prediction of secondary structure in and of itself does not necessarily yield much information on tertiary structure or protein topology. Secondary structure prediction is best utilized as a guide in model building or in selection of peptide targets in conjunction with other techniques, such as sequence homology, chemical modification, proteolytic digestion, site-specific mutagenesis, and so on.

Hydrophobicity Measurements
Given that soluble proteins fold in such a way as to minimize the exposure of hydrophobic side chains to the aqueous medium, it is possible to infer portions of the protein sequence that are likely to be buried or to be exposed on the protein surface by examining the relative hydrophilicity or hydrophobicity of the protein sequence. A comprehensive discussion of hydrophobicity methods with specific instructions can be found in Krystek et al., 1995a.

The method most widely available in protein sequence analysis software is that of Kyte and Doolittle (1982). They devised a set of values representative of the free energy change of each amino acid side chain going from a polar to a nonpolar environment. These values are averaged over a window (generally six to nine residues), and these average hydropathy values are plotted against the residue number of the first residue in the window. Positive values indicate hydrophobic segments of the protein, and negative values indicate hydrophilic regions. Segments of the protein chain with a strongly hydrophobic character correlate well with the interior regions of the protein, and those with a strongly hydrophilic character correlate well with regions on the surface. The Kyte–Doolittle hydrophobic propensity values are listed in table 2-6. An example of the graphical output from a Kyte–Doolittle analysis is shown in figure 2-14 (pages 44–45).

Hydrophilicity analysis of protein sequences has been used to predict antigenic determinants or epitopes in proteins. As one might intuitively expect, most antigenic determinants in proteins (though not all) are found in regions exposed on the surface of the protein molecule. Hopp and Woods (1981) modified a set of hydrophilicity values for amino acids ultimately derived from octanol-water partition coefficients, used them to calculate the hydrophilicity of protein sequences over a six-residue window, and plotted the results against the residue number. They found a good correlation between peaks of hydrophilicity and the known antigenic sequences of several proteins, although not all highly hydrophilic segments corresponded to known antigenic regions, and not all antigenic determinants corresponded to the highest hydrophilic peaks. The region of highest hydrophilicity did always correspond to an antigenic determinant in the set of 12 proteins examined. This method is useful for examining proteins of unknown structure, both to suggest peptides for epitope mapping studies or antibody production and also to look for other probable surface features, such as receptor binding sites, enzyme cleavage sites, or sites for post-translational modification.

From the earlier discussion on amphipathic secondary structures (see above, Amphipathic Structures), it is clear that sequences can exhibit an overall hydrophobic or hydrophilic profile that is rather flat and yet fold into structures that present very hydrophobic or hydrophilic faces. In order to evaluate this type of behavior, Eisenberg et al. (1982) developed the concept of a hydrophobic moment, which can be calculated for a protein sequence having a given secondary structure. For a helix, the hydrophobicity of a side chain is represented by a vector projecting outward from the helix, and its magnitude reflects the magnitude of the hydrophobicity. Its direction is determined by its position in the helix; successive side chains will be offset from each other by 100° in an α-helix. Hydrophilic side chains give rise to a hydrophobic vector of the same magnitude but pointing 180° back from the side chain projection. A hydrophobic moment can be calculated by averaging the magnitude and direction of these vectors over a window of residues. A large moment reflects the segregation of hydrophobic residues along one face of the helix and hydrophilic on the other. A similar calculation can be applied to β-sheet structure. In this case, the angular offset for succeeding residues is 160°.

Eisenberg et al. (1982) calculated hydrophobic moments for a series of helices in 26 different proteins of known structure and compared those values to the average hydrophobicity. The moment and average hydrophobicity were calculated for the entire helix sequence, so each different helix sequence gave a single value for moment and hydrophobicity. Buried or membrane-spanning helices tended to have large hydrophobicity and low hydrophobic moment, as one might intuitively expect, since they exist in a uniformly hydrophobic

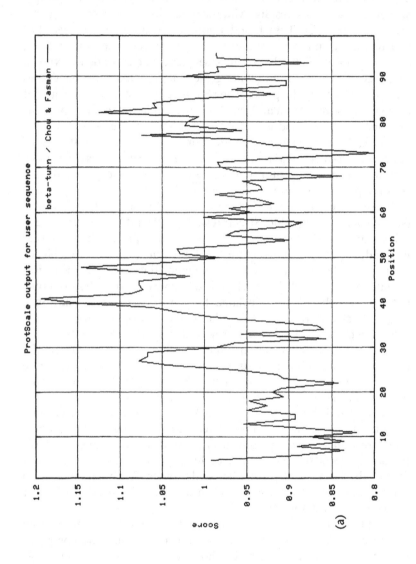

(a)

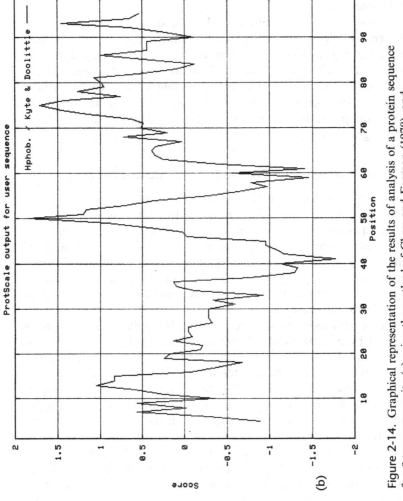

Figure 2-14. Graphical representation of the results of analysis of a protein sequence for β-turn propensity (a) using the method of Chou and Fasman (1978), and hydrophobicity (b) using the method of Kyte and Doolittle (1982).

environment. Amphiphilic helices have low overall average hydrophobicity but possess a large hydrophobic moment. They are found at interfaces between hydrophobic and hydrophilic environments, on the surface of proteins, or in peptides such as mellitin, which associate with plasma membranes. The asymmetric distribution of polar and nonpolar side chains allows these structures to interact favorably with both polar and nonpolar environments when they are situated at the interface.

In practice, with proteins of unknown structure, a hydrophobic moment is calculated over a small window, typically six residues, first assuming an α-helical structure and then assuming a β-sheet structure (Eisenberg et al., 1987). These are both plotted against residue number and compared with hydrophobicity and Chou–Fasman secondary structure calculations. In that way, one can identify probable amphipathic helices and β-structures and probable membrane-spanning helices.

Application of Sequence Analysis to Modeling and Design

Construction of detailed molecular models of peptide or protein structure based on sequence data is most fruitful when there is an X-ray crystal structure of a related protein available to use as a template. When the relatedness of the template protein to the protein of unknown structure is relatively distant, the use of secondary structure prediction, hydropathy measurements, and a certain amount of intuition also come into play. This convergent approach is illustrated in the following examples.

The complement protein C5a is a 74-residue protein that is released by proteolysis of a larger precursor during the complement activation cascade. It is thought to be involved in inflammatory responses by recruiting monocytes and polymorphonuclear leukocytes by chemotaxis and activating them to release free radicals and degradative enzymes. Inhibitors of C5a activity can have therapeutic application in the treatment of inflammatory diseases such as rheumatoid arthritis.

In an attempt to develop C5a antagonists, it would be very useful to have a model of the C5a structure. At the time, there was no X-ray crystal structure of C5a available, but there was a crystal structure of the related complement protein, C3a. Greer (1985) used this structure as a template on which to construct a model of C5a. The initial difficulty was that the sequences of C5a and C3a were only about 36% identical. Greer chose to begin the alignment with the central region of C3a, which was well defined in the crystal structure, and chose an alignment with C5a that would not cause any deletions or insertions in the helical regions. This alignment would not otherwise be considered the optimal one strictly on the basis of sequence. The model built using this alignment showed that the internal residues, which form the core of the structure, are, in fact, highly conserved, and that the residues on the surface are not conserved. Thus, while the overall similarity between the two proteins is only modest, the residues important for defining the structure are highly conserved. However, they do not appear adjacent to each other in the linear sequence, so sequence homology does not reveal this close relationship.

The second difficulty encountered was that the C3a amino terminus did not appear in the structure, possibly because it was too flexible in the crystal to give rise to substantial electron density. Thus, there was no template for the amino terminus of C5a. Secondary structural analysis suggested that the amino terminus of C5a had a strong potential for helical structure. When that region was folded into a helix, it was found to be strongly amphipathic, so it was reasonable that the hydrophobic face of this helix interacted with an accessible hydrophobic region elsewhere on the C5a molecule. One such hydrophobic patch was located on the surface in a region where the residues were highly conserved in human and porcine C5a and where the amino terminal helix could be docked.

This yielded a working model for the C5a structure where the core structure was identical to that of C3a, but the surface residues had considerable difference, which is consistent with the two proteins' distinct biological activities. The structure was also consistent with known pieces of structure–activity data. Porcine and human C5a will both bind to human C5a receptor, and there is a region on the surface of both molecules that is relatively conserved. This region is spatially near the C-terminal Arg residue, which is important for biological activity but not for binding to the receptor. When the three-dimensional solution structure of C5a was determined using 2-D NMR techniques and distance geometry methods (Zuiderweg et al., 1988), the proposed model was found to be in very good agreement with the deduced solution structure with respect to both the core structure, the helical conformation, and placement of the amino terminal sequence. The carboxyl terminus of the protein, however, seems to be highly mobile in solution and does not adopt a fixed conformation.

A similar approach was used to create initial structural models for the human immunodeficiency virus (HIV-1) protease. HIV-1 protease is responsible for the proteolytic maturation of the HIV structural proteins, reverse transcriptase, integrase, and RNase H, which are encoded by the *gag* and *pol* genes. It was initially known that many related retroviruses encode such a protease that processes the *gag* and *pol* translation products and that the HIV *gag* and *pol* proteins were indeed produced as polyprotein precursors. An approximate location in the genome for the protease was determined by protein sequence homology comparisons of the *gag* and *pol* regions of a number of retroviruses, considering especially two highly conserved tripeptide sequences (Asp-Thr-Gly and Ile-Ile-Gly) (Yasunaga et al., 1986). This was confirmed, and the exact location of the protease and its primary sequence were unambiguously determined when the gene was subsequently cloned and expressed (Debouck et al., 1987).

The tripeptide sequence, Asp-Thr-Gly, which is conserved in retroviral proteases, is also a conserved sequence in aspartyl proteases, and, in fact, represents the catalytically active aspartic acid residues of the active site. Aspartyl proteases are relatively well-studied enzymes, and a number of crystal structures are available. They consist of two domains, each of which contributes one Asp to the active site. The two domains have a fair degree of sequence homology, and both domains have a similar three-dimensional structure, suggesting that aspartyl proteases arose by gene duplication and fusion (Tang et al., 1978). The retroviral proteases are only about half the size of eukaryotic aspartyl proteases and contain only one Asp-Thr-Gly sequence. Pearl and Taylor (1987) hypothesized that the retroviral protease represents one lobe of an aspartyl protease, and that the protein forms a symmetrical dimer as the active protease. Using a combination of protein sequence comparisons among a large number of aspartyl proteases and secondary structure prediction, they were able to construct a model for the HIV-1 protease as a symmetrical dimer based on the crystal structure of the fungal aspartyl protease endothiapepsin. The dimeric nature of the HIV-1 protease and its identification as an aspartyl protease was later confirmed experimentally (Meek et al., 1989). The model predicted the general structure of the protease as four strands of β-sheet and a central hydrophobic cleft formed at the dimer interface that contains the substrate binding site and the active site aspartyl residues. This overall structure was confirmed when X-ray crystal structures of the HIV-1 protease were obtained (Lapatto et al., 1989; Wlodawer et al., 1989), although some important structural details varied. For example, the model was based on a sequence that incorrectly located the mature amino terminus of the protease and, therefore, failed to predict the intermolecular β-structure between the amino terminus of one monomer and the carboxyl terminus of the other monomer, which forms an important part of the dimer interface. Nevertheless, early recognition that the HIV-1 protease was likely to be a symmetrical dimer with a symmetrical active site led to the design of novel, symmetrical inhibitors of high potency (Erickson et al., 1990).

Peptide Design Based on Sequence

It has been a long-standing goal in peptide chemistry to be able to design short peptides that mimic some important aspect of protein structure or function. The peptides can be designed to be agonists of a biologically active protein by encompassing those residues necessary for activity. Antagonists can be designed using peptides containing only those residues required for binding to a receptor, a substrate, or a required protein subunit. (Fibrinogen receptor antagonists, which fall into this category, will be discussed in more detail later; see below, Cyclization.) When one has antibodies against the protein that neutralize the protein's activity, epitope mapping (or locating the antibody combining sites) can help identify protein sequence portions involved in the biological activity. Synthetic peptides based on the sequences of neutralizing antigens can be used as synthetic antigens in the preparation of antisera and vaccines. Conversely, antibodies can be produced with synthetic peptides that recognize specific proteins simply from a knowledge of the gene sequence without ever having isolated the protein.

Designing peptide agonists based on protein sequences is a formidable task. With increasing size of the polypeptide chain, it becomes increasingly likely that residues important for biological activity will be in close proximity in the three-dimensional structure but widely separated in the primary sequence. This makes design of a small peptide that can mimic that arrangement problematic, especially when there may not be any three-dimensional structural information available about the target protein. The case is often easier with peptides of moderate length, since there is often a minimum active fragment that retains a substantial portion of the biological activity. This can be illustrated with the case of gastrin-releasing peptide and bombesin.

Gastrin-releasing peptide (GRP) is a 27-amino-acid peptide originally isolated from porcine gut on the basis of its ability to cause release of gastrin (McDonald et al., 1979). Its C-terminus has a striking homology with bombesin, a 14-residue peptide isolated from frog skin (Anastasi et al., 1971), as well as a number of related peptides of amphibian origin. The sequences of GRP and bombesin are shown in table 2-7. Both GRP and bombesin produce a number of biological responses, including hypothermia or lowering of the core body temperature in rats exposed to cold. A series of truncated analogs was prepared and assayed for hypothermia activity in order to determine the minimum sequence necessary for agonist activity (table 2-7; Rivier and Brown, 1978; Märki et al., 1981). Deletions from the conserved C-terminus impaired activity. In contrast, the peptides could be shortened considerably from the amino terminus without loss of

Table 2-7 Gastrin-releasing peptide and bombesin analogs

Peptide	Sequence	Relative activity
GRP	A-P-V-S-V-G-G-T-V-L-A-K-M-Y-P-R-G-N-H-W-A-V-G-H-L-M-NH$_2$	0.28
Bombesin	<E-Q-R-L-G-N-Q-W-A-V-G-H-L-M-NH$_2$	1.00
	<E-Q-R-L-G-N-Q-W-A-V-G-H-L-M-OH	<0.01
	<E-Q-R-L-G-N-Q-W-A-V-G-H-NH$_2$	<0.01
	Ac-G-N-Q-W-A-V-G-H-L-M-NH$_2$	1.00
	Ac-N-H-W-A-V-G-H-L-M-NH$_2$	0.50
	Ac-H-W-A-V-G-H-L-M-NH$_2$	1.00
	Ac-W-A-V-G-H-L-M-NH$_2$	0.04

<E = pyroglutamic acid.

activity, and full activity was obtained with just the C-terminal octapeptide sequence.

Peptides derived from an internal protein sequence may require blocking groups at the amino and/or carboxyl terminus to mask the presence of charges that are not present in the native protein. For example, the HIV-1 protease mentioned earlier (see above, Application of Sequence Analysis to Modeling and Design) is an endopeptidase whose substrate is a viral polyprotein. One of the cleavage sites is encompassed by the heptapeptide Ser-Gln-Asn-Tyr-Pro-Val-Val, with cleavage occurring between the Tyr and Pro residues (Moore et al., 1989). A series of peptides of varying sizes were prepared either with N-acetyl and carboxamide blocking groups or free amino or carboxyl termini and their substrate activity compared. The results are shown in table 2-8, where K_m is the Michaelis binding constant, k_{cat} is the catalytic rate constant, and k_{cat}/K_m is a measure of catalytic efficiency. The fully blocked heptapeptide (4), the free amino peptide amide (5), and the acetyl peptide free acid (6) all have fairly similar activities. Removal of the C-terminal Val residue (peptide 7) causes a substantial loss in k_{cat} and catalytic efficiency, and removal of both Val residues (peptide 8) results in an inactive peptide. Truncation at the amino terminus causes a slight loss in activity in hexapeptide 9 and pentapeptide 11, with a complete loss of activity when tetrapeptide 13 is reached. The free amine hexapeptide 10 is as active as its acetylated parent, but the free amine pentapeptide 12 again shows a substantial loss in k_{cat} and catalytic efficiency. These data collectively suggest that the active site of the HIV-1 protease will accommodate a hexapeptide, stretching three residues on either side of the cleavage site. Charged groups and the presence of additional residues have little effect on substate activity since they extend out of the active site. The loss of activity in uncovering the free amino group in the pentapeptide 12 is probably due to the presence of a

Table 2-8 HIV-1 protease substrates

Compound	Sequence	K_m (mM)	k_{cat} (sec^{-1})	k_{cat}/K_m (mM$^{-1} \cdot$ s^{-1})
4	Ac-Ser-Gln-Asn-Tyr-Pro-Val-Val-NH$_2$	5.5	54	9.8
5	H-Ser-Gln-Asn-Tyr-Pro-Val-Val-NH$_2$	1.2	14.0	11.7
6	Ac-Ser-Gln-Asn-Tyr-Pro-Val-Val-OH	5.0	25.7	5.1
7	Ac-Ser-Gln-Asn-Tyr-Pro-Val-NH$_2$	4.9	0.49	0.090
8	Ac-Ser-Gln-Asn-Tyr-Pro-NH$_2$	neg.	neg.	neg.
9	Ac-Ala-Asn-Tyr-Pro-Val-Val-NH$_2$	13.5	9.3	0.69
10	H-Ala-Asn-Tyr-Pro-Val-Val-NH$_2$	18	29.7	1.6
11	Ac-Asn-Tyr-Pro-Val-Val-NH$_2$	17.1	37	2.2
12	H-Asn-Tyr-Pro-Val-Val-NH$_2$	12.8	0.88	0.068
13	Ac-Tyr-Pro-Val-Val-NH$_2$	neg.	neg.	neg.
14	H-Tyr-Pro-Val-Val-NH$_2$	neg.	neg.	neg.

charged group in the generally hydrophobic active-site cleft. This interpretation of the extent of the enzyme active site is confirmed by X-ray crystal structures of co-crystals of the protease containing substrate-based inhibitors bound in the active site (Miller et al., 1989; Wlodawer et al., 1991).

Peptide based protease substrates can be made more sensitive and easier to measure by the incorporation of fluorescent groups. Matayoshi et.al. (1990) made an HIV protease substrate that incorporated two fluorescent groups that internally quenched each other by resonance energy transfer. This substrate had the structure, DABCYL-Ser-Gln-Asn-Tyr-Pro-Ile-Val-Gln-EDANS. In this case, the fluorescence donor is EDANS (5-(2-aminoethylamino)-naphthalene sulfonic acid) and the quenching acceptor is DABCYL (4-dimethylamino-azobenzene 4'-carboxylic acid). When the substrate is cleaved at the Tyr-Pro bond by the protease, the two fluorescent groups are no longer held in close proximity so the quenching of the EDANS signal by DABCYL is significantly decreased. The resulting increase in fluorescence emission is used to monitor the course of the reaction. In this case, the fluorescence increased by 40-fold upon cleavage of the peptide and a K_m of 0.1 mM and a k_{cat} of 4.9 s^{-1} was determined for HIV protease. It thus appeared that the presence of the EDANS and DABCYL groups confered improved substrate binding properties over those found for the peptides in table 2-8. This is not always the case, however. Incorporation of fluorescent groups into a peptide substrate for poliovirus 3C protease (Weidner and Dunn, 1991) reduced the rate of cleavage by approximately 50-fold.

The addition of a charged moiety, remote enough from the active site that it does not interfere with catalytic activity, can be advantageous in terms of solubility. Succinyl groups are often added to the amino terminus of enzymes' substrates, for example, the chymotrypsin substrate succinyl-Ala-Ala-Pro-Phe-*p*-nitroanilide (DelMar et al., 1979). This modification not only improves solubility of hydrophobic peptides but also blocks potential aminopeptidase degradation at the amino terminus.

Epitope Mapping and Production of Protein-Specific Antibodies from Peptides

Epitope Mapping
One of the most common applications of peptides based on protein sequences is epitope mapping. Determination of antibody combining sites on proteins is used to study the mechanism of immunogenicity of proteins as well as to determine protein topology (on the assumption that most epitopes will be located at the protein surface). If there are monoclonal antibodies that block the protein's activity, epitope mapping can help determine the parts of the protein sequence involved in the expression of biological activity. If the protein is derived from a pathogenic organism, peptides based on neutralizing epitopes can be the basis of synthetic vaccines.

Determination of an antigenic epitope relies on preparation of a peptide representing a portion of the protein sequence and demonstrating competition with the protein in antibody binding. There are several approaches to determining which sequences to examine as potential epitopes. The complete approach would require synthesizing overlapping short peptides, generally hexapeptides, that cover the entire protein sequence. Several techniques for multiple solid-phase peptide synthesis make this approach possible (see chapter 3). Houghten (1985) devised a method in which small amounts of aminoacyl resin are sealed in small polystyrene mesh bags called "T-bags." The bags can be labeled with an indelible marker for identification. The common steps of washing, deprotection, and neutralization can be performed together for a number of these T-bags, which are then separated only for the coupling of the individual amino acids and at the end of the synthesis for HF (hydrogen fluoride) cleavage. While there is still a fair amount of labor involved, it is possible to turn out 50-mg quantities of a large number of peptides in a relatively short time. Houghten's method is advantageous because the peptides can be isolated, purified, and characterized, but characterization clearly becomes rate-limiting if a sufficiently large number of peptides is being prepared.

An alternative approach has been developed by Geysen et al. (1988). The peptides are synthesized on the tips of polystyrene pins arranged on a support such that the tips can be immersed in the wells of a standard 96-well microtiter plate. As with Houghten's approach, steps such as washing and deprotection are carried out in common while the couplings take place in the microliter wells. In the original conception, an ELISA (enzyme-linked immunosorbent assay) type assay was used so that the results could be read colorimetrically in microliter plates. The pins were incubated with the probe antibody, washed, incubated with the conjugate antibody (horseradish peroxidase coupled to anti-IgG for the appropriate species in which the probe antibody was raised), washed again, and then immersed in wells containing the enzyme substrate. Color developed only in those wells where the pins bound the original probe antibody. A large amount of labor is involved in this synthesis approach, also, but an extremely large number of peptides can be prepared in a short amount of time. The advantage to this approach is that, in theory, the pins can be reused with another probe antibody if the original peptide–antibody complex can be disrupted. The disadvantage is that the peptides remain attached to the pins and cannot be characterized; there is the possibility that the derivatized polystyrene matrix of the pin could affect the interaction of peptide and antibody. A recent modification to this method involves the attachment of a protected Lys-Pro to the pins and peptide assembly on the ε-amino group of the lysine. The peptide can be released from the pins by mild base treatment, giving the free peptide with a pendant diketopiperazine moiety (Bray et al., 1990).

One approach to limiting the number of peptides that must be prepared for epitope mapping has already been mentioned, namely, the use of a hydropathy profile of the protein sequence (see above, Hydrophobicity Measurements). This approach is based on the supposition that most epitopes will be found on the protein surface and that relatively hydrophilic stretches of the protein sequence are more likely to be on the surface. One disadvantage of this method is that the hydropathy profile does not give an unambiguous or very detailed indication of protein topology. It is also true that all hydrophilic regions of the protein do not necessarily correspond to epitopes, that all epitopes are not hydrophilic or even on the protein surface, and that the major antigenic regions are not necessarily the most hydrophilic.

The more typical approach to epitope mapping involves using protein biochemistry and molecular genetics techniques to create fragments of the protein, narrowing down the location of the epitope by determining successively smaller fragments that cross-react with the target antibody. With modem techniques and the judicious use of restriction endonucleases, it is often possible to clone and express the protein of interest along with a library of partial sequences.

Using sensitive methods of detection such as Western blotting, it is possible to rapidly identify fairly small regions of protein sequence as containing a particular epitope, which can then be refined further through the use of synthetic peptides. A typical case is the identification of the principal neutralizing epitope of the HIV envelope protein.

The HIV envelope protein, gp120, is the viral ligand for the CD4 receptor protein on T4 lymphocytes and is involved both in binding of the virus to target cells and in fusion of the cellular and viral membranes, allowing viral genetic material to enter the cell. Antisera to gp120 have been shown to block both infectivity and fusion (Allen et al., 1985; Veronese et al., 1985). Using protein and peptide fragments produced by recombinant techniques, limited proteolysis and peptide synthesis, the major neutralizing epitope was localized with both polyclonal antisera (Rusche et al., 1988) and a monoclonal antibody (Matsushita et al., 1988) to a 24-residue peptide which occurs in a 36-residue disulfide loop on the gp120 molecule. Synthetic peptides corresponding to portions of this loop were prepared, guided partially by sequence homology comparisons using the sequences of gp 120 from different isolates of HIV-1. An octapeptide sequence (Ile-Gln-Arg-Gly-Pro-Gly-Arg-Ala) located at the middle of the loop was shown to be the minimum sequence that could block the effects of neutralizing antibodies and that, when coupled to a carrier protein, could itself generate neutralizing antisera (Javaherian et al., 1989).

Producing Protein-Specific Antibodies
Antibodies that recognize intact proteins can be produced with synthetic peptides whose sequences are based on short stretches of the protein sequence. There are certain considerations, based on general properties of protein structure, that go into the design of appropriate synthetic peptides for this purpose. This section will concern itself mainly with these design considerations while more detailed treatment on the their production can be found in a number of good references (Grant, 1999; Tam and Spetzler, 1997). A synthetic peptide that will be successful for the production of antiserum that will recognize an intact protein should be based on a stretch of sequence that is exposed to solvent. Moreover, that sequence should be a continuous stretch of amino acids. Although discontinuous epitopes certainly exist, designing a synthetic peptide to recognize them is much more complex.

An inspection of known protein structures indicates that their surfaces are populated with a predominance of charged and polar residues (Creighton, 1993). Therefore, it is reasonable to assume that stretches of protein sequences that contain proportionately higher levels of these amino acids will be found on the surface of proteins.

Turns or loop structures tend to be more useful for antibody production because they are generally found on the surface of proteins connecting larger arrays of helices and sheets and they consist of continuous stretches of amino acids. Many amino acid residues in helices and sheets are also found exposed to solvent, but since these structures are also usually found in larger arrays such as entended sheets and helical bundles, they are less appealing for this purpose.

These simple considerations form the basis for predicting and designing synthetic peptides for this purpose. The main problem is determining which part of the protein satisfies these criteria. Many different indices have been devised which predict hydrophilicity, hydrophobicity, secondary structure, segmental mobility, side-chain accessibility, and sequence variability (see Van Regenmortel et al., 1988). All of these methods generally tend to yield similar results. A good general approach is to use an index of hydrophilicity (which is usually just an inversion of an index of hydrophobicity) along with a prediction of secondary structure. The results are then compared and areas of high turn propensity and hydrophilicity are identified. The ExPASy web site of the University of Geneva offers free access to a variety of different programs over the internet at http://expasy.hcuge.ch/tools/. Note that these procedures were developed for, and work best with, proteins that are water soluble and composed of a single globular structure. They may not be as successful with multi-subunit proteins or membrane proteins which may have large sections shielded from the solvent.

An effective immunogen can usually be produced by coupling the peptide to a carrier protein or by synthesizing a "multiple antigenic peptide" or MAP (see below). MAPs are assembled on a branched lysine core producing either four or eight identical peptides branching from the α and ε amines of the terminal lysines of the core.

General Procedure

A general procedure for selecting potential sequences is outlined here.

1. Using the selected algorithms, compute the hydropathy index and the tendency for β-turns of your protein sequence. Use a "window" of 7 or 9 and give equal weight to each amino acid. A window size of 9 includes four amino acids on each side of the central amino acid for which the value is computed and which is the simple average of the values for each amino acid in the window.

2. Compare the results of the two analyses and look for areas of sequence that are high in turn tendency and high in hydrophilicity (or low in hydrophobicity). These areas will generally be sequences high in Lys, Arg, His, Asp, Glu, Ser, Thr, Asn, Gln, Pro, and Tyr content.

3. Examine the sequences for glycosylation site motifs and discard any sequences which contain them unless you know that the protein is not glycosylated. Amino-linked carbohydrate chains can occur at Asn-X-Ser or Asn-X-Thr sequences. Hydroxyl-linked carbohydrate chains do not appear to have a set motif but there are now some programs available that aid in their prediction.

4. The immediate amino-terminal and carboxyl-terminal regions of proteins are often exposed to solvent. If these areas appear to be hydrophilic in nature, they should be considered as acceptable candidates.

5. Mimicking the charge state of the ends of the peptide may be helpful in certain instances. For instance, unless the peptide is at the immediate amino or carboxyl end of the protein, its ends will be in the peptide linkage to the rest of the protein and thus will not be protonatable. Synthesis of the peptide with blocked termini (acetyl-amino or carboxyl-amide) may be helpful in mimicking this situation by reducing the presence of potentially undesirable charge effects at the peptide termini. However, note that it is important to make sure the peptide has sufficient additional charge for solubility.

6. Peptides of 10 to 15 residues are usually adequate for the production of antibodies. Longer peptides are more difficult and expensive to make and usually unnecessary. Peptides do not necessarily need to be rigorously purified before conjugation, especially for polyclonal antibody production, since the antisera will be selected on the basis of specificity as well as titer.

7. Including a cysteine residue at either the amino or the carboxyl end of the peptide can be very helpful for cross-linking to carrier protein with heterobifunctional reagents.

An example of an analysis using the ExPASy web site is shown in figure 2-14 for the 99-residue HIV-1 protease. The figure shows an analysis using the Chou and Fasman β-turn propensity index (a) and the Kyte and Doolittle hydropobicity index (b). One area in particular, that around residue 41, shows both a high level of hydrophilicity (negative hydrophobicity) and high propensity for a β-turn. This sequence, which is NLPGKWKPK, contains a high level of the amino acids expected on surface turns (see step 2 above) and would be an excellent candidate for producing an antibody to this protein.

Designing the Immunogen

A synthetic peptide is seldom able to produce a sufficient immunogenic response on its own. A common solution to this is to chemically couple the synthetic peptide to a carrier protein which will boost the immune response. A less used but equally effective approach is the direct synthesis of a covalent multimer of the simple peptide sequence. This is called a "multiple antigenic peptide" or MAP (Tam and Spetzler, 1997). Unlike simple peptides, where additional chemical manipulations after synthesis are required for coupling to a carrier protein, a MAP is complete and ready for immunization at the conclusion of the synthetic protocol. However, MAPs are more difficult to produce as completely homogeneous synthetic products, more difficult to analyze post-synthetically, and they may also be more prone to insolubility problems. Both four- and eight-branched MAPS peptides have been found to be effective, but four-branched MAPS are less prone to synthesis problems and are easier to characterize.

Synthetic peptides produced as immunogens should be checked for homogeneity by analytical HPLC (high-performance liquid chromatography) and correct mass by mass spectrometry. The characterization of MAPs can be more problematic due to their multi-branched nature (Mints et al., 1997). If each of the four to eight peptide chains of a MAPS molecule have only a small percentage of modification at any particular residue, the aggregate will produce a relatively broad spectra or distribution. This can compromise HPLC and mass spectrometric analysis by producing broadly eluting peaks and heterogeneous spectra. However, this feature of MAPs usually does not tend to compromise their ability to form antigens to the proper peptide since the correct sequence is usually present in high enough concentration that a significant amount of the specific antibody is produced among the polyclonal population. Amino acid analysis, which is less sensitive to multiple small differences, may give a more reasonable assessment of the composition of the MAP.

Design and Use of Combinatorial Peptide Libraries

With the use of combinatorial techniques, it is now possible to screen peptide libraries potentially representing millions of peptide sequences to optimize peptide structure–activity relationships (SAR) for receptor or enzyme targets or even to discover ligands for targets whose natural ligands are unknown. Peptides are ideal molecules for the application of combinatorial technologies. Synthetic protocols are highly optimized and largely independent of peptide sequence. There are extremely sensitive and powerful analytical methods available for determining peptide composition and sequence. In fact, combinatorial chemistry was first conceived with peptides (Geysen et al., 1986, 1987) and most of the methodologies in use today were adapted from peptide technology.

With the ability to screen large libraries of diverse peptide sequences, design based on specific amino acid properties becomes much less important. The major factors influencing peptide library design derive from the specific techniques employed. While a comprehensive review of combinatorial chemistry is beyond the scope of this chapter, a brief discussion of some of the most widely used peptide combinatorial techniques and their underlying assumptions, limitations and utility will be useful.

Methods of Synthesis and Elucidation of Sequence

Because of the ease of handling, peptide combinatorial libraries are most often prepared using solid phase peptide synthesis, either on traditional polystyrene supports (beads, pins or crowns) or on hydrophilic supports such as polyethylene glycol (PEG)-polystyrene (TentaGel or ArgoGel) or special-purpose supports like PEG-polyacrylamide (PEGA). An extensive listing of supports is found in Hudson (1999a). While a number of highly ingenious approaches have been developed for combinatorial chemistry, the utility of combinatorial chemistry is largely in screening, and the most straightforward methods have generally proven adequate.

The simplest way to introduce a diverse set of amino acids at a given coupling step is to utilize an equimolar mixture of protected amino acids during the coupling step, and this is in fact how the first peptide combinatorial libraries were prepared. One can verify by sequencing or by amino acid composition that essentially equimolar amounts of amino acids can be incorporated at any given position in a peptide sequence (Ostresh et al., 1994; Rutter and Santi, 1989). It is important to note that an equimolar representation of amino acids at each position in a combinatorial mixture does not necessarily mean that each possible peptide sequence is also present in the mixture at equimolar amounts. Results from deconvolution of these mixtures really represent concensus sequences, and there is a basic implicit assumption that each position in a peptide sequence can be optimized independently of the other positions in the sequence. To the first approximation, this is not an unreasonable assumption for peptides, especially at the level of screening where combinatorial chemisty is most useful.

A more rigorous way to prepare mixtures in which every possible sequence is equally represented is the split-and-mix technique described independently by Furka (1991) and Lam et al (1991). The general scheme for split-and-mix synthesis is shown in figure 2-15. The process is conceptually straightforward and uses traditional solid-phase peptide synthesis. To create a library with 10 different amino acids, for example, the resin is split into 10 equal aliquots and each aliquot is coupled separately with one of the 10 different amino acids. The couplings can be monitored for completeness and recoupled as necessary. The 10 aliquots are then recombined, thoroughly mixed and then split again into 10 aliquots for the second round of coupling. Each aliquot now contains an equimolar mixture of the 10 different aminoacyl resins. The process can be repeated until the library is of the desired size.

There are two important things to bear in mind about split-and-mix libraries. The first is that the splitting or portioning of resin introduces

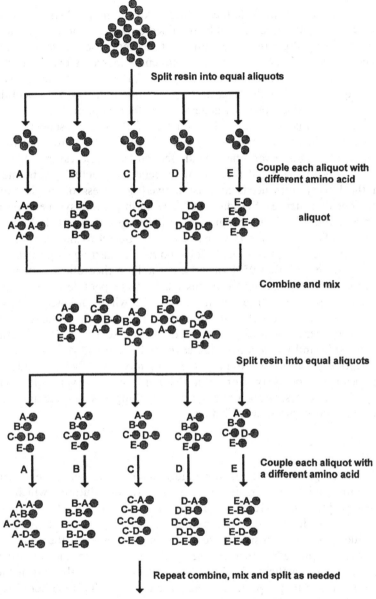

Figure 2-15. Split-and-mix approach to combinatorial library synthesis, illustrated for a 5 × 5 library. If the number of resin beads is large compared to the number of aliquots the resin is split into, each aliquot will contain an equimolar distribution of amino acids at each step. Note that all possible sequences are represented. At the end of the process, each resin aliquot contains a mixture of peptides with a common N-terminal residue. If the aliquot is cleaved, a mixture of peptides will be produced. Note also that each resin bead contains a unique peptide sequence. If the beads can be segregated and assayed individually, information can be obtained for individual sequences rather than for mixtures.

an element of statistics into the synthesis. An equimolar distribution of the various peptidyl resins will be obtained only if the number of beads is appreciably larger than the number of theoretical sequences present. Because the number of theoretical sequences increases exponentially with the length of the combinatorial peptide, this can become an issue quite quickly. Using all 20 naturally occurring amino acids, a peptide with four variable positions generates 160,000 theoretical sequences; for five variable positions, the number of theoretical sequences is 3,200,000. There have been a number of calculations as to what fold excess of beads over theoretical sequences is necessary; for a probability that 95% of theoretical sequences are actually contained in the library, one needs at least a five-fold excess of beads over theoretical sequences. Standard 100–200 mesh peptide synthesis resin contains about 106 beads per gram of resin (Burgess et al., 1994). For the example of a peptide library with five variable positions and using all 20 naturally occurring amino acids, it would require starting with 16 grams of resin to ensure that 95% of those 3.2 million theoretical sequences were actually represented! For this reason large peptide libraries are usually prepared as consensus libraries, which will be discussed below.

The second property of split-and-mix libraries is that each individual bead contains just a single peptide sequence. If the resin from a split-and-mix synthesis is cleaved in bulk, it will yield a mixture of peptides, just as libraries prepared using mixtures of amino acids. If the beads can be assayed on an individual basis, however, individual peptide sequences can be obtained directly. Single-bead screening will be discussed in more detail below.

Iterative Deconvolution

The most general method for determining active sequences from screening mixture libraries is iterative deconvolution, which is illustrated in figure 2-16. A set of mixtures in which one of the variable positions is defined in each mixture is assayed and the optimal residue(s) at that position are determined. Each of the variable positions is probed in turn by the synthesis of ever smaller mixtures in which each of the optimal residues is held fixed. At the end of the process, individual peptide sequences are obtained. This methodology was the initial deconvolution technique introduced in the first combinatorial libraries and has been used extensively and successfully (Berk et al., 1999).

There are a few points to consider about iterative deconvolution. The first is that, in the first round of screening, very large mixtures are being evaluated. If the activity has very stringent sequence requirements, a very small portion of the library components will be active and the assay will need to be sufficiently sensitive to detect what may

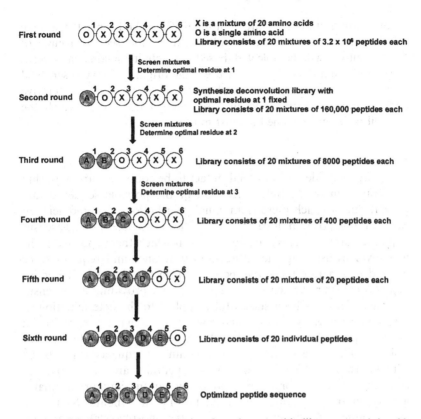

First round

1 2 3 4 5 6
O X X X X X

X is a mixture of 20 amino acids
O is a single amino acid
Library consists of 20 mixtures of 3.2 x 10⁶ peptides each

Screen mixtures
Determine optimal residue at 1

Second round

1 2 3 4 5 6
A O X X X X

Synthesize deconvolution library with
optimal residue at 1 fixed
Library consists of 20 mixtures of 160,000 peptides each

Screen mixtures
Determine optimal residue at 2

Third round

1 2 3 4 5 6
A B O X X X

Library consists of 20 mixtures of 8000 peptides each

Screen mixtures
Determine optimal residue at 3

Fourth round

1 2 3 4 5 6
A B C O X X

Library consists of 20 mixtures of 400 peptides each

Fifth round

1 2 3 4 5 6
A B C D O X

Library consists of 20 mixtures of 20 peptides each

Sixth round

1 2 3 4 5 6
A B C D E O

Library consists of 20 individual peptides

1 2 3 4 5 6
A B C D E F

Optimized peptide sequence

Figure 2-16. Iterative deconvolution for a hexapeptide library containing 20 different amino acids. At the first round each mixture is quite large, so the assay must be sensitive enough to detect what may be a weak activity in the mixture. Activity should increase at each successive round as optimized residues are held fixed and the number of peptides in each mixture decreases. The deconvolution process can be speeded up by holding two residues fixed at each round rather than one. This requires more robust parallel synthesis technology since in this example each library will now consist of 400 mixtures rather than 20, meaning that 400 individual syntheses rather than 20 are needed at each step. However the size of the mixtures also decreases by a factor of 20 at each round, so the assay sensitivity requirements are also reduced.

be very weak activity in the mixture. More seriously, deconvolution may also be confounded if the opposite is true and there are many weakly active peptides in the mixture, in which case the activity of the mixture is the sum of those weak activities (Terrett et al., 1995a, b). Specific activity does not increase with successive rounds of iterative deconvolution and optimized individual sequences cannot be identified. Finally, since resynthesis is required for iterative deconvolution, it does require additional effort. And, since the specific sublibraries which will be resynthesized in the second round of deconvolution will be selected based on activity against a specific target and are not likely to be of use if the original library is screened against a second target, the overhead of resynthesis makes iterative deconvolution less attractive in a screening paradigm in which the same libraries are screened against multiple targets.

Consensus Libraries

Since large peptide libraries tend in fact to be consensus libraries, that is, libraries in which only a fraction of the possible sequences are present but in which there is an equal distribution of amino acids at each variable position, it can be more efficient to determine consensus sequences directly. There are two ways in which this is possible. The first involves assaying the entire library at once and requires some method to separate the active peptides from the inactive. The active peptides can then be sequenced as a mixture, providing a consensus sequence. This has been successfully applied to the determination of kinase substrates by Cantley and coworkers (Songyang et al., 1994; Nair et al. 1995). A peptide library in which the phosphate acceptor residue is fixed at a specific position within the sequence is incubated with kinase and ATP. Phosphorylated peptides are then separated from the nonphosphorylated peptides by careful column chromatography on an iron (Songyang et al., 1994) or molybdate (Nair et al. 1995) column, which selectively retains the phosphorylated peptides. Careful elution of the retained peptides followed by sequencing of the mixture provides consensus sequences which, when resynthesized individually, have provided highly active substrates. A similar result is obtained with affinity selection, in which a library is passed through a target which is immobilized in some fashion (Chu et al., 1995) (covalently linked to a column or focused by isoelectric focusing, for example). Peptides which interact with the target elute later and can be collected and sequenced. The advantage of this method is that extremely large libraries can be probed in a single experiment, minimizing both effort and biological reagent. The disadvantages are that not every system is amenable to the separation of the active peptides, and the target must be sufficiently promiscuous to recognize a consensus sequence. The concentration of any one peptide in such a mixture is exceedingly small, so it is relatively unlikely to detect a peptide if the sequence requirements for activity are very stringent.

Scanning Libraries

A specific type of consensus library is the scanning library (Pinilla et al., 1992; 1994). In this configuration there is one sublibrary for each

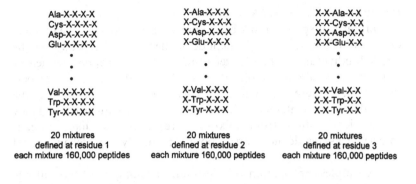

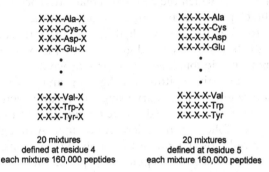

Figure 2-17. Scanning library for a pentapeptide sequence using the set of 20 naturally occurring amino acids. The library consists of $5 \times 20 = 100$ mixtures. Each mixture consists of $20 \times 20 \times 20 \times 20 = 160,000$ peptides. Screening the 100 mixtures identifies a consensus sequence. The complete library represents 3.2 million sequences and can be deconvoluted without resynthesis by assaying only 100 mixtures. The assay must be sufficiently sensitive to detect activity in a mixture of 160,000 peptides. Targets with extremely stringent sequence requirements may be less amenable to this approach.

amino acid at each variable position in the peptide, as shown in figure 2-17. Thus for a library using all 20 amino acids at each of five variable positions, there would be $20 \times 5 = 100$ sublibraries, each with a different amino acid fixed at one of the five variable positions. The efficiency of this arrangement is again readily apparent. A 20×5 library encodes 3.2×10^6 sequences, but they can be evaluated in 100 assays. The results of the assay would identify the best amino acid(s) at each of the various postions, allowing construction of a consensus sequence(s). The same caveats apply here as with other large mixtures (in the example each sublibrary would consist of 160,000 theoretical components) in that the target must be promiscuous enough to recognize a consensus sequence.

One-Bead, One-Compound Screening

The product of split-and-mix bead libraries is a mixture of beads each one of which bears a single peptide sequence. If the beads can be assayed individually, one can get results for single peptide sequences rather than results extrapolated to single peptides from assays performed on mixtures of sequences. This is an important point if the particular peptide interaction has a high degree of sequence stringency where identification from mixture screens is less likely. Bead-associated assays do however impose a relatively high technological barrier.

The simplest application of one-bead, one-compound screening is an on-bead assay in which the peptides remain attached to the bead, a technique first applied in probing antibody specificity (Lam et al., 1991). This requires the use of a hydrophilic support like TentaGel. In this case the beads were treated in bulk with a fluorescently labeled antibody, but any double-labeling technique would be appropriate. Beads bearing an active epitope will light up and can be physically picked with tweezers or a capillary using a dissecting microscope. The peptide sequence is determined by removing the antibody with a guanidine wash followed by microsequencing. Although PEG-PS resins are typically lower loading than polystyrene resins, each bead still typically contains 100–300 pmol of peptide, which is more than sufficient for solid-phase microsequencing, which generally requires less than 10 pmol of material.

On-bead analysis has also been used to probe enzyme specificity. In the case of kinases (Wu et al., 1994), beads are incubated in bulk with the kinase and ^{32}P-ATP to specifically label substrate-bearing beads. The beads are then washed and immobilized in a dilute gelatin matrix onto autoradiographic film or a phosphorimager to detect and pick the labeled beads. Typically this process is applied several times until the labeled beads are sufficiently dispersed to allow individual beads to be picked. In the case of proteases, fluorescence-quenched substrates have been prepared on resin (Singh et al., 1995). These substrates contain a fluor at the *C*-terminal end and a quencher at the *N*-terminal end, so that the intact peptides on the resin are not fluorescent. Endo proteolysis releases the quencher into solution and the *C*-terminal fragment which remains attached to the resin bead then fluoresces, allowing substrate containing beads to be picked. If proteolysis is not allowed to proceed to completion, microsequencing reveals both the complete substrate sequence as well as the cleavage site. Application of this method requires that there be no residual fluorescence on the resin support, which unfortunately is not the case with the PEG-PS resins, prompting Meldal (1994) to design a PEG-polyacrylamide (PEGA) resin specifically for this purpose.

The advantage of these on-bead assays is that they can be performed in a single experiment with bulk resin, since the labeled beads can be selected out of the mixture. One caveat with on-bead assays is that the support may influence the interaction with the target in unpredicatable ways. Another caveat is that while the peptide loading on the bead may be only 100–300 pmol, this can still translate into a concentration approaching millimolar levels, so it is quite possible to select low-affinity interactions which may not be reproduced when soluble versions of the same peptides are prepared.

Not every biological target is suitable for on-bead screening, so various techniques have been developed to remove the peptide from the bead before assaying. Ideally, this is accomplished in such a way that some residual peptide remains on the bead for identification by solid-phase sequencing. The trick is being able to perform the cleavage in such a way that the peptide still remains associated with the resin until it is released at the time of the assay. Cleavage of the resin in situ in the assay is usually not possible because of the harsh nature of most linkage cleavage conditions. Lerner (Quillan et al., 1995) has developed an interesting system in which peptides prepared on typical polystyrene resins are treated with neat gaseous trifluoroacetic acid (TFA), which effects a small amount of cleavage. Vacuum removal of the TFA leaves the cleaved peptide adsorbed to the resin beads. This technique has been used in melanophore-based assays for G-protein coupled receptor (GPCR) ligands. Pigmentation of frog skin melanophores is under the control of a GPCR system and transient expression of an exogenous GPCR into melanophores allows a simple visual read-out of ligand activity. Resin is adhered to plastic wrap, cleaved with TFA vapor and then layered over melanophores cultured in soft agar. Sufficient peptide diffuses out from the bead to cause a local response in melanophores immediately adjacent to beads containing an active ligand, allowing those beads to be picked and sequenced.

A more general approach to partial release of peptides from resin involves use of multiple selectively-cleavable linkers on the resin such that under some conditions a controlled amount of peptide can be released from the resin beads while retaining sufficient peptide covalently attached to allow for subsequent microsequencing (Patek and Lebl, 1999). With judicious choices of linkers, it is possible to design a resin from which peptides can be released under several different orthogonal conditions (Lebl et al., 1993). This allows small aliquots of beads to be assayed in one well. Beads from active wells can then be rearrayed as singles for a second round of cleavage to determine the individual active beads, which can then be selected for microsequencing.

It is also possible to effect total cleavage of the peptide from the resin if the resin beads are tagged in some way to encode the peptide sequence. The tags are some other type of molecule which is synthesized onto the resin and which can be detected in some orthogonal release and/or analysis. Chemical tags which are selectively cleaved and identified by electron capture gas chromatography (Ohlmeyer et al., 1993) or HPLC (Ni et al., 1996) have been described, as have polynucleotide tags, which are decoded by PCR (polymerase chain reaction) (Brenner and Lerner, 1992; Nielsen et al., 1993). Tagging adds an additional synthetic step for each peptide coupling in order for the tag to be introduced at each step and so increases the synthetic effort required. However tagging can allow identification of peptides which are otherwise not amenable to identification by sequencing (e.g., those containing blocked amino termini, unusual amino acids or D-amino acids, etc.)

The main disadvantage with bead libraries in general is that due to the statistical nature of such libraries it is necessary to sample many more beads than there are theoretical components in the library in order to be confident that the majority of theoretical sequences have in fact been assayed. This is not as big a problem with on-bead assays, which can be performed on bulk bead samples, but becomes a major logistical problem when beads must be arrayed individually so that released peptides may be assayed in solution but still associated with the bead of their origin. Arraying is very time-consuming and can be technically challenging depending on the exact format of the assay. There can also be an increased requirement for biological reagent, since the number of assays required is minimally three times the number of theoretical compounds in order to have a 95% chance that each peptide sequence has been sampled at least once, if individual beads are being arrayed and assayed. This requirement is often reduced by arraying multiple beads per well for the first round of assay and then rearraying individual beads only from the active wells. This can substantially reduce the number of assays required, but does require additional bead handling since active mixtures need to be rearrayed, and requires a mechanism for controlled partial release of peptides from the resin, adding to the synthetic overhead.

Parallel Synthesis

An alternative to mixture library synthesis is parallel synthesis of peptides. Geysen introduced synthesis on polystyrene pins which were arranged to match the spacing of standard 96-well microtiter plates (Geysen et al., 1984). Each individual well could contain a different amino acid during coupling and the pins could be washed and deprotected in parallel. This method has been used extensively to prepare overlapping peptide sequences for epitope mapping of proteins, for example. An alternative method for parallel synthesis which took advantage of the efficiency of split-and-mix synthesis was introduced by Houghten and involved the use of polypropylene mesh bags (T-bags) to hold the synthesis resin (Houghten, 1985). T-bags were labeled with india ink (the only marking system which could withstand repeated exposure to organic solvents used in peptide synthesis). T-bags could be combined in a single flask for common operations like deprotection and washing and could be segregated for coupling reactions. A variation on this which is more amenable to automation was introduced by Nicolaou et al. (1995) which relied on glass-encapsulated radiofrequency tags which were used to identify polypropylene mesh cannisters which contain the synthesis resin. An automated system for tracking and sorting these rf-tagged cannisters has been commercialized by Irori Quantum Microchemistry.[1]

The use of libraries of individually synthesized peptides offers advantages over mixture libraries in that the identity of peptides is already known from the synthesis and there is no need for deconvolution or identification by sequencing or reading chemical tags. In addition, there is the opportunity to characterize the peptides, verify sequence or structure, purify if necessary, and quantitate the biological response since exact weight or concentrations of the peptides can be determined. The disadvantage is that the number of samples to be processed multiplies exponentially as sites of diversity are added and can require considerable resources to accommodate large numbers of peptides. Parallel synthesis is probably best applied to situations where a relatively small number of positions are being varied.

Other Aspects of Peptide Design

Enzymatic Stability

For simple peptides, the serum half-life and biological activity is often limited by proteolytic hydrolysis of susceptible peptide bonds. To make these peptides pharmacologically or physiologically useful, it is necessary to increase their serum half-life by increasing their resistance to proteolysis. The exact pathway for proteolysis depends on the peptide in question, and determining the metabolic fate of a given peptide can be a difficult and tedious process. However, inspection of the peptide sequence can usually provide insight into the most likely sites for proteolytic degradation.

There are two types of proteolytic enzymes: exopeptidases, which cleave amino acids from either end of a peptide chain, and endopeptidases, which cleave peptide bonds in the middle of a peptide chain. Exopeptidases include aminopeptidases, which cleave the N-terminal amino acid from a peptide chain; carboxypeptidases, which cleave the C-terminal amino acid from a peptide chain; and dipeptidyl carboxypeptidases like angiotensin converting enzyme (ACE), which cleave the C-terminal dipeptide from a peptide chain. Their recognition sites include the charged amino or carboxyl termini of the peptide, and they may show side-chain specificities as well. Endopeptidases generally have specificities for a particular amino acid side chain or a discrete sequence. The most relevant endopeptidase activities to consider are trypsin-like, in which Lys-X or Arg-X bonds are cleaved; chymotrypsin-like, in which Phe-X, Trp-X or bulky aliphatic-X bonds are cleaved; and post-proline cleaving enzyme, in which Pro-X bonds are cleaved.

There are many ways to stabilize a peptide against specific proteolysis. One approach is as simple as acetylating the amino terminus to block aminopeptidase action because aminopeptidases require the presence of a free α-amino group. Another approach involves modifying or deleting an amino acid side chain or inverting its chirality (most proteolytic enzymes have greatly reduced activity against D-amino acid residues). The susceptible peptide bond itself may be altered as well. Introduction of steric bulk along the peptide backbone by substitution of methyl groups for α-hydrogens or amide hydrogens reduces the rate of enzymatic hydrolysis adjacent to those substitutions. Peptide bonds can also be substituted for by a variety of hydrolytically inert isosteric structures such as *trans* olefins. All of these approaches involve some modification to the peptide structure that can have effects on biological activity apart from any effects on proteolytic stabilization. Modifications that reduce susceptibility to proteolytic cleavage can, however, have a profound effect on the in vivo potency of peptides.

Arginine vasopressin (AVP) is a naturally occurring peptide hormone produced in the pituitary and exhibits antidiuretic activity in vivo. The antidiuretic activity from a bolus injection of AVP has a very short duration of action, suggestive of metabolic instability. Indeed, immunoreactive AVP disappears rapidly from the circulation after a bolus injection and is degraded rapidly upon exposure to serum in vitro (Edwards et al., 1973). The structure of arginine vasopressin is given in figure 2-18 (compound 15). From the structure, several potential sites for enzymatic degradation can be identified. The peptide has a free amino group and might be susceptible to aminopeptidase action. There is an arginine residue at position 8, which could be a site for tryptic-like cleavage. The amino-terminal hexapeptide is cyclized by means of a disulfide bond. Cyclic peptides are often resistant to proteolytic degradation due to the conformational constraints imposed by cyclization. Disulfides, however, can be opened by reduction or by interchange reactions with sulfhydryl containing molecules such as glutathione (γ-glutamyl-cysteinyl-glycine). In addition, most small cyclic peptides require the cyclic structure for biological activity since it folds them into conformations that are otherwise relatively unlikely for

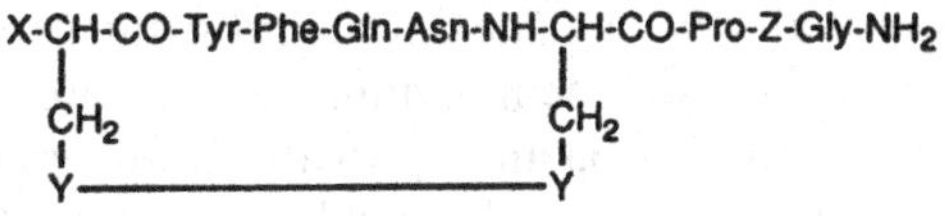

$$X\text{-CH-CO-Tyr-Phe-Gln-Asn-NH-CH-CO-Pro-Z-Gly-NH}_2$$

(structure showing X-CH-CO-Tyr-Phe-Gln-Asn-NH-CH-CO-Pro-Z-Gly-NH₂ with two CH₂ groups connected by Y—Y bridge)

Number	X	Y	Z	Compound
15	H_2N	S	Arg	AVP
16	H	S	Arg	dAVP
17	H_2N	S	D-Arg	DAVP
18	H	S	D-Arg	dDAVP
19	H	CH_2	Arg	[Asu[1,6]]-AVP

Figure 2-18. Structure of arginine vasopressin and analogs.

linear peptides. AVP is no exception, and the linear Ala-1,6 analog is
devoid of antidiuretic activity (Walter et al., 1967).

Figure 2-18 shows some AVP analogs that were modified in ways
that would expectedly increase the stability of the peptide. Analog 16
(dAVP) is the desamino version, which would be expected to be
resistant to any aminopeptidase activity. Analog 17 (DAVP) is the
D-Arg-8 peptide, which should be resistant to tryptic-like cleavages.
Analog 18 (dDAVP) contains both modifications together. Analog 19
is a dicarba analog, one in which the disulfide is replaced with a non-
reducible ethylene bridge; it is a desamino analog as well. The biological activities of these analogs are given in table 2-9. The ADH
(antidiuretic hormone) activity is a measure of antidiuretic potency
in vivo in the rat. AVP has 332 International Units of antidiuretic
activity per milligram of peptide (Sawyer et al., 1974). The desamino
peptide 16 exhibits a five-fold enhanced in vivo potency compared to
AVP (Bankowski et al., 1978), suggesting increased metabolic
stability. The D-Arg-8 analog 17 also retains good in vivo antidiuretic

Table 2-9 Biological activity of modified arginine vasopressin analogs

Number	Compound	ADH (IU mg^{-1})	K_{bind} (mM)	K_a (nM)
15	AVP	332	0.44	0.1
16	dAVP	1745	–	–
17	DAVP	114	1400	2800
18	dDAVP	955	3000	3450
19	[Asu[1,6]]-AVP	1274	1.4	0.4

potency (Zaoral et al., 1967a). The doubly substituted analog 18 shows roughly additive effects for the individual substitutions (Zaoral et al., 1967b). The desamino-dicarba analog 19 also shows enhanced potency in vivo, roughly equivalent to the disulfide analog 16 (Hase et al., 1972).

The peptide dDAVP (compound 18) not only exhibits enhanced in vivo antidiuretic activity but has a five-fold-longer serum half-life after bolus injection and is completely stable to incubation with plasma in vitro (Edwards et al., 1973), establishing the utility of enhancing metabolic stability. What is even more remarkable can be found by comparing the in vitro activity of these analogs. Vasopressin acts in the kidney by binding to specific receptors activating adenylate cyclase. Both the binding to rat kidney membrane preparations and the hormone-dependent activation of adenylate cyclase are listed in table 2-9. AVP has a subnanomolar binding affinity and activation constant (Hechter et al., 1978). In contrast, both DAVP and dDAVP show more than three orders of magnitude decrease in receptor affinity and activation constant (Roy et al., 1975; Butlen et al., 1978). The increased metabolic stability of dDAVP more than makes up for a tremendous loss in intrinsic activity at the receptor, allowing this analog to exhibit enhanced antidiuretic potency in vivo to the extent that it is clinically useful for the treatment of diabetes insipidis. This is a remarkable result and underscores the importance of metabolic stability in the expression of the in vivo biological activity of peptides.

There are more exotic ways to stabilize individual peptide bonds against proteolysis. These approaches involve replacing the amide linkage with a proteolytically resistant linkage. Some peptide bonds, like the trans double bond or the thioamide (C=SNH), are isosteric with the amide bond and preserve their geometry while others, like the ethylene (CH_2CH_2) or thiomethyl ether linkages (CH_2S), preserve the connectivity and chain length only. All of them alter the polarity and hydrogen bonding potential of the amide bond as well as imparting resistance to proteolysis.

A number of these substitutions were examined in the endogenous opiate peptide leucine enkephalin, which has the sequence H-Tyr-Gly-Gly-Phe-Leu-OH. Leu-enkephalin is susceptible to a number of enzymatic cleavages including aminopeptidase cleavage of the Tyr-Gly bond (Hambrook et al., 1976), carboxypeptidase cleavage of the Phe-Leu bond (Marks et al., 1977), and dipeptidyl carboxypeptidase cleavage of the Gly-Phe bond (Malfroy et al., 1978). Biological activity requires the free α-amino group and the L-tyrosine side chain at position 1 (Frederickson, 1977), so desamino or D-Tyr substitutions are not useful to decrease aminopeptidase degradation. Substitution of D-Ala for Gly-2 gives an analog with somewhat higher biological activity than the parent compound (Kosterlitz et al., 1980) and

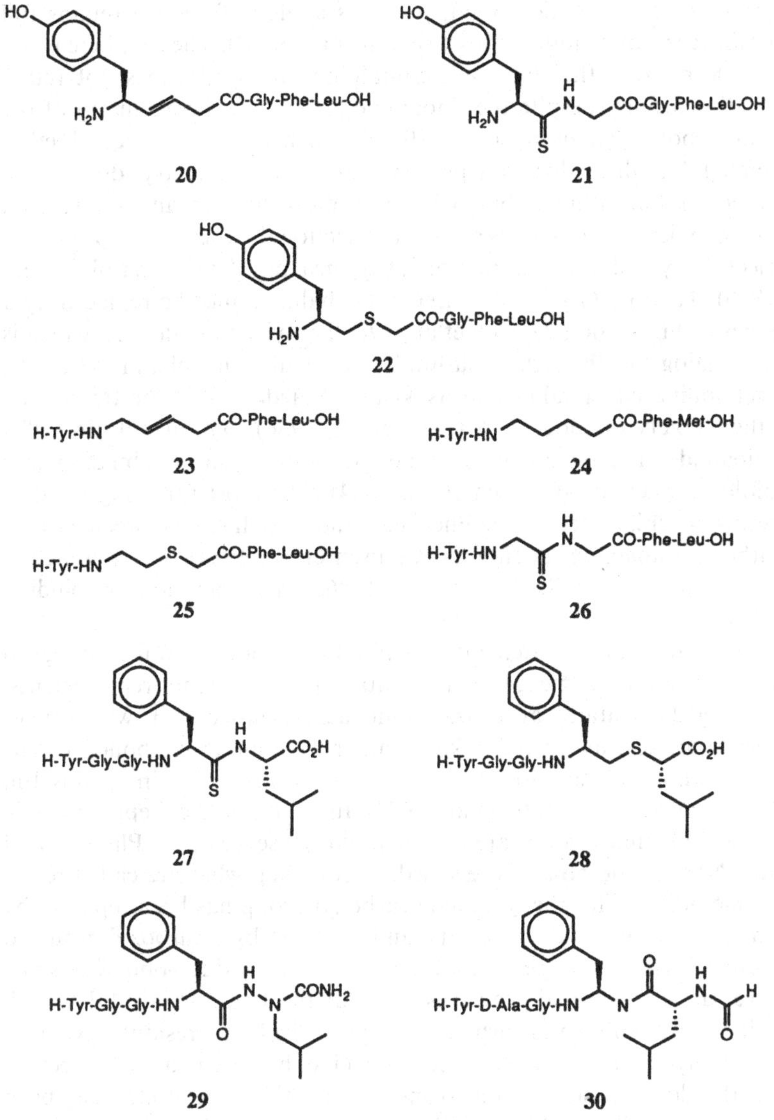

Figure 2-19. Enkephalin analogs with modified peptide bonds.

improved resistance to proteolysis (Miller et al., 1977). However, in the quest to further improve metabolic stability, a number of peptide bond replacements were also examined, some of which are summarized in figure 2-19. The results from amide-bond replacement differ with the particular amide bond being replaced. Replacement of the Tyr-Gly bond with a *trans* olefin isostere, which retains the geometry of the amide bond, provides analog 20, which retains full brain-receptor-binding affinity (Hann et al., 1982). The case here is not simple because the thioamide-containing analog 21 does not retain binding affinity despite the thioamide preserving the geometry of the amide bond (Clausen et al., 1984). Further, the thiomethyl ether analog 22, which does not preserve the amide geometry, does retain some binding affinity, but it behaves more like an antagonist than an agonist at low doses in a functional assay (inhibition of electrically induced contraction of guinea pig ileum; Spatola et al., 1986). The Gly-Gly bond in Leu-enkephalin cannot be replaced by a *trans* olefin (analog 23; Cox et al., 1980), ethylene (analog 24, which is an analog of the other naturally occurring enkephalin sequence, methionine enkephalin; Kawasaki and Maeda, 1982), or thiomethyl ether moiety (analog 25; Spatola et al., 1986), but substitution of a thioamide (analog 26) gives a peptide with enhanced binding and biological activity (Clausen et al., 1984). The most forgiving bond in terms of replacement is the Phe-Leu bond, which can be replaced with either a thioamide (analog 27; Clausen et al., 1984) or a thiomethyl ether (analog 28; Spatola et al., 1986) with retention of binding affinity.

The lack of a requirement of a Phe-Leu amide bond for biological activity led to some more innovative peptide bond replacements. Analog 29 contains an α-aza-amino acid (aza-leucine), which gives rise to an acyl hydrazide linkage rather than an amide bond. The α-aza-amino acid has no chiral center. This analog again retains full biological activity (Dutta et al., 1977). In the D-Ala-2 enkephalinamide analog 30, amide bonds are retained, but the sense of the Phe-Leu and Leu-NH$_2$ amide bonds is reversed, introducing what are called retro-amide bonds. The phenylalanine carboxyl group has been replaced by an amino group and the leucine amino group by a carboxyl group to form a NHCO linkage rather than a CONH peptide bond. The same reversal of functional groups is made at the Leu-NH$_2$ peptide bond. (Since both amino and carboxyl groups of the Leu-5 residue have been "exchanged," it is now in reality D-leucine, but the relative projection of the leucine side chain from the peptide backbone has been preserved.) The D-Ala substitution should block aminopeptidase action, and the retro-peptide bonds should block chymotryptic-like cleavages. The modified peptide exhibits both enhanced potency in vitro and prolonged activity in vivo compared to D-Ala-2 enkephalin-amide (Chorev et al., 1979).

Conformational Design and Constraint

Conformational considerations in peptide synthesis remain one of the greatest challenges in the design of biologically active peptides. In addition to the nature of the side chains, the three-dimensional shape assumed by a peptide when interacting with its target, be it receptor, antibody, or enzyme, determines its biological activity. Unlike proteins, peptides of under 15 or so residues in length tend not to exhibit a stable or even a preferred solution conformation. There is generally too little hydrophobic character in a short peptide that can be sequestered from the polar environment by folding. The peptide backbone and polar side chains will have a driving force to be solvated, and entropic considerations will generally favor solvation by the bulk solvent. Thus, there will usually be an ensemble of conformational states in solution. If biological activity involves only one discrete conformer, this conformational ensemble essentially represents a dilution of the biologically active species. The problem is most acute for peptides designed to mimic a portion of a protein structure. In their native environment, these peptide sequences can rely on the protein's structural rigidity to hold them in a particular conformation, while as free peptides they have no such constraining influence. Even for analogs of linear peptide hormones, which have evolved to express high biological activity despite conformational mobility, favorable conformational constraints can impart an appreciable increase in biological activity.

Conformational constraints can be divided conceptually into the three following categories: Local constraints involve restricting the conformational mobility of a single residue in a peptide; regional constraints are those which affect a group of residues that form some secondary structural unit, such as a β-turn; global constraints involve the entire peptide structure.

Peptides that are long enough to adopt stable solution conformations can have conformational preferences optimized globally by modifying the peptide sequence in accordance with the empirical rules of secondary structure prediction such as those devised by Chou and Fasman (1978; see above, Secondary Structure Prediction). This approach does not "freeze" a peptide into a particular secondary structure, but it stabilizes that structure relative to random coil or other structures, increasing the number of peptide molecules that would be found in that conformation at any given time.

Peptides that are too small to adopt stable conformations on their own require covalent modifications to introduce local or regional conformational constraints. These constraints typically involve addition of sterically bulky substituents adjacent to a rotatable bond to restrict its mobility or the incorporation of cyclic structures. It is impossible to cover all the approaches that have been taken to introduce conformational constraints into peptides, but some of the more successful approaches will be discussed below.

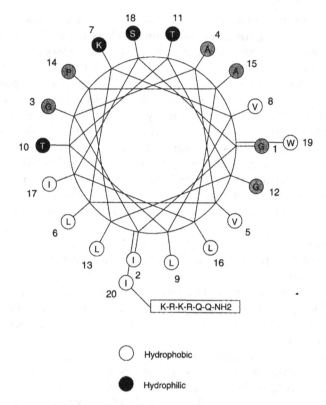

K-R-K-R-Q-Q-NH2

○ Hydrophobic

● Hydrophilic

Figure 2-20. (a) Helix wheel projection of a mellitin
1–20 helix. There is some segregation of hydrophilic and
hydrophobic residues, but the arrangement is interrupted
by Gly and Ala residues and by proline-14, which
destabilizes the helix. The basic hexapeptide tail hangs off
the helix.

Globally Optimized Secondary Structure
It is possible to design peptides of 15 to 20 residues in length that will
exhibit a preference for adopting helical structures in solution. The
most successful applications involve the design of peptides containing
amphiphilic helices, such as the bee venom peptide mellitin. Mellitin is
a 26-residue peptide with the sequence Gly-Ile-Gly-Ala-Val-Leu-Lys-
Val-Leu-Thr-Thr-Gly-Leu-Pro-Ala-Leu-Ile-Ser-Trp-Ile-Lys-Arg-Lys-
Arg-Gln-Gln-NH$_2$ with the biological activity of binding to and lysing
erythrocytes (Habermann, 1972). The lytic activity requires the basic
C-terminal hexapeptide, but the amino-terminal 20-residue peptide
can still interact with phospholipid bilayers and membranes
(Schröder et al., 1971). This suggested to Kaiser and coworkers that

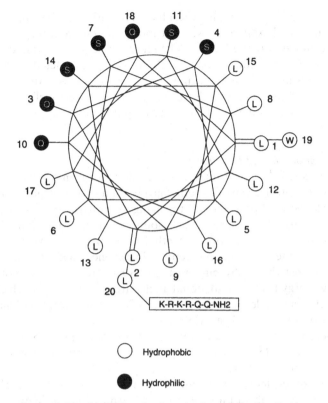

Figure 2-20. (b) Helix wheel projection of the idealized amphiphilic lytic peptide based on the mellitin structure. The hydrophobic face is composed solely of leucine residues while the hydrophilic face is composed of serine and glutamine residues. There is little sequence homology in the two helical sequences 1–20, but the length of the amphipathic helix has been preserved while the amphiphilicity has been optimized. Tryptophan-19 and the basic hexapeptide tail have been retained.

the amino-terminal 20 residues might have formed an amphiphilic helix that bound at the membrane interface and held the basic hexa-peptide tail responsible for the lytic activity close to the membrane surface where it could interact (DeGrado et al., 1981). A helical wheel projection of mellitin 1–20 is shown in figure 2-20(a). There are several problems with this as an ideal amphiphilic helix. The Gly residues and especially the Pro-14 residue tend to reduce the helical propensity of the peptide. Also, several of the Gly residues are found on the hydro-phobic face. DeGrado et al. (1981) designed an idealized sequence for the 1–20 region in which the hydrophobic residues were replaced with Leu, and the hydrophilic residues were either Ser or Gln; the C-terminal basic hexapeptide was not modified. The resulting idealized helix is shown in figure 2-20(b). The idealized peptide had an identical profile of activity compared to mellitin, but it interacted with lipid monolayers more strongly, formed tetramers at a lower concentation, and lysed erythrocytes at a lower concentration than mellitin, which agrees with the formation of a better amphiphilic helix than mellitin.

Local Conformational Constraints

The simplest local constraints that can be placed on a given residue involve the substitution of a methyl group for a hydrogen adjacent to a rotatable bond. For example, replacing the α-hydrogen on alanine with a methyl group gives aminoisobutyric acid (Aib; figure 2-21). The greater steric bulk of the methyl group reduced the rotational freedom of the two adjacent peptide backbone angles ϕ and ψ (see figure 2-6; see above, Protein Secondary Structure). In the case of Aib, the low energy backbone torsional angles are $\phi = \psi \approx \pm 60°$, which are the backbone angles associated with helical structure (Marshall and Bosshard, 1972). α, α-Diethylglycine, with its bulkier substitution at the α-carbon, tends to constrain the backbone angles ϕ and ψ to approximately $180°$, angles typical of a fully extended chain (Benedetti et al., 1988).

Some other local constraint-producing amino acids are shown in figure 2.21. Substitution of a methyl group for the hydrogen on the amino group, to give N-methyl-alanine for example, has an effect on the *cis–trans* ratio of the amide bond, lowering the relative energy of the *cis* isomer. It has only a modest constraining effect on its own torsional angle ϕ, but it affects the angle ψ of the preceding residue, constraining it to large positive values typical of those found in extended or β-structure (Marshall and Bosshard, 1972). Proline is a special case of N-methyl amino acids because its own angle ϕ is constrained to about $-80°$ by being contained in the five-membered pyrrolidine ring. Proline is often found in reverse turn structures, occupying the $i + 1$ position in Type II and III$'$ β-turns, the $i + 2$ position in Type II$'$ and III β-turns, and in inverse γ-turn conformations (Smith and Pease, 1980). An interesting proline congener is 5,5-dimethylthiazolidine-4-carboxylic acid (Dtc), a β,β-disubstituted thioproline analog. Dtc is conformationally similar to proline, except that the γ-turn is disfavored in Dtc because of steric interaction between the β-methyl groups and the carbonyl carbon. Also, it was suggested as a probe for proline-containing inverse γ-turns (Samanen et al., 1990). Constraint-producing amino acid substitutions are typically used as probes for the biologically active conformation. If the constrained peptide retains good biological activity, then it can be

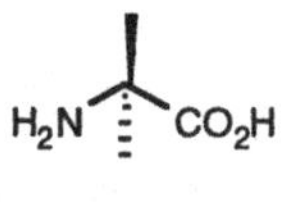

Aminoisobutyric acid
(Aib)

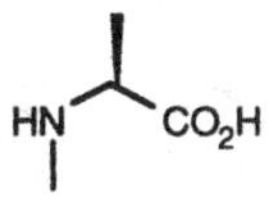

N-Methyl-L-alanine
(MeAla)

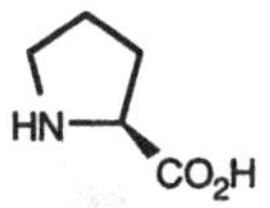

Proline
(Pro)

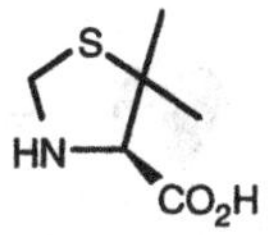

Dimethylthiazolidine
carboxylic acid (Dtc)

Figure 2-21. Some constraint-producing amino acids.

inferred that the particular backbone constraint is compatible with the biologically active conformation. Negative results are more difficult to interpret because the additional steric bulk producing the constraint may also interfere with the peptide binding to its target molecule regardless of conformational effects. This approach has been applied to a number of peptide hormones including arginine vasopressin (Moore et al., 1985), bradykinin (Turk et al., 1975), and angiotensin (Samanen et al., 1989). In most cases, the constrained analogs do not exhibit any better potency than their unconstrained analogs, in spite of the reduction in conformational mobility. This may be due in part to the fact that these modifications do not eliminate conformational mobility but only reduce it to some extent. The results of incorporation of constraint-producing amino acids can also be somewhat capricious. *N*-methyl or α-methyl substitutions, for example, have converted angiotensin agonists into antagonist analogs (Peña et al., 1974; Turk et al., 1976). In addition to providing conformational constraints, *N*-methyl and α-methyl amino acids can impart other properties to peptides, such as increased resistance to proteolysis, and have been successfully used in the design of metabolically stable and orally active renin inhibitors (Pals et al., 1986; Thaisrivongs et al., 1987).

Regional Constraints

Turns are a prominent feature in peptide secondary structure that have often been implicated as important elements for biological activity, as discussed previously (see above, Turns). Turns or chain reversals must

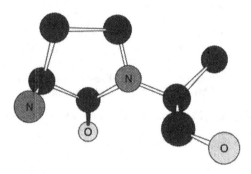

Figure 2-22. γ-Lactam turn mimetic.
Ball-and-stick of γ-lactam mimetic showing
β-turn conformation; compare with
figure 2-11(d).

occur in cyclic peptides or in peptides that occur as short loops in proteins, but linear peptides can fold into conformations containing turns as well. There are steric constraints placed on the side chains and backbone torsions of the corner residues in a turn. Gly and Pro residues most readily accommodate these constraints, and their appearance in a linear sequence is suggestive of a potential turn structure. A number of investigators have designed turn mimetics, cyclic moieties designed to replace the internal residue(s) of a turn while maintaining the overall geometry associated with the turn (Nagai and Sato, 1985; Huffman et al., 1988; Kahn et al., 1988). The mimetic most frequently applied to peptide design is the γ-lactam of Freidinger (1981; figure 2-22), which induces a Type II′ β-turn conformation.

The γ-lactam has successfully been applied to the peptide hormone, luteinizing-hormone-releasing hormone (LHRH), whose sequence is shown in figure 2-23. LHRH contains a Gly residue at position 6. When substituted by a D-Ala, the biological activity increased, while the L-Ala analog exhibited low activity (Monahan et al., 1973). This outcome suggests that the Gly residue might be occupying the $i + 1$ position of a Type II′ β-turn consisting of the sequence Tyr-Gly-Leu-Arg. As discussed previously (see above, Protein Amino Acids), a

<Glu-His-Trp-Ser-Tyr-Gly-Leu-Arg-Pro-Gly-NH₂

LHRH

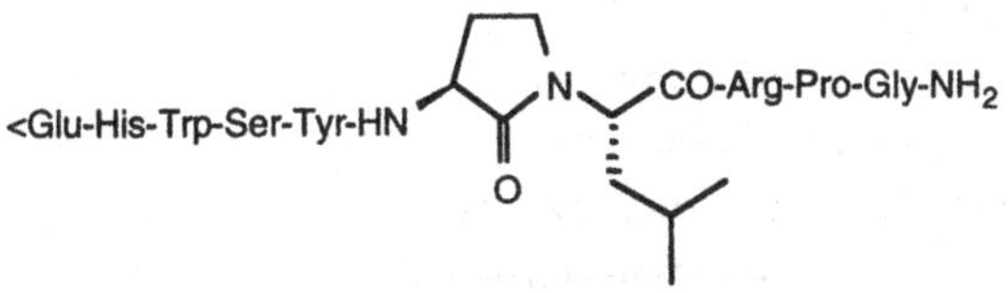

γ-Lactam analog

Figure 2-23. LHRH and γ-lactam-containing analog.

D-amino acid at that position would have an equatorial placement of the side chain, while an L-amino acid would have the less favorable axial placement of the side chain. Freidinger et al. (1980) prepared the γ-lactam-containing peptide shown in figure 2-23, in which the corner Gly-Leu residues were replaced by the lactam structure. The constrained peptide proved to be more potent than the parent LHRH, both in vivo and in an in vitro assay, suggesting that conformational stabilization is indeed responsible for the improved activity.

Cyclization

The simplest way to introduce a conformational constraint into a linear peptide is by cyclization, either through disulfide bond formation or by coupling the α- or Lys ε-amino groups with the carboxyl terminus of an Asp or Glu side chain. Cyclization usually requires peptides of moderate length to adopt some sort of folded conformation consistent with bringing the two ends together, but the exact nature of the conformation(s) induced may not be predictable. As mentioned previously, even a cyclic disulfide hexapeptide still has a great deal of conformational mobility (Hagler et al., 1985). Nevertheless, cyclization greatly reduces the number of accessible conformations compared to the linear peptide and, in favorable cases, can substantially improve or alter peptide biological activity.

One case in point involves the preparation of small peptide fibrinogen receptor antagonists. Platelets express a cell surface glycoprotein, gpIIb/IIIa, which binds to the extracellular matrix protein fibronectin and to the plasma protein fibrinogen. Fibrinogen expresses multiple platelet-binding domains and causes aggregation of activated platelets. This adhesive process is important in the formation of blood clots, and inhibitors of this process are potentially useful as

Table 2-10 Fibrinogen receptor binding peptides

Compound	Sequence	Antiaggregator IC_{50} (μM)	gpIIb/IIIa binding (μM)
	Fibrinogen		0.043
31	Ac-Arg-Gly-Asp-Ser-NH$_2$	91	4.2
32	Ac-Cys-Arg-Gly-Asp-Ser-Cys-NH$_2$	33	5.3
33	Ac-Cys-Arg-Gly-Asp-Cys-NH$_2$	16	0.78
34	Ac-Cys-Arg-Gly-Asp-Pen-NH$_2$	4.1	N.D.
35	Ac-Cys-MeArg-Gly-Asp-Pen-NH$_2$	0.36	0.027

Pen

antithrombotic agents. Both fibrinogen and fibronectin are extremely large proteins, and their three-dimensional structures are not known. The binding domain on fibronectin was localized to a tetrapeptide sequence Arg-Gly-Asp-Ser by the use of a proteolytic fragment of fibronectin and a series of synthetic peptides based on secondary structure predictions for the sequence of the fragment (Pierschbacher and Ruoslahti, 1984). Small peptides containing this Arg-Gly-Asp-Ser sequence were shown to block binding of platelets both to fibronectin and to fibrinogen, but at concentrations in the 100 mM range (Plow et al., 1985).

In order to try to improve binding affinity, Samanen and coworkers (1991) prepared the series of cyclic analogs shown in table 2-10. Both inhibition of aggregation of activated platelets and binding affinity to the gpIIb/IIIa receptor reconstituted in liposomes were measured. The smallest peptide containing the binding sequence is Ac-Arg-Gly-Asp-Ser-NH$_2$ (31), which has about 100 μM antiaggregatory activity and a binding affinity for the reconstituted receptor of 4 μM, 100-fold less potent than fibrinogen itself. Enclosing this sequence in a disulfide loop (32) has only a modest effect on activity, and removing the Ser from the sequence (33) enhances activity and binding affinity further. Replacing the C-terminal cysteine with penicillamine (Pen, β, β-dimethyl-cysteine; compound 34), which introduces an additional conformational constraint to the side chain and disulfide, increases the activity still further. The addition of an N-methyl group to the Arg residue (35), which adds yet another conformational constraint, now gives a compound with submicromolar antiaggregatory activity and binding affinity, which is now comparable to that of fibrinogen itself. The end result is a suitably constrained pentapeptide with the binding properties and affinity of a very large protein.

Summary

We have seen examples of peptides designed to mimic protein epitopes or binding sites, and we have seen examples of small peptides that have been modified for enhanced proteolytic stability or for conformational constraints. The exact nature of the design of biologically active peptides depends to a large extent on their intended use. However, several general principles should be kept in mind. Peptides are generally much more flexible and much less soluble than proteins. Smaller peptides do not tend to adopt any preferred solution structure, and longer peptides, which can exhibit strong tendencies to fold into a particular secondary structure, are still in equilibrium with partially folded structures. Peptides are less likely to fold in such a way as to shield hydrophobic regions from solvent exposure, contributing to both the lack of stable solution structures and to insolubility. Conformation, solubility, proteolytic resistance, and other properties are modified by altering the nature of the peptide backbone, the side chains, or both. Even small modification of one residue in a peptide may have a large effect on biological activity, conformation, solubility, or proteolytic susceptibility. Finally, because peptides are synthesized chemically rather than ribosomally, one is not necessarily limited to the use of the 20 naturally occurring amino acids in designing analogs. There are hundreds of synthetically available amino acids with a wide variety of structure, functional groups, charge, polarity, hydrophobicity, or chemical reactivity that can be incorporated to tailor a given peptide to a specific property or use.

Note

1. Irori Quantum Microchemistry, 9640 Towne Centre Drive, San Diego, CA 92121. (619) 546-1300.

References

Alderton, G., and Fevold, H. L. (1951). Lanthionine in subtilin. J. Am. Chem. Soc. 73:463–464.

Allen, J. S., Coligan, J. E., Barin, F., McLane, M. F., Sodroski, J. G., Rosen, C. A., Haseltine, W. A., Lee, T. H., and Essex, M. (1985). Major glycoprotein antigens that induce antibodies in AIDS patients are encoded by HTLV-III. Science 228:1091–1093.

Anastasi, A., Erspamer, V., and Bucci, M. (1971). Isolation and structure of bombesin and alytensin, two analoguous active peptides from the skin of the European amphibians Bombina and Alytes. Experentia 27:166–167.

Baker, E. N., and Hubbard, R. E. (1984). Hydrogen bonding in globular proteins. Prog. Biophys. Mol. Biol. 44:97–179.

Bankowski, K., Manning, M., Haldar, J., and Sawyer, W. H. (1978). Design of potent antagonists of the vasopressor response to arginine-vasopressin. J. Med. Chem. 21:850–853.

Benedetti, E., Barone, V., Bavoso, A., Blasio, B. D., Lelj, F., Pavone, V., Pedone, C., Toniolo, C., Leplawy, M. T., Kaczmarek, K., and Redlinski, A. (1988). Structural versatility of peptides from $C^{\alpha,\alpha}$-dialkylated glycines. I. A conformational energy computation and X-ray diffraction study of homo-peptides from $C^{\alpha,\alpha}$-diethylglycine. Biopolymers 27:357–371.

Berk, S. C., Rohrer, S. P., Degrado, S. J., Birzin, E. T., Mosley, R. T., Hutchins, S. M., Pasternak, A., Schaeffer, J. M., Underwood, D. J., and Chapman, K. T. (1999). A combinatorial approach toward the discovery of non-peptide, subtype-selective somatostatin receptor ligands. J Combinatorial Chem. 1:388–391.

Berridge, N. J., Newton, G. G. F., and Abraham, E. P. (1952). Purification and nature of the antibiotic nisin. Biochem. J. 52:529–535.

Bray, A. M., Maeji, N. J., and Geysen, H. M. (1990). The simultaneous multiple production of solution phase peptides; assessment of the Geysen method of simultaneous peptide synthesis. Tet. Lett. 31:5811–5814.

Brenner, S., and Lerner, R. A. (1992). Encoded combinatorial chemistry. Proc. Natl. Acad. Sci. USA 89:5381–5383.

Burley, S. K., and Petsko, G. A. (1985). Aromatic-aromatic interaction: a mechanism of protein structure stabilization. Science 229:23–28.

Burley, S. K., and Petsko, G. A. (1988). Weakly polar interactions in proteins. Adv. Protein Chem. 39:125–189.

Burgess, K., Liaw, A. I., and Wang, N. (1994). Combinatorial technologies involving reiterative division/coupling/recombination: statistical considerations. J. Med. Chem. 37:2985–2987.

Butlen, D., Guillon, G., Rajerison, R. M., Jard, S., Sawyer, W. H., and Manning, M. (1978). Structural requirements for activation of vasopressin-sensitive adenylate cyclase, hormone binding, and antidiuretic actions: effects of highly potent analogues and competitive inhibitors. Mol. Pharmacol. 14:1006–1017.

Chorev, M., Shavitz, R., Goodman, M., Minick, S., and Guillemin, R. (1979). Partially-modified retro-inverso enkephalinamides. Science 204:1210–1212.

Chou, P. Y., and Fasman, G. D. (1978). Prediction of the secondary structure of proteins from their amino acid sequence. Advances Enzymol. 47:45–148.

Chu, Y.-H., Kirby, D. P., and Karger, B. L. (1995). Free solution identification of candidate peptides from combinatorial libraries by affinity capillary electrophoresis mass-spectrometry. J. Am. Chem. Soc. 117:5419–5420.

Clausen, K., Spatola, A. F., Lemieux, C., Schiller, P. W., and Lawesson, S.-O. (1984). Evidence of a peptide backbone contribution toward selective receptor recognition for leucine enkephalin thioamide analogs. Biochem. Biophys. Res. Comm. 120:305–310.

Cox, M. T., Gormley, J. J., Hayward, C. F., and Peter, N. F. (1980). Incorporation of *trans*-olefininc dipeptide isosteres into enkephalin and substance P analogs. J. Chem. Soc. Chem. Comm. 800–802.

Creighton, T. E. (1993). *Proteins: Structure and Molecular Properties*, 2nd ed., New York, W. H. Freeman.

Dayhoff, M. O., Eck, R. V., Chang, M. A., and Sochard, M. R. (1965). *Atlas of Protein Sequence and Structure*, Silver Spring, Md., National Biomedical Research Foundation.

Dayhoff, M., Schwartz, R. M., and Orcutt, B. C. (1978). *Atlas of Protein Sequence and Structure*, Silver Spring, Md., National Biomedical Research Foundation.

Debouck, C., Gorniak, J. G., Strickler, J. E., Meek, T. D., Metcalf, B. W., and Rosenberg, M. (1987). Human immunodeficiency virus protease expressed in *Escherichia coli* exhibits autoprocessing and specific maturation of the gag precursor. Proc. Natl. Acad. Sci. USA 84:8903–8906.

DeGrado, W. F., Kézdy, F. J., and Kaiser, E. T. (1981). Design, synthesis and characterization of a cytotoxic peptide with mellitin-like activity. J. Am. Chem. Soc. 103:679–681.

DelMar, E. G., Largman, C., Brodrick, J. W., and Goekas, M. C. (1979). A sensitive new substrate for chymotrypsin. Anal. Biochem. 99:316–320.

Dutta, A. S., Gormley, J. J., Hayward, C. F., Morley, J. S., Shaw, J. S., Stacey, G. J., and Turnbull, M. T. (1977). Enkephalin analogues eliciting analgesia after intravenous injection. Life Sci. 21:549–562.

Edwards, C. R. W., Kitau, M. J., Chard, T., and Besser, G. M. (1973). Vasopressin analogue DDAVP in diabetes insipidis: clinical and laboratory studies. Br. Med. J. 3:375–378.

Eisenberg, D. (1984). Three-dimensional structure of membrane and surface proteins. Ann. Rev. Biochem. 53:595–623.

Eisenberg, D., Weiss, R. M., and Terwilliger, T. C. (1982). The helical hydrophobic moment: a measure of the amphilhilicity of a helix. Nature 299:371–374.

Eisenberg, D., Wilcox, W., and Eshita, S. (1987). Hydrophobic moments as tools for analysis of protein sequences and structures. In *Proteins, Structure and Function*, J. J. L'Italien, ed., New York, Plenum Press, pp. 425–436.

Erickson, J., Neidhart, D. J., VanDrie, J., Kempf, D. J., Wang, X. C., Norbeck, D. W., Plattner, J. J., Rittenhouse, J. W., Turon, M., Wideburg, N., et al. (1990). Design, activity, and 2.8 Å crystal structure of a C2 symmetric inhibitor complexed to HIV-1 protease. Science 249:527–533.

Frederickson, R. C. A. (1977). Enkephalin pentapeptides: a review of current evidence for a physiological role in vertebrate neurotransmission. Life Sci. 21:23–42.

Freidinger, R. M. (1981). Computer graphics and chemical synthesis in the study of conformation of biologically active peptides. In *Peptides, Synthesis, Structure, Function*, Proceedings of the Seventh American Peptide Symposium, D. H. Rich and E. Gross, eds., Rockford, Ill., Pierce Chemical Co.

Freidinger, R. M., Veber, D. F., Perlow, D. S., Brooks, J. R., and Saperstein, R. (1980). Bioactive conformation of luteinizing hormone-releasing hormone: evidence from a conformationally constrained analog. Science 210:656–658.

Frömmel, C., and Preissner, R. (1990). Prediction of prolyl residues in *cis*-conformation in protein structures on the basis of amino acid sequence. FEBS Lett. 277:159–163.

Fukushima, D., Kaiser, E. T., Kézdy, F. J., Kroon, D. J., Kupferberg, J. P., and Yokoyama, S. (1980). Rational design of synthetic models for lipoproteins. Ann. N.Y. Acad. Sci. 348:365–373.

Furka, A., Sebestyen, F., Asgedom, M., and Dibo, G. (1991). General method for rapid synthesis of multicomponent peptide mixtures. Int. J. Peptide Protein Res. 37:487–493.

Garnier, J., Osguthorpe, D. J., and Robson, B. 1978. Analysis of the accuracy and implications of simple methods for predicting the secondary structure of globular proteins. J. Biol. Chem. 120:97–120.

Geysen, H. M., Meloen, R. H., and Barteling, S. J. (1984). Use of peptide synthesis to probe viral antigens for epitopes to a resolution of a single amino acid. Proc. Natl. Acad. Sci. USA. 81:3998–4002.

Geysen, H. M., Rodda, S. J., and Mason, T. J. (1986). A priori delineation of a peptide which mimics a discontinuous antigenic determinant. Mol. Immunol. 23:709–715.

Geysen, H. M., Rodda, S. J., Mason, T. J., Tribbick, G., and Schoofs, P. G. (1987). Strategies for epitope analysis using peptide synthesis. J. Immunol. Methods 102:259–274.

Geysen, H. M., Rodda, S. J., and Mason, T. J. (1988). A synthetic strategy for epitope mapping. In *Peptides, Chemistry and Biology*, Proceedings of the Tenth American Peptide Symposium, G. R. Marshall, ed., Leiden, Escom.

Grant, G. A. (1999). Synthetic peptides for the production of antibodies that recognize intact proteins. *Current Protocols in Protein Science*, 18.3.1–18.3.14, New York, John Wiley and Sons.

Gray, T. M., and Matthews, B. W. (1984). Intrahelical hydrogen bonding of serine, threonine and cysteine residues within α-helices and its relevance to membrane-bound proteins. J.. Mol. Biol. 175:75–81.

Greer, J. (1985). Model structure for the inflammatory protein C5a. Science 228:1055–1060.

Habermann, E. (1972). Bee and wasp venoms. Science 177:314–322.

Hagler, A. T., Osguthorpe, D. J., Dauber-Osguthorpe, P., and Hempel, J. C. (1985). Dynamics and conformational energetics of a peptide hormone: vasopressin. Science 227:1309–1315.

Hambrook, J. M., Morgan, B. A., Rance, M. J., and Smith, C. F. C. (1976). Mode of deactivation of the enkephalins by rat and human plasma and rat brain homogenates. Nature 262:782–783.

Hancock, W. S., Prescott, D. T., Marshall, G. R., and Vagelos, P. R. 1972. Acyl carrier protein. XVIII. Chemical synthesis and characterization of a protein with acyl carrier protein activity. J. Biol. Chem. 247:6224–6233.

Hann, M. M., Sammes, P. G., Kennewell, P. D., and Taylor, J. B. (1982). On the double bond isostere of the peptide bond: preparation of an enkephalin analogue. J. Chem. Soc. Perkin I:307–314.

Harrison, R. K., and Stein, R. L. (1990). Substrate specificities of the peptidyl prolyl *cis*-trans isomerase activities of cyclophilin and FK-506 binding protein: evidence for the existence of a family of distinct enzymes. Biochemistry 29:3813–3816.

Hase, S., Sakakibara, S., Wahrenburg, M., Kirchberger, M., Schwartz, I. L., and Walter, R. (1972). 1,6-Aminosuberic acid analogs of lysine- and arginine- vasopressin and -vasotocin. Synthesis and biological properties. J. Am. Chem. Soc. 94:3590–3600.

Hechter, O., Terada, S., Spitsberg, V., Nakahara, T., Nakagawa, S., and Flouret, G. (1978). Neurohypophyseal hormone-responsive renal adenylate cyclase. III. Relationship between affinity and intrinsic activity in neurohypophyseal hormones and structural analogs. J. Biol. Chem. 253:3230–3237.

Hirschmann, R., Nutt, R. F., Veber, D. F., Vitali, R. A., Varga, S. L., Jacob, T. A., Holly, F. W., and Denkewalter, R. G. (1969). Studies on the total synthesis of an enzyme. V. The preparation of enzymatically active material. J. Am. Chem. Soc. 91:507–508.

Hopp, T. P., and Woods, K. R. (1981). Prediction of protein antigenic determinants from amino acid sequences. Proc. Natl. Acad. Sci. USA 78:3824–3828.

Houghten, R. A. (1985). General method for the rapid solid phase synthesis of large numbers of peptides: specificity of antigen–antibody interaction at the level of individual amino acids. Proc. Natl. Acad. Sci. USA 82:5131–5135.

Howard, J. B. (1981). Reactive site in human α_2-macroglobulin: circumstantial evidence for a thiolester. Proc. Natl. Acad. Sci. USA 78:2235–2239.

Hudson, D. (1999a). Matrix assisted synthetic transformations: a mosaic of diverse contributions. I. The pattern emerges. J. Comb. Chem. 1:333–360.

Hudson, D. (1999b). Matrix assisted synthetic transformations: A mosaic of diverse contributions. II. The pattern is completed. J. Comb. Chem. 1:403–457.

Huffman, W. F., Callahan, J. F., Eggleston, D. S., Newlander, K. A., Takata, D. T., Codd, E. E., Walker, R. F., Schiller, P. W., Lemieux, C., Wire, W. S., and Burks, T. F. (1988). Reverse turn mimics. In *Peptides, Chemistry and Biology*, Proceedings of the Tenth American Peptide Symposium, G. R. Marshall, ed., Leiden, Escom.

Hunt, S. (1985). The non-protein amino acids. In *Chemistry and Biochemistry of the Amino Acids*, G. C. Barrett, ed., London, Chapman and Hall.

Ito, N., Phillips, S. E. V., Stevens, C., Ogel, Z. B., McPherson, M. J., Keen, J. N., Yadav, K. D. S., and Knowles, P. F. (1991). Novel thioether bond revealed by a 1.7Å crystal structure of galactose oxidase. Nature 350:87–90.

Javaherian, K., Langlois, A. J., McDanal, C., Ross, K. L., Eckler, L. I., Jellis, C. L., Profy, A. T., Rusche, J. R., Bolognesi, D. P., Putney, S. D., and Matthews, T. J. (1989). Principal neutralizing domain of the human immunodeficiency virus type 1 envelope protein. Proc. Natl. Acad. Sci. USA 86:6768–6772.

Kahn, M., Lee, Y.-H., Wilke, S., Chem, B., Fujita, K., and Johnson, M. E. (1988). The design and synthesis of mimetics of peptide β-turns. J. Mol. Recognition 1:75–79.

Kaiser, E. T., and Kézdy, F. J. (1984). Amphiphilic secondary structure: design of peptide hormones. Science 223:249–255.

Kauzmann, W. (1959). Some factors in the interpretation of protein denaturation. In *Advances in Protein Chemistry*, C. B. Anfinsen, Jr., M. L. Anson, J. T. Edsall and F. M. Richards, eds., New York, Academic Press.

Kawasaki, K., and Maeda, M. (1982). Amino acids and peptides II. Modification of glycylglycine bond in methionine enkephalin. Biochem. Biophys. Res. Comm. 106:113–116.

Kellis, J. T., Jr., Nyberg, K., Sali, D., and Fersht, A. R. (1988). Contribution of hydrophobic interactions to protein stability. Nature 333:784–786.

Kim, P. S., and Baldwin, R. L. (1982). Specific intermediates in the folding reactions of small proteins and the mechanism of protein folding. Ann. Rev. Biochem. 51:459–489.

Knowles, J. R. 1991. Enzyme catalysis: not different, just better. Nature 350:121–124.

Kosterlitz, H. W., Lord, J. A. H., Paterson, S. J., and Waterfield, A. A. (1980). Effects of changes in the structure of enkephalins and narcotic analgesic drugs on their interactions with μ- and δ-receptors. Br. J. Pharmacol. 68:333–342.

Krystek, S. R. Jr., Metzler, W. J., and Novotny, J. (1995a). Hydrophobicity profiles for protein sequence analysis. In *Current Protocols in Protein Science*, 2.2.1–2.2.13, New York, John Wiley and Sons.

Krystek, S. R. Jr., Metzler, W. J., and Novotny, J. (1995b). Protein secondary structure prediction. In *Current Protocols in Protein Science*, 2.3.1–2.3.20, New York, John Wiley and Sons.

Kyte, J., and Doolittle, R. F. (1982). A simple method for displaying the hydropathic character of a protein. J. Mol. Biol. 157:105–132.

Lam, K. S., Salmon, S. E., Hershg, E. M., Hruby, V. J., Kazmierski, W. M., and Knapp, R. J. (1991). A new type of synthetic peptide library for identifying ligand-binding activity. Nature 354:82–84.

Landry, S. J., and Gierasch, L. M. (1991). Recognition of nascent polypeptide for targetin and folding. TIBS 159–163.

Landschutz, W. H., Johnson, P. F., and McKnight, S. L. (1988). The leucine zipper: a hypothetical structure common to a new class of DNA binding proteins. Science 240:1759–1764.

Lapatto, P., Blundell, T., Hemmings, A., Overington, J., Wilderspin, A., Wood, S., Merson, J. R., Whittle, P. J., Danley, D. E., Geoghegan, K. F., Hawrylik, S. J., Lee, S. E., Scheld, K. G., and Hobart, P. M. (1989). X-ray analysis of HIV-1 proteinase at 2.7 Å resolution confirms structural homology among retroviral enzymes. Nature 342:299–302.

Lebl, M., Patek, M., Kocis, P., Krchnak, V., Hruby, V. J., Salmon, S. E., and Lam, K. S. 1993. Multiple release of equimolar amounts of peptides from a polymeric carrier using orthogonal linkage-cleavage chemistry. Int. J. Peptide Protein Res. 41:201–203.

Lipman, D. J., and Pearson, W. R. (1985). Rapid and sensitive protein similarity searches. Science 227:1435–1441.

Malfroy, B., Swerts, J. P., Guyon, A., Roques, B. P., and Schwartz, J. C. (1978). High affinity enkephalin-degrading peptidase in brain is increased after morphine. Nature 276:523–526.

Märki, W., Brown, M., and Rivier, J. E. (1981). Bombesin analogs: effects on thermoregulation and glucose metabolism. Peptides 2, Suppl. 2:169–177.

Marks, N., Grynbaum, A., and Neidle, A. (1977). On the degradation of enkephalins and endorphines by rat and mouse brain extracts. Biochem. Biophys. Res. Comm. 74:1552–1559.

Marshall, G. R., and Bosshard, H. E. (1972). Angiotensin II. Biologically active conformation. Circ. Res., Suppl. II 30–31:II-143–150.

Matayoshi, E. D., Wang, G. T., Krafft, G. A., and Erickson, J. (1990). Novel fluorogenic substrates for assaying retroviral proteases by resonance energy transfer. Science 247:954–958.

Matsushita, S., Robert-Guroff, M., Rusche, J., Koito, A., Hattori, T., Hoshino, H., Javaherian, K., Takatsuki, K., and Putney, S. (1988). Characterization of a human immunodeficiency virus neutralizing monoclonal antibody and mapping of the neutralizing epitope. J. Virol. 62:2107–2114.

McDonald, T. J., Jornvall, H., Nilsson, G., Vagne, M., Ghatei, M., Bloom, S. R., and Mutt, V. (1979). Characterization of a gastrin releasing peptide from porcine non-antral gastric tissue. Biochem. Biophys. Res. Comm. 90:227.

Meek, T. D., Dayton, B. D., Metcalf, B. W., Dreyer, G. B., Strickler, J. E., Gorniak, J. G., Rosenberg, M., Moore, M. L., Magaard, V. W., and Debouck, C. (1989). Human immunodeficiency virus 1 protease expressed in *Escherichia coli* behaves as a dimeric aspartic protease. Proc. Natl. Acad. Sci. USA 86:1841–1845.

Meldal, M., Svendsen, I., Breddam, K., and Auzanneau, F.-I. (1994). Portion-mixing peptide libraries of quenched fluorogenic substrates for complete subsite mapping of endoprotease specificity. Proc. Natl. Acad. Sci. USA. 91:3314–3318.

Merrifield, B. (1963). Solid phase peptide synthesis. I. The synthesis of a tetrapeptide. J. Am. Chem. Soc. 85:2149–2154.

Michaels, G., and Garian, R. (1995). Computational methods for protein sequence analysis. in *Current Protocols in Protein Science*, 2.1.1–2.1.18, New York, John Wiley and Sons.

Miller, M., Schneider, J., Sathyanarayana, B. K., Toth, M. V., Marshall, G. R., Clawson, L., Selk, L., Kent, S. B. H., and Wlodawer, A. (1989). Structure of complex of synthetic HIV-1 protease with a substrate-based inhibitor at 2.3 Å resolution. Science 246:1149–1152.

Miller, R. J., Chang, K.-J., Cuatrecasas, P., and Wilkinson, S. (1977). The metabolic stability of enkephalins. Biochem. Biophys. Res. Comm. 74:1311–1317.

Mints, L., Hogue Angeletti, R., and Nieves, E. (1997). Analysis of MAPS Peptides. In *ABRF News*, Vol. 8, J. Rush, ed., Bethesda, Md., Association of Biomolecular Resource Facilities, pp. 22–26.

Monahan, M. W., Amoss, M. S., Anderson, H. A., and Vale, W. (1973). Synthetic analogs of the hypothalamic luteinizing hormone releasing factor with increased agonist or antagonist properties. Biochemistry 12:4616–4620.

Moore, M. L., Huffman, W. F., Bryan, W. M., Silvestri, J., Chang, H.-L., Marshall, G. R., Stassen, F., Stefankiewicz, J., Sulat, L., Schmidt, D.,

Kinter, L., McDonald, J., and Ashton-Shue, D. (1985). Vasopressin antagonist analogs containing α-methyl amino acids at position 4. In *Peptides: Structure and Function*, Proceedings of the Ninth American Peptide Symposium, C. M. Deber, V. J. Hruby and K. D. Kopple, eds., Rockford, Ill., Pierce Chemical Co.

Moore, M. L., Bryan, W. M., Fakhoury, S. A., Magaard, V. W., Huffman, W. F., Dayton, B. D., Meek, T. D., Hyland, L., Dreyer, G. B., Metcalf, B. W., Strickler, J. E., Gorniak, J. G., and Debouck, C. (1989). Peptide substrates and inhibitors of the HIV-1 protease. Biochem. Biophys. Res. Comm. 159:420–425.

Morgan, B. P., Scholtz, J. M., Ballinger, M. D., Sipkin, I. D., and Bartlett, P. A. (1991). Differential binding energy: a detailed evaluation of the influence of hydrogen bonding and hydrophobic groups on the inhibition of thermolysin by phosphorus-containing inhibitors. J. Am. Chem. Soc. 113:297–307.

Morishima, H., Takita, T., Aoyagi, T., Takeuchi, T., and Umezawa, H. (1970). The structure of pepstatin. J. Antibiot. 23:263–265.

Nagai, U., and Sato, K. (1985). Synthesis of a bicyclic dipeptide with the shape of a β-turn central part. Tet. Lett. 26:647–650.

Nair, S. A., Kim, M. H., Warren, S. D., Choi, S., Songyang, Z., Cantley, L. C., and Hangauer, D. G. (1995). Identification of efficient pentapeptide substrates for the tyrosine kinase pp60c-src. J. Med. Chem. 38:4276–4283.

Needleman, S., and Wunsch, C. (1970). A general method applicable to the search for similarities in the amino acid sequence of two proteins. J. Mol. Biol. 48:443–453.

Ni, Z.-J., Maclean, D., Holmes, C. P., Gallop, M. A. (1996). Encoded combinatorial chemistry: binary coding using chemically robust secondary amine tags. Methods Enzymol. 267:261–272.

Nicolaou, K. C., Xiao, X.-Y., Parandoosh, Z., Senyei, A., and Nova, M. P. (1995). Radiofrequency encoded combinatorial chemistry. Angew. Chem. Int. Ed. Eng. 34:2289–2291.

Nielsen, J., Brenner, S., and Janda, K. J. (1993). Synthetic methods for the implementation of encoded combinatorial chemistry J. Am. Chem. Soc. 115:9812–9813.

Nutt, R. F., Brady, S. F., Darke, P. L., Ciccarone, T. M., Colton, C. D., Nutt, E. M., Rodkey, J. A., Bennett, C. D., Waxman, L. H., Sigal, I. S., Anderson, P. S., and Veber, D. F. (1988). Chemical synthesis and enzymatic activity of a 99-residue peptide with a sequence proposed for the human immunodeficiency virus protease. Proc. Natl. Acad. Sci. USA 85:7129–7133.

Ohlmeyer, M. H. J., Swanson, R. N., Dillard, L. W., Reader, J. C., Asouline, G., Kobayashi, R., Wigler, M., and Still, W. C. (1993). Complex synthetic chemical libraries indexed with molecular tags. Proc. Natl. Acad. Sci. USA 90:10922–10926.

O'Shea, E. K., Rutkowski, R., and Kim, P. S. (1989). Evidence that the leucine zipper is a coiled coil. Science 243:538–542.

Ostresh, J. M., Winkle, J. H., Hamashin, V. T., and Houghten, R. A. (1994). Peptide libraries: determination of relative reaction rates of protected amino acids in competitive couplings. Biopolymers 34:1681–1689.

Pals, D. T., Thaisrivongs, S., Lawson, J. A., Kati, W. M., Turner, S. R., DeGraaf, G. L., Harris, D. W., and Johnson, G. A. (1986). An orally active inhibitor of renin. Hypertension 8:1105–1112.

Pandey, R. C., Cook, Jr., J. C., and Rinehart, Jr., K. L. (1977). High resolution and field desorption mass spectrometry studies and revised structures of alamethicins I and II. J. Am. Chem. Soc. 99:8469–8483.

Patek, M., and Lebl, M. (1999). Safety-catch and multiply cleavable linkers in solid-phase synthesis Biopolymers 47:353–363.

Payne, J. W., Jakes, R., and Hartley, B. S. (1970). The primary structure of alamethicin. Biochem. J. 117:757–766.

Pearl, L. H., and Taylor, W. R. (1987). A structural model for the retroviral proteases. Nature 329:351–354.

Pearson, W. R., and Lipman, D. J. (1988). Improved tools for biological sequence comparison. Proc. Natl. Acad. Sci. USA 85:2444–2448.

Peña, C., Stewart, J. M., and Goodfriend, T. C. (1974). A new class of angiotensin inhibitors: N-methylphenylalanine analogs. Life Sci. 14:1331.

Pierschbacher, M. D., and Ruoslahti, E. (1984). Cell attachment of fibronectin can be duplicated by small synthetic fragments of the molecule. Nature 309:30–33.

Pinilla, C., Appel, J. R., Blanc, P., and Houghten, R. A. (1992). Rapid identification of high affinity peptide ligands using positional scanning synthetic peptide combinatorial libraries. BioTechniques 13:901–905.

Pinilla, C., Appel, J. R., Blondelle, S. E., Dooley, C. T., Eichler, J., Ostresh, J. M., and Houghten, R. A. (1994). Versatility of positional scanning synthetic combinatorial libraries for the identification of individual compounds. Drug Dev. Res. 33:133–145.

Plow, E. F., Pierschbacher, M. D., Ruoslahti, E., Marguerie, G. A., and Ginsberg, M. H. (1985). The effect of Arg-Gly-Asp-containing peptides on fibrinogen and von Willebrand factor binding to platelets. Proc. Natl. Acad. Sci. USA 82:8057–8061.

Polácek, I., Krejcí, I., Nesvada, H., and Rudinger, J. (1970). Action of [1,6-di-alanine]-oxytocin and [1,6-di-serine]-oxytocin on the rat uterus and mammary gland in vitro. Eur. J. Pharmacol. 9:239–248.

Quillan, J. M., Jayawickreme, C. K., and Lerner, M. R. (1995). Combinatorial diffusion assay used to identify topically active melanocyte-stimulating hormone receptor antagonists. Proc. Natl. Acad. Sci. USA 9:2894–2898.

Rinehart, Jr., K. L., Gloer, J. B., Cook, Jr., J. C., Miszak, S. A., and Scahill, T. A. (1981). Structures of the didemnins, antiviral and cytotoxic depsipeptides from a Caribbean tunicate. J. Am. Chem. Soc. 103:1857–1859.

Rivier, J. E., and Brown, M. R. (1978). Bombesin, bombesin analogues, and related peptides: effects on thermoregulation. Biochemistry 17:1766–1771.

Rivier, J., Brown, M., and Vale, W. (1976). D-Trp8 somatostatin: an analog of somatostatin more potent than the native molecule. Biochem. Biophys. Res. Comm. 65:746–751.

Rose, G. D., and Roy, S. (1980). Hydrophobic basis of packing in globular proteins. Proc. Natl. Acad. Sci. USA 77:4643–4647.

Roy, C., Barth, T., and Jard, S. (1975). Vasopressin-sensitive kidney adenylate cyclase. J. Biol. Chem. 250:3149–3156.

Rüegger, A., Kuhn, M., Lichti, H., Loosli, H.-R., Huguenin, R., Quiquerez, C., and Wartburg, A. V. (1976). Cyclosporin A, ein immunosuppressiv wirksamer Peptidmetabolit aus *Trichoderma polysporem* Rifai. Helv. Chim. Acta 59:1075–1092.

Rusche, J. R., Javaherian, K., McDanal, C., Petro, J., Lynn, D. L., Grimaila, R., Langlois, A., Gallo, R. C., Arthur, L. O., Fischinger, P. J., Bolognesi, D. P., Putney, S. D., and Matthews, T. J. (1988). Antibodies that inhibit fusion of human immunodeficiency virus-infected cells bind a 24-amino acid sequence of the viral envelope, gp120. Proc. Natl. Acad. Sci. USA 85:3198–3202.

Rutter, W. J., and Santi, D. V. (1989). Preparation of peptide mixtures and selecting peptides with specific properties. PCT Int. Appl., 57 pp.

Samanen, J., Narindray, D., Cash, T., Brandeis, E., Adams, J., W., Yellin, T., Eggleston, D., Debrosse, C., and Regoli, D. (1989). Potent angiotensin II antagonists with non-β-branched amino acids in position 5. J. Med. Chem. 32:466–472.

Samanen, J., Zuber, G., Bean, J., Eggleston, D., Romoff, T., Kopple, K., Saunders, M., and Regoli, D. (1990). 5,5-Dimethylthiazolidine-4-carboxylic acid (Dtc) as a proline analog with restricted conformation. Int. J. Peptide Protein Res. 35:501–509.

Samanen, J., Ali, F. E., Romoff, T., Calvo, R., Sorenson, E., Bennett, D., Berry, D., Kostler, P., Vasko, J., Powers, D., Stadel, J., and Nichols, A. (1991). SK&F 106760: Reinstatement of high receptor affinity an a peptide fragment (RGDS) of a large glycoprotein (fibrinogen) through conformational constraints. In *Peptides 1990: Proceedings of the 21st European Peptide Symposium*, E. Giralt and D. Andreu, eds., Leiden, Escom.

Sarantakis, D., McKinley, W. A., and Grant, N. H. (1973). The synthesis and biological activity of Ala[3,14]-somatostatin. Biochem. Biophys. Res. Comm. 55:538–542.

Sawyer, W. H., Acosta, M., Balaspiri, L., Judd, J., and Manning, M. (1974). Structural changes in the arginine vasopressin molecule that enhance antidiuretic activity and specificity. Endocrinology 94:1106–1115.

Schröder, E., Lübke, K., Lehmann, M., and Beetz, I. (1971). Peptide syntheses. XLVII. Mellitin. 3. Hemolytic acitivty and action on surface tension of aqueous solutions on synthetic mellitins and their derivatives. Experientia 27:764–765.

Singh, J., Allen, M. P., Ator, M. A., Gainor, J. A., Whipple, D. A., Solowiej, J. E., Treasurywala, A. M., Morgan, B. A., Gordon, T. D., and Upson, D. A. (1995). Validation of screening immobilized peptide libraries for discovery of protease substrates. J. Med. Chem. 38:217–219.

Smith, J. A., and Pease, L. G. (1980). Reverse turns in peptides and proteins. In *CRC Critical Reviews in Biochemistry*, G. D. Fasman, ed., Boca Raton, Fla., CRC Press.

Songyang, Z., Blechner, S., Hoagland, N., Hoekstra, M. F., Piwnica-Worms, H., and Cantley, L. H. (1994). Use of an oriented peptide library to determine the optimal substrates of protein kinases. Curr. Biol. 4:973–982.

Sothrup-Jensen, L., Petersen, T. E., and Magnuson, S. (1980). A thiol-ester in α_2-macroglobulin is cleaved during proteinase complex formation. FEBS Lett. 121:275–279.

Spatola, A. F., Saneii, H., Edwards, J. V., Bettag, A. L., Anwer, M. K., Rowell, P., Browne, B., Lahti, R., and von Voightlander, P. (1986). Structure-activity relationships of enkephalins containing serially replaced thiomethylene amide bond surrogates. Life. Sci. 38:1243–1249.

Tack, B. F., Harrison, R. A., Janatova, J., Thomas, M. L., and Prahl, J. W. (1980). Evidence for the presence of an internal thiolester bond in third component of human complement. Proc. Natl. Acad. Sci. USA 77:5764–5768.

Tam, J. P., and Spetzler, J. C. (1997). Multiple antigen peptide system. Methods Enzymol. 289:612–637.

Tanford, C. (1973). *The Hydrophobic Effect*, New York, John Wiley & Sons.

Tang, J., James, M. N. G., Hsu, I. N., Jenkins, J. A., and Blundell, T. L. (1978). Structural evidence for gene duplication in the evolution of the acid proteases. Nature 271:618–621.

Terrett, N. K., Gardner, M., Gordon, D. W., Kobylecki, R. J., and Steele, J. (1995a). Combinatorial synthesis—the design of compound libraries and their application to drug discovery. Tetrahedron 51:8135–73.

Terrett, N. K., Bojanic, D., Brown, D., Bungay, P. J., Gardner, M., Gordon, D. W., Mayers, C., and Steele, J. (1995b). The combinatorial synthesis of a 30,752-compound library—discovery of sar around the endothelin antagonist, FR-139,317. Bioorg. Med. Chem. Lett. 5:917–22.

Thaisrivongs, S., Pals, D. T., Lawson, J. A., Turner, S. R., and Harris, D. W. (1987). α-Methylproline-containing renin inhibitory peptides: In vivo evaluation in an anesthetized, ganglion-blocked, hog renin infused rat model. J. Med. Chem. 30:536–541.

Thanki, N., Thornton, J. M., and Goodfellow, J. M. (1990). Influence of secondary structure on the hydration of serine, threonine and tyrosine residues in proteins. Prot. Eng. 3:495–508.

Turk, J., Needleman, P., and Marshall, G. R. (1975). Analogs of bradykinin with restricted conformational freedom. J. Med. Chem. 18:1139–1142.

Turk, J., Needleman, P., and Marshall, G. R. (1976). Analogs of angiotensin II with restricted conformational freedom including a new antagonist. Mol. Pharmacol. 12:217–224.

Van Regenmortel, M. H. V., Briand, J. P., Muller, S., and Plaué, S. (1988). Synthetic polypeptides as antigens. In *Laboratory Techniques in Biochemistry and Molecular Biology*, Vol. 19, R. H. Burdon and P. H. van Knippenberg, eds., Amsterdam, Elsevier.

Veronese, F. D., DeVico, A. L., Copeland, T. D., Oroszlan, S., Gallo, R. C., and Sarngadharan, M. G. (1985). Characterization of gp41 as the transmembrane protein coded by the HTLV-III/LAV envelope gene. Science 229:1402–1405.

Walter, R., Rudinger, J., and Schwartz, I. L. (1967). Chemistry and structure–activity relations of the antidiuretic hormones. Am. J. Med. 42:653–677.

Walter, R., Smith, C. W., Mehta, P. K., Boonjarern, S., Arruda, J. A. L., and Kurtzman, N. A. (1977). Conformational considerations of vasopressin as a guide to development of biological probes and therapeutic agents. In *Disturbances in Body Fluid Osmolality*, T. E. Andreoli, J. J. Grantham and F. C. Rector Jr., eds., Bethesda, Md., American Physiological Society.

Weidner, J. R., and Dunn, B. M. (1991). Development of synthetic peptide substrates for the poliovirus 3C proteinase. Arch. Biochem. Biophys. 286:402–408.

Wickner, W. T., and Lodish, H. F. (1985). Multiple mechanisms of protein insertion into and across membranes. Science 230:400–407.

Wieland, T. (1968). Poisonous principles of mushrooms of the genus Amanita. Science 159:946–952.

Wlodawer, A., Miller, M., Jaskólski, M., Sathyanarayana, B. K., Baldwin, E., Weber, I. T., Selk, L. M., Clawson, L., Schneider, J., and Kent, S. B. H. (1989). Conserved folding in retroviral proteases: crystal structure of a synthetic HIV-1 protease. Science 245:616–621.

Wlodawer, A., Miller, M., Swain, A. L., and Jaskólski, M. (1991). Structure of three inhibitor complexes of HIV-1 protease. In *Methods in Protein Sequence Analysis*, Jörnvall, Höög and Gustavsson, eds., Basel, Birkhauser Verlag.

Wolfsberg, T. G., and Madden, T. L. (1995). Sequence similarity searching using BLAST. In *Current Protocols in Protein Science*, 2.5.1–2.5.29, New York, John Wiley and Sons.

Wu, J., Ma, Q. N., and Lam, K. S. (1994). Identifying substrate motifs of protein kinases by a random library approach. Biochemistry 33:14825–14833.

Yajima, H., and Fujii, N. (1981). Studies on peptides. 103. Chemical synthesis of a crystalline protein with the full enzymic activity of ribonuclease A. J. Am. Chem. Soc. 103:5867–5871.

Yasunaga, T., Sagata, N., and Ikawa, Y. (1986). Protease gene structure and *env* gene variability of the AIDS virus. FEBS Lett. 199:145–150.

Zaoral, M., Kolc, J., and Sorm, F. (1967a). Amino acids and peptides. LXX. Synthesis of D-Arg8- and D-Lys8 vasopressin. Coll. Czech. Chem. Comm. 32:1242–1248.

Zaoral, M., Kolc, J., and Sorm, F. (1967b). Amino acids and peptides. LXXI. Synthesis of 1-Deamino-8-D-γ-aminobutyrine-vasopressin, 1-Deamino-8-D-lysine-vasopressin and 1-Deamino-8-D-arginine-vasopressin. Coll. Czech. Chem. Comm. 32:1250–1257.

Zuiderweg, E. R. P., Henkin, J., Mollison, K. W., Carter, G. W., and Greer, J. (1988). Comparison of model and nuclear magnetic resonance structures for the human inflammatory protein C5a. Proteins: Struct. Funct. Genet. 3:139–145.

3

Principles and Practice of Solid-Phase Peptide Synthesis

Gregg B. Fields
Janelle L. Lauer-Fields
Rong-qiang Liu
George Barany

Peptides play key structural and functional roles in biochemistry, pharmacology, and neurobiology, and are important probes for research in enzymology, immunology, and molecular biology. The amino acid building blocks can be among the 20 genetically encoded L-residues, or else unusual ones, and the sequences can be linear, cyclic, or branched. It follows that rapid, efficient, and reliable methodology for the chemical synthesis of these molecules is of utmost interest. A number of synthetic peptides are significant commercial or pharmaceutical products, ranging from the sweet dipeptide L-Asp-L-Phe-OMe (aspartame) to clinically used hormones such as oxytocin, adrenocorticotropic hormone, calcitonin, and gonadotropin releasing hormone (GnRH) super-agonists. Synthesis can lead to potent and selective new drugs by judicious substitutions that change functional groups and/or conformations of the parent peptide. These include introduction of N- or C-alkyl substituents, unnatural or D-amino acids, side-chain modifications including sulfate or phosphate groups or carbohydrate moieties, and constraints such as disulfide bridges between half-cystines or side-chain lactams between Lys and Asp or Glu. Commercially important products that evolved from such studies include protease inhibitors, such as captopril and other angiotensin converting enzyme (ACE) inhibitors, peptidomimetic HIV protease inhibitors, and the somatostatin analog lanreotide. Most of the biologically or medicinally important peptides which are the targets for useful structure–function studies by chemical synthesis comprise under 50 amino acid residues, but occasionally a synthetic approach can lead to important conclusions about small proteins (full or domains) in the 100–200 residue size range.

Methods for synthesizing peptides are divided conveniently into two categories: solution (classical) and solid-phase peptide synthesis (SPPS). The classical methods have evolved since the beginning of the twentieth century, and they are described amply in several reviews and books (Wünsch, 1974; Finn and Hofmann, 1976; Bodanszky and Bodanszky, 1984; Goodman et al., 2001). The solid-phase alternative was conceived and elaborated by R. B. Merrifield beginning in 1959, and has also been covered comprehensively (Erickson and Merrifield, 1976; Birr, 1978; Barany and Merrifield, 1979; Stewart and Young, 1984; Merrifield, 1986; Barany et al., 1987, 1988; Kent, 1988; Atherton and Sheppard, 1989; Fields and Noble, 1990; Barany and Albericio, 1991; Fields et al., 1992; Gutte, 1995; Fields, 1997; Lloyd-Williams et al., 1997; Chan and White, 2000; Kates and Albericio, 2000). Solution synthesis retains value in large-scale manufacturing, and for specialized laboratory applications. However, the need to optimize reaction conditions, yields, and purification procedures for essentially every intermediate (each of which has unpredictable solubility and crystallization characteristics) renders classical methods time consuming and labor intensive. Consequently, most workers now requiring peptides for their research opt for the more accessible solid-phase approach.

In this chapter, revised with respect to an edition published a decade ago, we discuss critically the scope and limitations of the best available procedures for solid-phase synthesis of peptides. Developments and trends in the field considered new and mentioned briefly in the previous edition have now faced the test of time, and text has been adjusted accordingly. Literature citations are weighted toward detailed reports with full experimental descriptions, with some bias toward those describing procedures with which we and our collaborators have laboratory experience.

Overview of Solid-Phase Strategy

The concept of SPPS (figure 3-1) is to retain chemistry proven in solution (*protection scheme, reagents*), but to add a covalent attachment step (*anchoring*) that links the nascent peptide chain to an insoluble *polymeric support*. Subsequently, the anchored peptide is extended by a series of addition (*deprotection/coupling*) cycles, which are required to proceed with exquisitely high yields and fidelities. It is the essence of the solid-phase approach that reactions are driven to

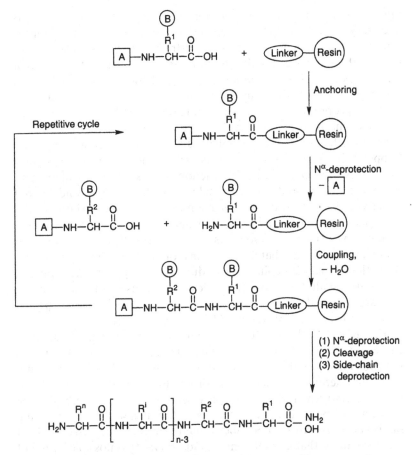

Figure 3-1. Stepwise solid-phase synthesis of linear peptides.
An N^α-derivatized amino acid is attached to an insoluble support (*resin*) via
a *linker* moiety. The N^α-protecting group (A) is then removed selectively
(deprotection). The next amino acid, which is again N^α-protected, is coupled
to the amino acid-linker support. These deprotection/coupling cycles are
repeated until the desired sequence of amino acids is generated. Finally, the
peptide-linker support is cleaved to obtain the peptide as a free acid or an
amide, depending on the chemical nature of the linker. Ideally, the cleavage
reagent also removes the amino acid side-chain-protecting groups (B), which
were chosen to be stable to the repetitive N^α-deprotection conditions. See
text for further details, discussion, and variations.

completion by the use of *excess* soluble reagents, which can be
removed by simple *filtration* and *washing* without manipulative losses.
Because of the speed and simplicity of the repetitive steps, which are
carried out in a single reaction vessel at ambient temperature, the
major portion of the solid-phase procedure is readily amenable to

automation. Once chain elaboration has been accomplished, it is necessary to release (*cleave*) the crude peptide from the support under conditions that are minimally destructive towards sensitive residues in the sequence. Finally, there must follow prudent *purification* and scrupulous *characterization* of the synthetic peptide product, in order to verify that the desired structure is indeed the one obtained.

An appropriate polymeric support (resin) must be chosen that has adequate mechanical stability, as well as desirable physicochemical properties that facilitate solid-phase synthesis (see below, Polymeric Support). In practice, such supports include those that exhibit significant levels of swelling in useful reaction/wash solvents. Swollen resin beads are reacted and washed batchwise with agitation, and filtered either with suction or under positive nitrogen pressure. Alternatively, solid-phase synthesis may be carried out in a continuous-flow mode, by pumping reagents and solvents through resins that are packed into columns. The usual batchwise resins often lack the rigidity and strength necessary for column procedures. More appropriate supports, which are usually, but not always, lower in terms of functional capacity, are obtained when mobile polymer chains are chemically grafted onto, or physically embedded within, an inert matrix.

Regardless of the structure and nature of the polymeric support chosen, it must contain appropriate functional groups onto which the first amino acid can be anchored. In early schemes which still have considerable popularity, chloromethyl groups are introduced onto a polystyrene resin by a direct Friedel–Crafts reaction, following which an N^α-protected amino acid as its triethylammonium or cesium salt is added to provide a polymer-bound benzyl ester. Subsequently, it was recognized that greater control and generality is possible by use of "handles," which are defined as bifunctional spacers that on one end incorporate features of a smoothly cleavable protecting group. The other end of the handle contains a functional group, often a carboxyl, that can be activated to allow coupling to functionalized supports, for example ones containing aminomethyl groups. Particularly advantageous, though more involved to prepare, are "preformed" handles, which serve to link the first amino acid to the resin in two discrete steps, and thereby provide maximal control over this essential step of the synthesis (see below, Attachment to Support).

The next stage of solid-phase synthesis is the systematic elaboration of the growing peptide chain. In the vast majority of solid-phase syntheses, suitably N^α- and side-chain-protected amino acids are added *stepwise* in the $C \rightarrow N$ direction. A particular merit of this strategy is that the best practical realizations have been shown experimentally to proceed with only negligible levels of racemization. A "temporary" protecting group is removed quantitatively at each step to liberate the N^α-amine of the peptide-resin, following which the next incoming protected amino acid is introduced with its carboxyl group suitably activated (see below, Formation of Peptide Bond). It is frequently worthwhile to verify that the coupling has gone to completion by some monitoring technique (see below, Monitoring).

Once the desired linear sequence has been assembled satisfactorily on the polymeric support, the anchoring linkage must be cleaved. Depending on the chemistry of the original handle and on the cleavage reagent selected, the product from this step can be a C-terminal peptide acid, amide, or other functionality. The cleavage can be conducted so as to retain "permanent" side-chain-protecting groups, and thus yield protected segments which, once purified, are suitable for further condensation. Alternatively, selected "permanent" groups can be retained on sensitive residues for later deblocking in solution. However, the most widely used approach involves final deprotection carried out essentially concurrent to cleavage; in this way, the released product is directly the free peptide. Finally, in what is formally a hybrid approach that has come to the fore quite recently, free peptide segments—themselves prepared by solid-phase synthesis—can be modified with suitable activating groups and then ligated selectively, in solution or even on the solid phase, to create larger structures.

Protection Schemes

The preceding section outlined the key steps of the solid-phase procedure, but dealt only tangentially with combinations of "temporary" and "permanent" protecting groups, and the corresponding methods for their removal. The choice and optimization of protection chemistry is perhaps the key factor in the success of any synthetic endeavor. Even when a residue has been incorporated safely into the growing resin-bound polypeptide chain, it may still undergo irreversible structural modification or rearrangement during subsequent synthetic steps. The vulnerability to damage is particularly pronounced at the final deprotection/cleavage step, since these are usually the harshest conditions. At least two levels of protecting group stability are required, in so far as the "permanent" groups used to prevent branching or other problems on the side chains must withstand repeated applications of the conditions for quantitative removal of the "temporary" N^{α}-amino protecting group. However, structures of "permanent" groups must be such that conditions can be found to remove them with minimal levels of side reactions that might affect the integrity of the desired product. The necessary stability is often approached by kinetic "fine-tuning," which is a reliance on quantitative rate differences whenever the same chemical mechanism (usually acidolysis) serves to remove both classes of protecting groups. An often-limiting consequence of such schemes based on graduated lability is that they force adoption of relatively severe final deprotection conditions. Alternatively, *orthogonal* protection schemes can be used. These involve two or more classes of groups that are removed by differing chemical mechanisms, and therefore can be removed in any order and in the presence of the other classes. Orthogonal schemes offer the possibility of substantially milder overall conditions, because selectivity can be attained on the basis of differences in chemistry rather than in reaction rates.

"Temporary" Protection of N^α-Amino Groups

Boc Chemistry

The so-called standard Merrifield system is based on graduated acid lability (figure 3-2, in a modern improved version). The acidolyzable "temporary" N^α-*tert*-butyloxycarbonyl (Boc) group is introduced onto amino acids with either di-*tert*-butyl dicarbonate (Boc$_2$O) or 2-(*tert*-butoxycarbonyloxyimino)-2-phenylacetonitrile (Boc-ON), in aqueous 1,4-dioxane containing NaOH or triethylamine (Et$_3$N) (Bodanszky and Bodanszky, 1984). The Boc group is stable to alkali and nucleophiles, and removed rapidly by inorganic and organic acids (Barany and Merrifield, 1979). Boc removal is usually carried out with trifluoroacetic acid (TFA) (20–50%) in dichloromethane (DCM) for 20–30 min, and, for special situations, HCl (4 N) in 1,4-dioxane for 35 min. Deprotection with neat (100%) TFA, which offers enhanced peptide-resin solvation compared to TFA–DCM mixtures, proceeds in as little as 4 min (Kent and Parker, 1988; Wallace et al., 1989). Following acidolysis, a rapid diffusion-controlled neutralization step with a tertiary amine, usually 5–10% Et$_3$N or N,N-diisopropylethylamine (DIEA) in DCM for 3–5 min, is interpolated to release the free N^α-amine. Alternatively, Boc-amino acids may be coupled without prior neutralization by using "in situ" neutralization, that is, coupling in the presence of DIEA or NMM (Suzuki et al., 1975; Schnölzer et al., 1992; Sueiras-Diaz and Horton, 1992; Alewood et al., 1997; Miranda and Alewood, 1999). "Permanent" side-chain-protecting groups are ether, ester, and urethane derivatives based on benzyl alcohol, suitably "fine-tuned" with electron-donating methoxy or methyl groups or electron-withdrawing halogens for the proper level of acid stability/lability. Alternatively, ether and ester derivatives based on cyclopentyl or cyclohexyl alcohol are sometimes applied, as their use mitigates the levels of certain side reactions. These "permanent" groups are sufficiently stable to repeated cycles of Boc removal, yet cleaved cleanly in the presence of appropriate scavengers by use of liquid anhydrous hydrogen fluoride (HF) at 0 °C or trifluoromethanesulfonic acid (TFMSA) at 25 °C (see below, Cleavage). The 4-methylbenzhydrylamine (PAM) or 4-(hydroxymethyl)phenylacetic acid (MBHA)

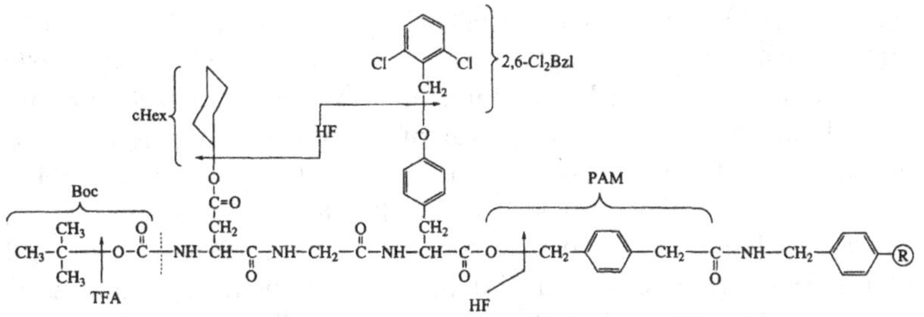

Figure 3-2. Merrifield protection scheme for solid-phase synthesis, based on graduated acidolysis. Temporary N^α-amino protection is provided by the Boc group, removed at each step by the moderately strong acid TFA. Permanent Bzl and cHex type side-chain-protecting groups, and the PAM linkage, are then cleaved simultaneously by HF or other strong acids, with a free peptide acid being formed in high yield.

anchoring linkages are similarly "fine-tuned" to be cleaved at the same time (see below, Attachment to Support).

Fmoc Chemistry
A mild orthogonal alternative is constructed using Carpino's base-labile "temporary" N^α-9-fluorenylmethyloxycarbonyl (Fmoc) group (figure 3-3). The optimal reagent for preparation of Fmoc-amino acids is fluorenylmethyl succinimidyl carbonate (Fmoc-OSu), applied in a partially aqueous/organic mixture in the presence of base; the alternative procedure involving derivatization by Fmoc-chloride is

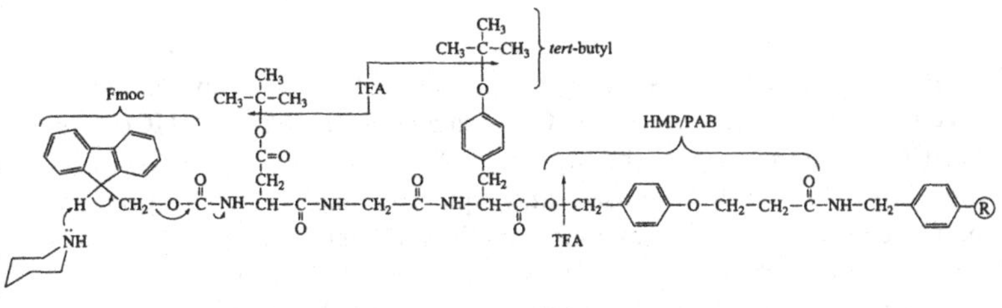

Figure 3-3. A mild two-dimensional orthogonal protection scheme for solid-phase synthesis. Temporary N^α-amino protection is provided by the Fmoc group, removed by the indicated base-catalyzed β-elimination mechanism. Permanent *t*Bu-based side-chain-protecting groups and the HMP/PAB ester linkage are both cleaved by treatment with TFA to yield the free peptide acid. A third dimension of orthogonality may be added with an acid-stable, photolabile anchoring linkage (details in text).

accompanied by unacceptable levels (2–20%) of Fmoc-dipeptide formation (Pacquet, 1982; Sigler et al., 1983; Lapatsanis et al., 1983; Tesser et al., 1983; Ten Kortenaar et al., 1986; de Lisle Milton et al., 1987; Fields et al., 1989). Removal of the Fmoc group is achieved usually with 20–55% piperidine in N,N-dimethylformamide (DMF) or N-methylpyrrolidone (NMP) for 10–18 min (Atherton et al., 1978a,b; Chang et al., 1980a; Albericio et al., 1990a; Fields and Fields, 1991); piperidine in DCM is not recommended, as an amine salt precipitates after relatively brief standing. The base abstracts the acidic proton at the 9-position of the fluorene ring system; β-elimination follows to give a highly reactive dibenzofulvene intermediate which is trapped by excess secondary amine to form a stable, harmless adduct (Carpino and Han, 1972). The Fmoc group may also be removed by 2% 1,8-diazabicyclo[5.4.0]undec-7-ene (DBU) in DMF; however, this reagent is recommended for continuous-flow syntheses only since the dibenzofulvene intermediate does not form an adduct with DBU and thus must be washed rapidly from the peptide-resin (Wade et al., 1991). For batch syntheses, a solution of DBU–piperidine–DMF (1:1:48) is effective, as the piperidine component scavenges the dibenzofulvene (C. G. Fields et al., 1993b; Kates et al., 1996). After Fmoc removal mediated by base, the liberated N^α-amine of the peptide-resin is free and ready for immediate acylation without an intervening neutralization step (compare to previous paragraph on Boc chemistry). "Permanent" protection compatible with N^α-Fmoc protection is provided primarily by ether, ester, and urethane derivatives based on *tert*-butanol. These derivatives are cleaved at the same time as appropriate anchoring linkages, by use of TFA at 25 °C. Scavengers must be added to the TFA to trap the reactive carbocations which form under the acidolytic cleavage conditions.

A problem that occurs during preparation of Fmoc-amino acids is the precipitation of either the Fmoc-OSu reagent or the base (NaHCO$_3$ or Na$_2$CO$_3$) upon mixing of the organic (1,4-dioxane, acetone, DMF, and acetonitrile have been proposed) and aqueous co-solvents. This problem is best overcome by use of aqueous dimethoxyethane with Na$_2$CO$_3$ (Fields et al., 1989; Nagase et al., 1994). A solution of Fmoc-OSu (3.0 mmol) in dimethoxyethane (10 mL) is added slowly to the amino acid (2.0 mmol) dissolved in 10% aqueous Na$_2$CO$_3$ (10 mL); final yields of Fmoc-amino acids after workup are in the 75–95% range. Another successful procedure (Han et al., 1996b) involves addition of a solution of Fmoc-OSu (1.1 mmol) in acetonitrile (2.0 mL) to a solution of amino acid (1.0 mmol) plus triethylamine (2.5 mmol) in water (2.0 mL); yields are in the 80% range.

The Fmoc group has been shown to be completely stable to treatment with TFA, HBr in acetic acid (HOAc), or HBr in nitromethane for 1–2 days (Carpino and Han, 1972). Somewhat less stability was found in dipolar aprotic solvents (Atherton et al., 1979). Fmoc-Gly was deprotected after 7 days in dimethylacetamide (DMA), DMF, and NMP to the extent of 1%, 5%, and 14% respectively. Although these low levels of decomposition are considered relatively insignificant, it is nevertheless prudent to purify (by distillation at reduced pressure) the aforementioned solvents just before use. With NMP, Fmoc group removal is attributed directly to the presence of methylamine as an impurity (Otteson et al., 1989). The addition of HOBt (0.01–0.1 M) greatly reduces the detrimental effect of methylamine (Albericio and Barany, 1987); Fmoc-Gly-HMP-resin was <0.05% deprotected after 12 h in NMP containing 0.01 M HOBt (Fields et al., 1990). Fmoc-amino acids can be stored in purified or "synthesis-quality" NMP with little decomposition for 6–8 weeks, in the dark at 25 °C (J. Kent et al., 1991).

"Permanent" Protecting Groups for Reactive Amino Acid Side Chains

Once the means for N^α-amino protection has been selected, compatible protection for the side chains of trifunctional amino acids must be specified. These choices are made in the context of potential side reactions, which should be minimized. Problems may be anticipated either during the coupling steps or at the final deprotection/ cleavage. For certain residues (e.g., Cys, Asp, Glu, Lys), side-chain protection is absolutely essential, whereas for others, an informed decision should be made depending upon the length of the synthetic target and other considerations. Most solid-phase syntheses follow maximal rather than minimal protection strategies. Almost all of the useful N^α-Boc and N^α-Fmoc protected derivatives can be manufactured in bulk, and are found in the catalogues of the major suppliers of peptide synthesis chemicals. The most widely used "permanent" protecting groups for the trifunctional amino acids have been listed (table 3-1), together with information on how derivatives are prepared, conditions for their intentional deblocking, and conditions under which the indicated side-chain protection is either entirely stable, or prematurely cleaved by reagents used for peptide synthesis.

A useful resource to find amino acid derivatives is the Peptide Synthesis Database located on the World Wide Web at http://ChemLibrary. BRI.NRC.CA/home.html; this site is searchable and contains links to commercial suppliers.

Table 3-1 Amino acid side-chain protection for solid-phase peptide synthesis[a]

Side-chain protecting group	Protected amino acid derivatives	Stability	Removal[a]	Reagent(s) for introduction	Reference(s)[a]
Compatible with Boc chemistry					
Benzyl	Asp/Glu(OBzl)	TFA	strong acid	Bzl-OH + H[+]	Barany & Merrifield, 1979[#] Bodanszky & Bodanszky, 1984[*,b] Tam & Merrifield, 1987[#]
	Ser/Thr/Tyr(Bzl)	TFA base	strong acid	Bzl-Br + base	Yajima et al., 1988[#] Kiso et al., 1989[*]
	Hyp(Bzl)	TFA	strong acid	Bzl-Br	Weber & Nitschmann, 1978
2-Adamantyl	Asp(O-2-Ada)	TFA piperidine 1 M HCl	strong acid	Ada-2-OH + DCC	Okada & Iguchi, 1988[*,#]
Cyclohexyl	Asp(OcHex)	TFA	strong acid	cHex-OH + carbodiimide + DMAP or cHex-OH + H[+]	Tam & Merrifield, 1987[#] Tam et al., 1988[*,#] Yajima et al., 1988[#] Penke & Tóth, 1989[*] Kiso et al., 1989[*]
9-Fluorenylmethyl	Cys(Fm)	TFA HF	piperidine[d]	Fm-OTos + base	Albericio et al., 1990c[*,#,c]
	Asp/Glu(OFm)	TFA HF	piperidine TBAF	Fm-OH + DCC/DMAP	Albericio et al., 1990c[*,#,c]

Protecting group	Structure	Derivative			Reagent	References
2,6-Dichlorobenzyl 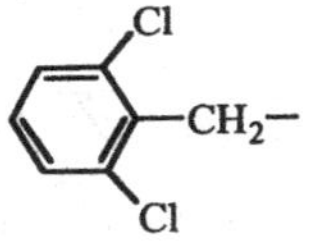		Tyr(2,6-Cl$_2$Bzl)	TFA	strong acid	2,6-Cl$_2$Bzl-Br + base	Erickson & Merrifield, 1973a[*,#]; Tam & Merrifield, 1987[#]; Yajima et al., 1988[#]; Kiso et al., 1989[#]
2-Bromobenzyloxycarbonyl 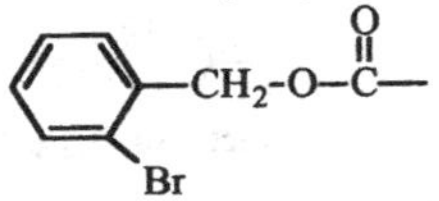		Tyr(2-BrZ)	TFA	strong acid	2-BrZ-ONp + base	Yamashiro & Li, 1973[*,#]; Tam & Merrifield, 1987[#]
2-Chlorobenzyloxycarbonyl 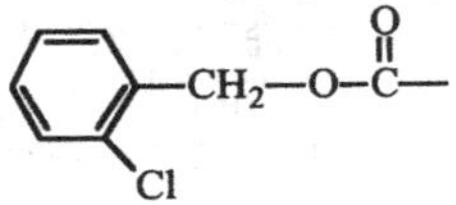		Lys(2-ClZ)	TFA	strong acid	2-ClZ-OSu + base	Erickson & Merrifield, 1973b[*,#]; Bodanszky & Bodanszky, 1984[*]; Tam & Merrifield, 1987[#]; Kiso et al., 1989[#]
9-Fluorenylmethyloxycarbonyl 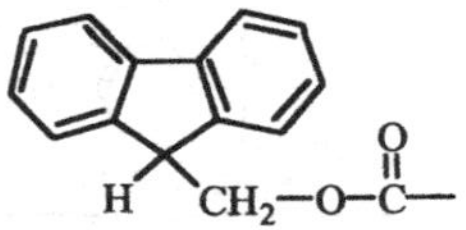		Lys(Fmoc)	TFA Pd(O)	piperidine TBAF	Fmoc-N$_3$ + MgO	Albericio et al., 1990c[*,#,c]

(continued)

103

Table 3-1 (*Continued*)

Side-chain protecting group	Protected amino acid derivatives	Stability	Removal[a]	Reagent(s) for introduction	Reference(s)[a]
4-Toluenesulfonyl (structure)	His(Tos)	TFA	strong acid Ac_2O/pyridine HOBt[@]	Tos-Cl + base	Stewart et al., 1972[#] Barany & Merrifield, 1979[#] van der Eijk et al., 1980[#] Bodanszky & Bodanszky, 1984[*]
	Arg(Tos)	HBr/TFA TFA		Tos-Cl + base	Tam & Merrifield, 1987[#]
Mesitylene-2-sulfonyl (structure)	Arg(Mts)	TFA	strong acid	Mts-Cl + base	Yajima et al., 1978[*,#] Tam & Merrifield, 1987[#] Yajima et al., 1988[#]
2,4-Dinitrophenyl (structure)	His(Dnp)	acids	thiophenol	1-fluoro-2,4-dinitrobenzene + base	Chillemi & Merrifield, 1969[*,#] Stewart et al., 1972[#] Tam & Merrifield, 1987[#] Applied Biosystems, Inc., 1989a[#]
Benzyloxymethyl $-CH_2-O-CH_2-$	His(Bom)	TFA	strong acid	$ClCH_2OBzl$	Brown et al., 1982[*,#] Tam & Merrifield, 1987[#] Kiso et al., 1989[#]
Formyl (on N^{in})	Trp(CHO)	TFA HF[c]	TFMSA piperidine HF[c]	HCO_2H	Bodanszky & Bodanszky, 1984[*] Tam & Merrifield, 1987[#] Yajima et al., 1988[#]

Protecting group	Residue	Removal conditions		Introduction	References
Cyclohexyloxycarbonyl	Trp(Hoc)	TFA DIEA piperidine	HF + p-cresol HF + 1,4 butanedithiol + p-cresol	Hoc-Cl + tetra-n-butyl ammonium hydrogensulfate + NaOH	Nishiuchi et al., 1996[*,#]
Sulfoxide (on thioether)	Met(O)	TFA base HF[c]	MMA DMF–SO₃ HF[c] NH₄I TMSBr + thioanisole	H₂O₂	Houghton & Li, 1979[#] Bodanszky & Bodanszky, 1984[*] Tam & Merrifield, 1987[#] Fujii et al., 1987[#] Beck & Jung, 1994[#] Ferrer et al., 1999[#]
4-Methylbenzyl	Cys(Meb)	TFA	strong acid Tl(Tfa)₃	Meb-Br + base	Erickson & Merrifield, 1973a[*,#] Barany & Merrifield, 1979[#] Tam & Merrifield, 1987[#] Fujii et al., 1987[#]
3-Nitro-2-pyridinesulfenyl	Cys(Npys)	TFA HF[c]	thiols HF[c] HOBt[@] piperidine[@]	Npys-Cl	Matsueda & Walter, 1980[*] Albericio et al., 1989b[#] Rosen et al., 1990[#]

(continued)

Table 3-1 (*Continued*)

Side-chain protecting group	Protected amino acid derivatives	Stability	Removal[a]	Reagent(s) for introduction	Reference(s)[a]
Compatible with Fmoc chemistry					
tert-Butyl	Asp/Glu(O*t*Bu)	base, Pd(0)	TFA	C_4H_8/H^+	Chang *et al.*, 1980b[*]; Bodanszky & Bodanszky, 1984[*]
	Ser/Thr/Tyr(*t*Bu)	base, Pd(0)	TFA	C_4H_8/H^+	Meienhofer, 1985[#]; Loffet *et al.*, 1989[*]; Lajoie *et al.*, 1990[*,e]
	Hyp(*t*Bu)	base, Pd(0)	TFA	C_4H_8/H^+	Greene, 1991[#]
4,4-Dimethyl-2,6-dioxocyclohexylidene-3-methylbutyl-aminobenzyl ester	Asp/Glu(Dmab)	piperidine, DBU, Pd(0), TFA	hydrazine	Dmab-OH + DCC	Chan *et al.*, 1995[*,#]
tert-Butyloxycarbonyl	Lys(Boc)	base	TFA	Boc₂O or Boc-ON	Bodanszky & Bodanszky, 1984[*]; Meienhofer, 1985[#]
	His(Boc)	piperidine	TFA	Boc₂O[f]	Atherton & Sheppard, 1989[*,#]
	Trp(Boc)	base	TFA	Boc₂O	White *et al.*, 1992[*]; Riniker *et al.*, 1993[#]; Fields & Fields, 1993[#]

Name	Abbreviation	Removal	Cleavage	Introduction	References
4,4-Dimethyl-2,6-dioxocyclohex-1-ylidene-ethyl CH_3 CH_3 CH_3	Lys(Dde)	piperidine TFA	hydrazine	2-acetyl-dimedone +Fmoc-lys	Bycroft et al., 1993[*,#]; Bloomberg et al., 1993[#]; C.G. Fields et al., 1993a[#]; Rohwedder et al., 1998[#]
4,4-Dimethyl-2,6-dioxocyclohex-1-ylidene-3-methylbutyl $CH_2CH(CH_3)_2$ CH_3 CH_3	Lys(ivDde)	piperidine DBU Pd(0) TFA	hydrazine	dimedone + isopentanoic acid + DCC/DMAP +Fmoc-Lys	Chhabra et al., 1998[*,#]
4-Methyltrityl CH_3	His(Mtt)	piperidine hydrazine Pd(0)	dilute TFA	Mtt-Cl + Et_3N +$(CH_3)_2$ $SiCl_2$	Barlos et al., 1991d[*,#]
	Lys(Mtt)	piperidine hydrazine Pd(0)	dilute TFA	Mtt-Cl + Et_3N + $(CH_3)_3$ SiCl	Aletras et al., 1995[*,#]
4-Methoxy-2,3,6-trimethyl-benzenesulfonyl CH_3O CH_3 CH_3 CH_3	Arg(Mtr)	piperidine	TFA[g]	2,3,5-trimethyl-anisole[e] + $ClSO_3H$	Fujino et al., 1981[*,#]; Atherton & Sheppard, 1989[#]

(continued)

Table 3-1 (*Continued*)

Side-chain protecting group	Protected amino acid derivatives	Stability	Removal[a]	Reagent(s) for introduction	Reference(s)[a]
2,2,5,7,8-Pentamethyl-chroman-6-sulfonyl	Arg(Pmc)	piperidine	TFA	2,2,5,7,8-pentamethyl-chroman + ClSO₃H	Ramage & Green, 1987[*,#]
2,2,4,6,7-Pentamethyldihydrobenzo-furan-5-sulfonyl	Arg(Pbf)	piperidine	TFA	2,2,4,6,7-pentamethyl-dihydrobenzo-furan + ClSO₃H	Carpino et al., 1993[*,#] Fields et al., 1993[#]
Triphenylmethyl	His(Trt)	piperidine 1 M HCl	TFA	Trt-Cl +(CH₃)₂SiCl₂	Sieber & Riniker, 1987[#] Barlos et al., 1982[*]
	Cys(Trt)	piperidine	HOAc TFA Hg(II) I₂	Trt-OH + BF₃·Et₂O	Photaki et al., 1970[#] Bodanszky & Bodanszky, 1984[*] Meienhofer, 1985[#] Greene, 1991[#]
	Asn/Gln(Trt)	piperidine 1 M HCl	TFA	Trt-OH + Ac₂O/H⁺	Sieber & Riniker, 1991[*,#]
	Ser/Thr	piperidine	TFA + TIS	Trt-Cl + Et₃N	Barlos et al., 1987[*,#] Barlos et al., 1991b[*] Barlos et al., 1998[#]

tert-Butoxymethyl CH_3 $CH_3—C—O—CH_2—$ CH_3	His(Bum)	piperidine	TFA	ClCH$_2$O*t*Bu	Colombo *et al.*, 1984[*,#]
2,4,6-Trimethoxybenzyl OCH$_3$ —CH$_2$— CH$_3$O— —OCH$_3$	Asn/Gln(Tmob)	piperidine	TFA	Tmob-NH$_2$ + DCC/HOSu	Weygand *et al.*, 1968a[*] Hudson, 1988b[*]
	Cys(Tmob)	piperidine	TFA	Tmob-OH + TFA	Munson *et al.*, 1992[*,#]
Methoxytrityl	Cys(Mmt)	piperidine hydrazine Pd(0)	dilute TFA/TES	Mmt-Cl	Barlos *et al.*, 1996[*,#]

(*continued*)

Table 3-1 (*Continued*)

Side-chain protecting group	Protected amino acid derivatives	Stability	Removal[a]	Reagent(s) for introduction	Reference(s)[a]
Compatible with both Boc and Fmoc chemistries					
Allyl $CH_2=CH-CH_2-$	Asp/Glu(OAl)	acids base	Pd(0)	Al-OH $+(CH_3)_3SiCl$ or H_2SO_4	Belshaw et al., 1990[*] Greene, 1991[*,#] Lyttle & Hudson, 1992[*,#]
Allyloxycarbonyl $CH_2=CH-CH_2-O-\overset{O}{\overset{\|}{C}}-$	Lys(Aloc)	acids base	Pd(0)	Aloc-Cl	Lyttle & Hudson, 1992[*,#]
9-Xanthenyl 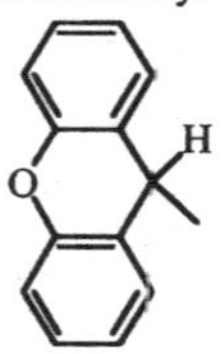 	Asn/Gln(Xan)	base	strong acid TFA⊕	Xan-OH + HOAc	Dorman et al., 1972[*,#] Stewart & Young, 1984[#] Tam & Merrifield, 1987[#] Han et al., 1997b[*,#]

110

Protecting group	Derivative				References
Acetamidomethyl $CH_3-\overset{\overset{\displaystyle O}{\|}}{C}-NH-CH_2-$	Cys(Acm)	piperidine acids	Hg(II) I_2 Tl(Tfa)$_3$	Acm-OH/TFA	Veber et al., 1972[*,#] Bodanszky & Bodanszky, 1984[*] Atherton et al., 1985a[#] Albericio et al., 1987a[*,c] Tam & Merrifield, 1987[#] Albericio et al., 1987[*,g] Fujii et al., 1987[#] Brady et al., 1988[#] McCurdy, 1989[#] Andreu et al., 1994[#] Sakakibara, 1999[#]
Trimethylacetamidomethyl $(CH_3)_3C-\overset{\overset{\displaystyle O}{\|}}{C}-NH-CH_2-$	Cys(Tacm)	piperidine acids	Hg(II) I_2 AgBF$_4$	Tacm-OH/TFA	Yoshida et al., 1990[#] Kiso et al., 1990[*,#]
tert-Butylsulfenyl $CH_3-\overset{\overset{\displaystyle CH_3}{\|}}{\underset{\underset{\displaystyle CH_3}{\|}}{C}}-S-$	Cys(StBu)	piperidine acids	thiols phosphines	tBu-SH tBu-SCN	Wünsch & Spangenberg, 1971[*] Atherton et al., 1985a[#] Romani et al., 1987[#]

[a] The protecting group structure drawn in the left column does not include the functional group that is being protected through a point of attachment at the far right of the structure. The order corresponds, by category, to first mention in the text. Abbreviations are provided in the second column. Conditions or reagents listed under "Removal" are intended for quantitative deprotection (see "Cleavage"), unless marked by "@", in which case the removal is an unacceptable side reaction that indicates an incompatibility with the protecting group. "Strong acid" means HF, TFMSA, or equivalent reagents (see "Cleavage"). TFA cleavages are best carried out in the presence of appropriate scavengers (see "Cleavage"). References marked by " * " refer to preparation of the side-chain derivative; references marked by "#" refer to stability and lability of each derivative.

[b] Boc-Thr(Bzl) is best prepared by the procedure of Chen et al., 1989.

[c] Stable to "high" (90%) HF, but removed by "low" HF-scavengers; see "Cleavage".

[d] Complete deprotection requires 4 h treatment by 50% piperidine in DMF.

[e] One-pot synthesis of the Fmoc-amino acid side-chain protected derivative.

[f] Fmoc-His(Boc) is prepared by reacting (Boc)₂O with commercially available Fmoc-His(Fmoc).

[g] Arg(Mtr) deprotection by TFA can be slow, especially in multiple Arg(Mtr)-containing peptides. See "Cleavage".

[h] Tmob-amine is reacted with a carboxyl protected Asp or Glu to produce Asn(Tmob) or Gln(Tmob), respectively.

The side-chain carboxyls of Asp and Glu are protected as benzyl (OBzl) esters for Boc chemistry and as *tert*-butyl (O*t*Bu) esters for Fmoc chemistry. A sometimes serious side reaction with protected Asp residues involves an intramolecular elimination to form an aspartimide, which can then partition in water to the desired α-peptide and the undesired by-product with the chain growing from the β-carboxyl (Bodanszky and Kwei, 1978; Barany and Merrifield, 1979; Tam et al., 1988). Aspartimide formation is sequence dependent, with Asp(OBzl)-Gly, -Ser, -Thr, -Asn, and -Gln sequences showing the greatest tendency to cyclize under basic conditions (Bodanszky et al., 1978; Bodanszky and Kwei, 1978; Nicolás et al., 1989); the same sequences are also quite susceptible in strong acid (Barany and Merrifield, 1979; Fujino et al., 1981; Tam et al., 1988). For models containing Asp(OBzl)-Gly, the rate and extent of aspartimide formation was substantial both in base (100% after 10 min treatment with piperidine–DMF [1 : 4], 50% after 1–3 h treatment with Et₃N or DIEA) and in strong acid (a typical value is 36% after 1 h treatment with HF at 25°C). By comparison, sequences containing Asp(O*t*Bu)-Gly are somewhat susceptible to base-catalyzed aspartimide formation (11% after 4 h treatment with piperidine–DMF [1 : 4]) (Nicolás et al., 1989), but do not rearrange at all in acid (Kenner and Seely, 1972). Sequence dependence studies of Asp(O*t*Bu)-X peptides revealed that piperidine could induce aspartimide formation when X = Arg(Pmc), Asn(Trt), Asp(O*t*Bu), Cys(Acm), Gly, Ser, Thr, and Thr(*t*Bu) (Yang et al., 1994; Lauer et al., 1995), and also showed that aspartimide formation can be conformation-dependent (Dölling et al., 1994). This side reaction can be minimized by including 0.1 M HOBt in the piperidine solution (Lauer et al., 1995), or eliminated by using an amide backbone protecting group (e.g., 2-hydroxy-4-methoxybenzyl) for the residue in the X position of an Asp-X sequence (Quibell et al., 1994).

To minimize the acid-catalyzed imide/$\alpha \to \beta$ rearrangement side reaction, Boc-Asp may be protected with either the 2-adamantyl (O-2-Ada) (Okada and Iguchi, 1988) or cyclohexyl (OcHex) (Tam et al., 1988) groups. The base-labile 9-fluorenylmethyl (OFm) group offers orthogonal side-chain protection for Boc-Asp/Glu (Bolin et al., 1989; Albericio et al., 1990c; Al-Obeidi et al., 1990), while the palladium-sensitive allyl (OAl) group (Belshaw et al., 1990; Lyttle and Hudson, 1992) offers orthogonal side-chain protection for both Boc- and Fmoc-Asp/Glu. The *N*-[1-(4,4,-dimethyl-2,6-dioxocyclo-hexylidene)-3-methylbutyl]aminobenzyl (Dmab) group can be used for quasi-orthogonal protection of Asp during Fmoc chemistry, as Dmab is stable to piperidine–DMF (1 : 4) and can be removed selectively with 2% hydrazine in DMF (Chan et al., 1995).

The side-chain hydroxyls of Ser, Thr, and Tyr are protected as Bzl ethers for Boc and either *t*Bu or Trt ethers for Fmoc SPPS. In strong acid, the Bzl protecting group blocking the Tyr phenol can migrate to the 3-position of the ring (Erickson and Merrifield, 1973a). This side reaction is decreased greatly when Tyr is protected by the 2,6-dichlorobenzyl (2,6-Cl$_2$Bzl) (Erickson and Merrifield, 1973a) or 2-bromobenzyloxycarbonyl (2-BrZ) (Yamashiro and Li, 1973) group; consequently, the latter two derivatives are much preferred for Boc SPPS. No corresponding C-alkylation occurs in Fmoc chemistry.

The ε-amino group of Lys is best protected by the 2-chlorobenzyl-oxycarbonyl (2-ClZ) or Fmoc group for Boc chemistry, and recipro-cally by the Boc group for Fmoc chemistry. The 2-ClZ group offers the desired acid stability for Boc chemistry, by comparison to the benzyl-oxycarbonyl (Z) and other ring-chlorinated Z groups. Branching due to premature side-chain deprotection by TFA is avoided, but the 2-ClZ group is still readily removable by strong acids (Erickson and Merrifield, 1973b). Orthogonal side-chain protection for both Boc- and Fmoc-Lys is provided by the palladium-sensitive allyloxycarbonyl (Aloc) group (Lyttle and Hudson, 1992; Kates et al., 1993b). In addi-tion, 1-(4,4-dimethyl-2,6-dioxocyclohex-1-ylidene)ethyl (Dde) side-chain protection of Lys during Fmoc chemistry allows for selective deprotection with 2% hydrazine in DMF (Bycroft et al., 1993). Lys(Dde) has been used successfully for the synthesis of branched peptides and peptide templates (Bycroft et al., 1993; C. G. Fields et al., 1993a; Grab et al., 1996; Xu et al., 1996). The isovaleryl derivative of Dde, that is, the ivDde group, may prove advantageous because of its enhanced stability to piperidine (Chhabra et al., 1998). A further selectively removable side-chain protecting group for Lys is 4-methyl-trityl (Mtt), which is labile to 1% TFA–triisopropylsilane in DCM (Aletras et al., 1995).

The highly basic trifunctional guanidino side-chain group of Arg may be protected or unprotected (i.e., protonated). Appropriate benzenesulfonyl derivatives are the 4-toluenesulfonyl (Tos) or mesityl-ene-2-sulfonyl (Mts) groups in conjunction with Boc chemistry, and either 4-methoxy-2,3,6-trimethylbenzenesulfonyl (Mtr), 2,2,5,7,8-pen-tamethylchroman-6-sulfonyl (Pmc), or 2,2,4,6,7-pentamethyldihydro-benzofuran-5-sulfonyl (Pbf) with Fmoc chemistry. These groups most likely block the ω-nitrogen of Arg, and their relative acid lability is Pbf > Pmc > Mtr $\gg$ Mts > Tos (Fujino et al., 1981; Green et al., 1988; Carpino et al., 1993). The Tos and Mts groups are removed by the same strong acids that cleave Bzl-type groups. The Mtr group may require extended TFA–thioanisole treatment (2–8 h) for removal, while Arg(Pmc) and Arg(Pbf) are deprotected readily by 50% TFA (< 2 h). A number of other Arg protecting groups have been proposed, particularly N^{ω}-mono or $N^{\delta,\omega}$-bis-urethane deriva-tives. However, based on current information, the aforementioned benzenesulfonyl derivatives seem to offer the best prospects of clean incorporation of Arg without the production of contaminating ornithine (Orn) at a later stage (Rink et al., 1984).

Due to the high pK_a of the guanidino group (~12.5), it can be protected selectively by protonation with HCl or HBr. Successful SPPS using protonated Arg requires a proton source (i.e., HOBt) for all subsequent coupling steps. This protocol is recommended to suppress intermolecular acylation of the guanidino group, which would lead to Orn formation (Atherton et al., 1984; Atherton and Sheppard, 1989).

A common side reaction of most Boc/Fmoc-Arg derivatives is δ-lactam formation (Barany and Merrifield, 1979; Rzeszotarska and Masiukiewicz, 1988). During carbodiimide activation, δ-lactam formation (intramolecular aminolysis) competes with peptide bond formation (intermolecular aminolysis). Acylations in the presence of HOBt are commonly used to inhibit δ-lactam formation. N^α-protected Arg derivatives may also be coupled as preformed esters, from which δ-lactam side products have been separated from the desired ester prior to use (Atherton et al., 1988b). However, preformed esters will undergo conversion to the δ-lactam relatively shortly after dissolving in DMF (D. Hudson, unpublished results).

Activated His derivatives are uniquely prone to racemization during stepwise SPPS, due to an intramolecular abstraction of the proton on the optically active α-carbon by the imidazole π-nitrogen (Jones et al., 1980). Racemization could be suppressed by either reducing the basicity of the imidazole ring, or by blocking the base directly (Riniker and Sieber, 1988). Consequently, His side-chain-protecting groups can be categorized depending on whether the τ- or π-imidazole nitrogen is blocked. The Tos group blocks the N^τ of Boc-His, and is removed by strong acids. However, the Tos group is also lost prematurely during SPPS steps involving HOBt; this allows acylation or acetylation (during capping) of the imidazole group, followed by chain termination due to $N^{im} \rightarrow N^\alpha$-amino transfer of the acyl or acetyl group (Ishiguro and Eguchi, 1989; Kusunoki et al., 1990). Therefore, HOBt should never be used during couplings of amino acids once a His(Tos) residue has been incorporated into the peptide-resin. An HF-stable, orthogonally removable, N^τ-protecting group for Boc strategies is the 2,4-dinitrophenyl (Dnp) function. Final Dnp deblocking is best carried out at the peptide-resin level prior to the HF cleavage step, by use of thiophenol in DMF (see box below). The τ-nitrogen of Fmoc-His can be protected by the Boc and triphenylmethyl (Trt) groups. When His is N^τ-protected by the Boc group, the basicity of the imidazole ring is reduced sufficiently so that acylation by the preformed symmetrical anhydride (PSA) method proceeds with little racemization (Atherton and Sheppard, 1989). His(Boc) is reasonably stable to repetitive base treatment (Atherton and Sheppard, 1989). His(Trt) is completely stable to piperidine, but removed with TFA (Sieber and Riniker,

Dnp removal from His(Dnp)-containing peptide-resins is achieved by use of 20 mmol thiophenol per mmol His(Dnp) residue. The peptide-resin is suspended in DMF (5 mL gram^{-1} of resin), thiophenol is added, and the reaction proceeds for 1 h at 25 °C. After thorough washing of the Boc-peptide-resin with DMF, H_2O, ethanol, and DCM, the N^{α}-Boc group is removed and the peptide is cleaved with HF or TFMSA. If Dnp groups still remain, the peptide is dissolved in 6 M guanidine·HCl, 50 mmol Tris acetate, pH 8.5 (10–20 mg peptide per mL solution), then deprotected by adding 2-mercaptoethanol to 20% (v/v) and treating for 2 h at 37 °C. The peptide should then be purified immediately by gel filtration or HPLC (Applied Biosystems, Inc., 1989a).

1987). The Trt group reduces the basicity of the imidazole ring (the pK_a decreases from 6.2 to 4.7), although racemization by the PSA method is not eliminated completely (Sieber and Riniker, 1987). Since Dnp and Trt N^{τ}-protection do not allow PSA coupling with low racemization, it is recommended that the appropriate derivatives be coupled as preformed esters or in situ with carbodiimide in the presence of HOBt (Sieber and Riniker, 1987; Riniker and Sieber, 1988). Boc-His(Tos) is coupled efficiently using benzotriazolyl N-oxy-trisdimethylaminophosphonium hexafluorophosphate (BOP, 3 equiv.) in the presence of DIEA (3 equiv.); these conditions minimize racemization and avoid premature side-chain deprotection by HOBt (Forest and Fournier, 1990).

Blocking of the π-nitrogen of the imidazole ring has been shown to be effective in reducing His racemization (Fletcher et al., 1979). The N^{π} of His is protected by the benzyloxymethyl (Bom) and tert-butoxymethyl (Bum) groups for Boc and Fmoc chemistry, respectively. Couplings using N^{π}-protected Boc-His(Bom) or Fmoc-His(Bum) PSAs result in racemization-free incorporation of His (Brown et al., 1982; Colombo et al., 1984). HF deprotection of His(Bom) and TFA deprotection of His(Bum) liberates formaldehyde, which can modify susceptible side-chains (see below, Cleavage).

The carboxamide side chains of Asn and Gln are often left unprotected in SPPS, but this approach leaves open the danger of dehydration to form nitriles upon activation with in situ reagents. On the other hand, acylations by activated esters result in minimal side-chain dehydration (Barany and Merrifield, 1979; Mojsov et al., 1980; Gausepohl et al., 1989b) (see below, Formation of Peptide Bond). Nitrile formation is also inhibited during in situ carbodiimide acylations when HOBt is added (Mojsov et al., 1980; Gausepohl et al., 1989b) (see below, Formation of Peptide Bond). However, the presence of HOBt does not effectively inhibit N^{α}-protected Asn dehydration during BOP in situ acylations (Gausepohl et al., 1989b).

Boc- and Fmoc-His are prepared by the general procedures described earlier for Boc- and Fmoc-amino acids. Crude Boc-His is dissolved in methanol (MeOH)–0.1 M pyridinium acetate buffer (pH 3.8) (1:1) and purified by ion-exchange chromatography over Dowex 50 (H^+ form) using a pH gradient from 3.8 to 5.8 (Kawasaki et al., 1989). Crude Fmoc-His is purified by washing with H_2O and hot MeOH (Kawasaki et al., 1989). N^τ-protected Boc- and Fmoc-His derivatives are prepared by straightforward reactions of appropriate side-chain derivatizing reagents with either N^α-protected or free His (see references in table 3-1). N^π-protected derivatives are prepared by protecting His-OMe at the N^α-amino position (by the Z or Boc group) and N^τ (by the Boc group), derivatizing N^π with the appropriate chloromethyl ether (simultaneously removing the N^τ-Boc group), and saponifying the methyl ester with NaOH (Brown et al., 1982; Colombo et al., 1984). The N^α-Z group is removed by hydrogenolysis and replaced by the Fmoc group (Colombo et al., 1984).

At the point where an N^α-amino-protecting group is removed from Gln, the possibility exists for an acid-catalyzed intramolecular aminolysis which displaces ammonia and leads to pyroglutamate formation (Barany and Merrifield, 1973, 1979; DiMarchi et al., 1982; Orlowska et al., 1987). Cyclization occurs primarily during couplings, as N^α-protected amino acids and HOBt promote this side reaction (DiMarchi et al., 1982). Consequently, it is recommended that the incoming residue that is to be incorporated onto Gln be activated as a nonacidic species, for example, PSA or a preformed ester (see below, Formation of Peptide Bond).

Although conditions are available for the safe incorporation of Asn and Gln with free side chains during SPPS, there are compelling reasons for their protection. Side-chain-protecting groups such as 9H-xanthen-9-yl (Xan), 2,4,6-trimethoxybenzyl (Tmob), and Trt minimize the occurrence of dehydration (Mojsov et al., 1980; Hudson, 1990c; Gausepohl et al., 1989b; Sieber and Riniker, 1990; Han et al., 1996a) and pyroglutamate formation (Barany and Merrifield, 1979), and may also inhibit hydrogen bonding that otherwise leads to secondary structures which substantially reduce coupling rates. Unprotected Fmoc-Asn and -Gln have poor solubility in DCM and DMF; solubility is improved considerably by Tmob, Trt, or Xan side-chain protection. The Xan group does not entirely survive the TFA deprotection conditions in Boc chemistry (Dorman et al., 1972; Stewart and Young, 1984).

The highly sensitive side-chain of Trp is best protected by the N^{in}-formyl (CHO) or cyclohexyloxycarbonyl (Hoc) group for Boc chemistry and the N^{in}-Boc group for Fmoc chemistry. Trp(CHO) is deprotected at the peptide-resin level by treatment with piperidine–DMF (9:91), 0 °C, 2 h, *prior* to HF cleavage (Fields et al., 1993);

In peptides where several Asn, Gln, His, and Cys residues are close in sequence, it may be worthwhile to limit the global use of Trt side-chain protection. Interspersing side-chain unprotected Asn and/or Gln residues in such congested sequences should limit difficult couplings due to steric hindrance. In addition, Asn or Gln adjacent to Trp should be left unprotected, since the Tmob, Trt, and Xan side-chain-protecting groups can modify Trp during TFA deprotection/cleavage (Southard, 1971; see also Cleavage, below, for additional references).

the formyl group is also removed by 20–25% HF in the presence of dimethyl sulfide and 4-thiocresol (see below, Cleavage). The Hoc group is removed by HF and does not require the presence of thiols (Nishiuchi et al., 1996). Unprotected Trp may be incorporated by Boc chemistry when 2.5% anisole plus either 2% dimethyl phosphite or indole are added to the TFA deprotection solution (Stewart and Young, 1984; Hudson et al., 1986). During the TFA cleavage procedure in Fmoc chemistry, the Boc group protecting the side chain of Trp is removed partially, with the indole tied up as an N-carbamate involving a molecule of carbon dioxide. Complete deprotection occurs in aqueous solution (Franzén et al., 1984; White, 1992). The intermediate carboxy-indole is relatively resistant to alkylation (White, 1992; Riniker et al., 1993; Johnson et al., 1993) and sulfonation (White, 1992; Choi and Aldrich, 1993; Fields and Fields, 1993).

The side chain of Met generally survives cycles of Fmoc chemistry, but protection during Boc chemistry by the reducible sulfoxide function is often advisable. Smooth deblocking of Met(O) occurs in 20–25% HF in the presence of dimethyl sulfide (see below, Cleavage), or by N-methylmercaptoacetamide (MMA) (10 equiv.) in 10% aqueous HOAc at 37 °C for 12–36 h (Houghten and Li, 1979), NH_4I–dimethyl sulfide (30 equiv. each) in TFA at 0 °C for 3 h (Fujii et al., 1987, Ferrer et al., 1999), trimethylsilyl bromide (TMSBr) and 1,2-ethanedithiol (EDT) under anhydrous conditions (Beck et al., 1994), or $DMF \cdot SO_3$–EDT (5 equiv. each) in pyridine–DMF (1 : 4) at 20°C for 1 h (Futaki et al., 1990b). $DMF \cdot SO_3$–EDT treatment of Met(O) can be carried out only while hydroxyl residues are side-chain protected, as free hydroxyls will be sulfated (Futaki et al., 1990a).

The most challenging residue to manage in peptide synthesis is Cys, which for some applications is required in the free sulfhydryl form and for others as a contributor to a disulfide linkage (Barany and Merrifield, 1979; Andreu et al., 1994; Annis et al., 1997). Another issue is the selective formation of multiple disulfides by the concurrent use of two or more classes of Cys protecting groups (see below, Post-Translational Modifications and Unnatural Structures). Compatible with Boc chemistry are the 4-methylbenzyl (Meb), 4-methoxybenzyl

(Mob), acetamidomethyl (Acm), trimethylacetamidomethyl (Tacm), *tert*-butylsulfenyl (S*t*Bu), 3-nitro-2-pyridinesulfenyl (Npys), and 9-fluorenylmethyl (Fm) β-thiol protecting groups; compatible with Fmoc chemistry are the Acm, 4-methoxytrityl (Mmt), Tacm, S*t*Bu, Trt, Tmob, and Xan groups. The Mmt, Trt, Tmob, and Xan groups are labile in TFA; due to the tendency of the resultant stable carbonium ions to realkylate Cys (Photaki et al., 1970; Munson et al., 1992; Han et al., 1996b), effective scavengers are needed (see below, Cleavage). The Meb group is optimized for removal by strong acid (Erickson and Merrifield, 1973a); Cys(Meb) residues may also be converted directly to the oxidized (cystine) form by thallium (III) trifluoroacetate [Tl(Tfa)$_3$], although some cysteic acid forms at the same time. Cys(Npys) and Cys(Fm) are stable to acid, and cleaved respectively by thiols and base. The Acm and Tacm groups are acid- and base-stable, and removed by mercuric (II) acetate [Hg(OAc)$_2$] in aqueous acidic media (Veber et al., 1972; Sakakibara, 1999), or TFA solutions of either silver tetrafluoroborate (Fujii et al., 1989) or silver trifluoromethanesulfonate (Tamamura et al., 1998) in the presence of anisole. Metal-mediated deprotections are followed by treatment with H$_2$S or excess mercaptans to free the β-thiol. In multiple Cys(Acm)-containing peptides, mercuric (II) acetate may not be a completely effective removal reagent (Kenner et al., 1979). Mercuric (II) acetate can modify Trp, but can be used safely with Trp-containing peptides when an acidic solvent, for example, 50% aqueous HOAc, is used (Nishio et al., 1994). Alternatively, Cys(Acm) residues are converted directly to disulfides by treatment with I$_2$ or Tl(Tfa)$_3$ in solution (Kamber 1971; Fujii et al., 1987; Andreu et al., 1994) or on-resin (Albericio et al., 1991a; Munson and Barany, 1993; Andreu et al., 1994; Edwards et al., 1994; Annis et al., 1997). Finally, the acid-stable S*t*Bu group is removed by reduction with thiols or phosphines (Romani et al., 1987).

C-Terminal esterified (but not amidated) Cys residues are racemized by repeated piperidine deprotection treatments during Fmoc SPPS. Following 4 h exposure to piperidine–DMF (1 : 4), the extent of racemization found was 36% D-Cys from Cys(S*t*Bu), 12% D-Cys from Cys(Trt), and 9% D-Cys from Cys(Acm) (Atherton et al., 1991). Racemization of esterified Cys(Trt) was reduced from 12% with piperidine–DMF (1 : 4) to only 2.6% with 1% DBU–DMF after 4 h treatment (Atherton et al., 1991; Wade et al., 1991). Additionally, the steric hindrance of the 2-chlorotrityl linker has been shown to minimize racemization of C-terminal Cys residues (Fujiwara et al., 1994). More recently, a novel side-chain-anchoring strategy that combines *S*-Xan protection and xanthenyl (XAL) handle for anchoring amides has been applied to prevent *C*-terminal Cys racemization. Detachment of completed peptides anchored through the Cys side chain can be carried out under mild conditions, either by acidolysis or by application of oxidative reagents (Barany, 1999). The above-mentioned approaches also circumvent formation of 3-(1-piperidinyl)alanine byproducts, a sometimes serious side reaction with C-terminal cysteine (Lukszo et al., 1996) depending on the Cys protecting group ($StBu$ is particularly bad; this explains in retrospect a previously mysterious finding by Eritja et al. [1987]) and anchoring linkage (HMP/PAB-type is problematic; 2-ClTrt and XAL are relatively safe).

Cys is also susceptible to racemization during Fmoc SPPS. When applying protocols for Cys incorporation which include phosphonium and aminium salts as coupling agents (BOP, HBTU, HATU, and PyAOP), as well as preactivation in the presence of suitable additives (HOBt and HOAt) and tertiary amine bases (DIEA and NMM), 5–33% racemization is observed (Han et al., 1997a). These high levels of racemization are generally reduced by avoiding preactivation, using a weaker base (such as collidine), and switching to the solvent mixture DMF–DCM (1:1). Couplings with less than 1% racemization were performed successfully using either BOP plus HOBt, HBTU plus HOBt, or HATU plus HOAt, in all cases in the presence of collidine, without preactivation, and in DMF–DCM. Alternatively, the Pfp ester of a suitable Fmoc-Cys derivative can be used. Under the aforementioned optimized conditions, Cys protecting groups Tmob, Trt, Xan, and Acm do not enhance the risk of racemization (Han et al., 1997a).

N-Terminal Cys residues are modified covalently by formaldehyde, liberated during HF deprotection of His(Bom) residues (see below, Cleavage). Additional difficulties, often poorly understood, have arisen with a range of protected Cys derivatives in a variety of applications (Barany and Merrifield, 1979; Atherton et al., 1985b; Andreu et al., 1994).

As is clear from the preceding discussion, the Boc and Fmoc groups have risen to the fore as the most widely used and commercially viable N^{α}-amino-protecting groups for SPPS. A plethora of other N^{α}-amino-protecting groups, some illustrating remarkably creative organic chemistry, have been proposed over the years (figure 3-4). Among these, the 2-(4-biphenyl)propyl[2]oxycarbonyl (Bpoc) (Wang and Merrifield 1969; Kemp et al., 1988), 2-(3,5-dimethoxyphenyl)propyl[2]-oxycarbonyl (Ddz) (Birr et al., 1972; Voss and Birr, 1981), and 4-methoxybenzyloxycarbonyl (Moz) (Wang et al., 1987; Chen et al., 1987) groups are removed in dilute TFA, the dithiasuccinoyl (Dts) (Barany and Merrifield, 1977; Barany and Albericio, 1985; Albericio and Barany, 1987; Zalipsky et al., 1987; Planas et al., 1999) and 3-nitro-2-pyridinesulfenyl (Npys) (Matsueda and Walter, 1980; Wang et al., 1982; Ikeda et al., 1986; Hahn et al., 1990) groups are removed by thiolysis, the 6-nitroveratryloxycarbonyl (Nvoc) group is removed by photolysis (Patchornik et al., 1970; Fodor et al., 1991), and

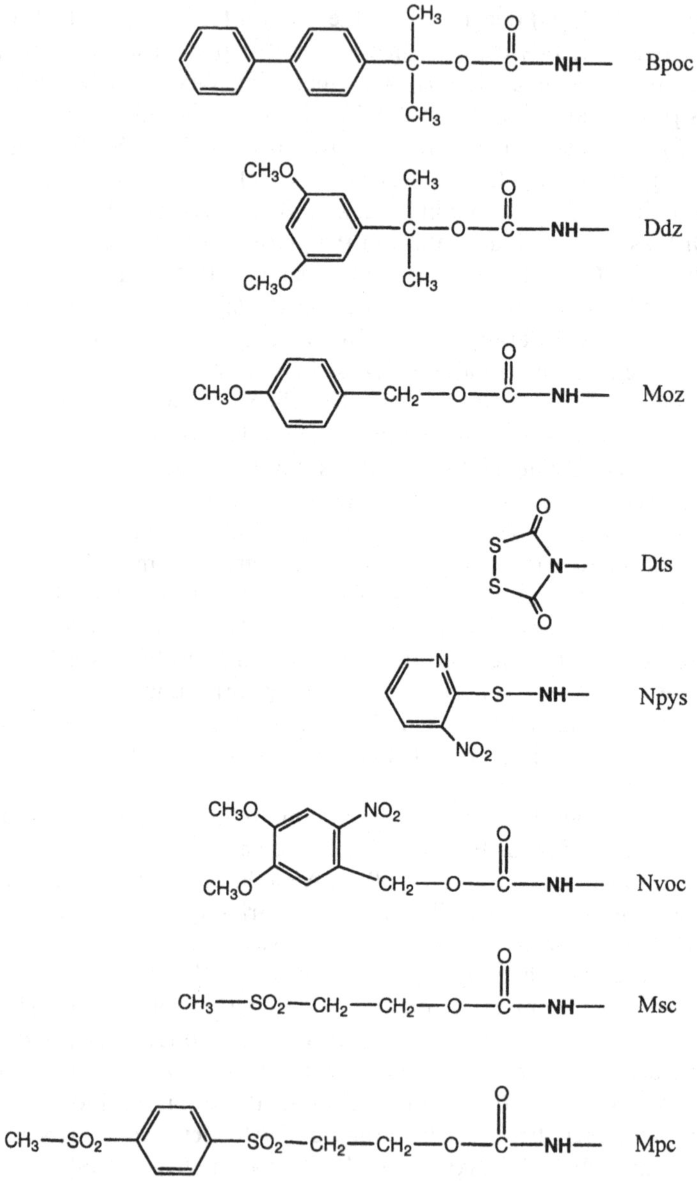

Figure 3-4. Alternative N^{α}-amino protecting groups for SPPS. The protected nitrogen which is part of the amino acid is shown in boldface.

2-(methylsulfonyl)ethoxycarbonyl (Msc) (Tesser and Balvert-Geers, 1975; Camarero et al., 1998; Canne et al., 1999) and the 2-[4-(methyl-sulfonyl)phenylsulfonyl]ethoxycarbonyl (Mpc) (Schielen et al., 1991) groups are base-labile. Chemistries relying on these protecting groups are implicit in the cited references, but otherwise are beyond the scope of the present article.

Polymeric Support

The term "solid phase" often conjures misleading images among the uninitiated. Supports that lead to successful results for macromolecule synthesis are far from static, and due to the need for reasonable capa-cities, it is rare for solid-phase chemistry to take place exclusively on surfaces. The resin support is quite often a polystyrene suspension polymer cross-linked with 1% of 1,3-divinylbenzene; the level of func-tionalization is typically 0.2–1.0 mmol/g. Dry polystyrene beads have an average diameter of about 50 μm, but with the commonly used solvents for peptide synthesis, namely DCM and DMF, they swell 2.5- to 6.2-fold in volume (Sarin et al., 1980). Thus, the chemistry of solid-phase synthesis takes place within a well-solvated gel containing mobile and reagent-accessible chains (Sarin et al., 1980; Live and Kent, 1982). Polymer supports have also been developed based on the concept that the insoluble support and peptide backbone should be of comparable polarities (Atherton and Sheppard, 1989). A resin of copolymerized dimethylacrylamide, N,N'-bisacryloylethylenediamine, and acryloylsarcosine methyl ester (typical loading 0.3 mmol/g), com-mercially known as polyamide or Pepsyn, was synthesized to satisfy this criteria (Arshady et al., 1981). Under the best solvation conditions for both polystyrene and polyamide supports, reaction rates approach, but generally do not reach, those attainable in solution. It has been shown for a polystyrene carrier that macroscopic dimensions of both the dry and solvated beads change dramatically once an appreciable level of peptide has been built up (Sarin et al., 1980). Thus, for this specific case, reactions continued to occur efficiently throughout the interior of a peptide-resin that was four-fold the weight of the starting support.

A fertile area of inquiry has been the testing of supports with macroscopic physical properties and possibly other characteristics dif-fering from 1% cross-linked polystyrene and polyamide gel beads. These include membranes (Bernatowicz et al., 1990), cotton and other appropriate carbohydrates (Frank and Döring, 1988; Lebl and Eichler, 1989; Eichler et al., 1989), controlled-pore silica glass (Büttner et al., 1988), linear polystyrene chains covalently grafted onto dense Kel-F particles (Tregear, 1972; Kent and Merrifield, 1978; Albericio et al., 1989a), polyethylene sheets (Berg et al., 1989), or "augmented surface polyethylene prepared by chemical transformation" (ASPECT) (Cook and Hudson, 1996, 1997; Adams et al., 1998a; Hudson, 1999). Supports developed specifically to withstand the back-pressures that arise during continuous-flow procedures have been low-density, highly permeable inorganic matrices with polyamide embedded within. These embedded matrices include polyamide-kieselguhr (known commercially as Pepsyn K) (Atherton et al., 1981b) and polyamide-Polyhipe (Small and Sherrington, 1989). Pepsyn K has a typical loading of $0.1 \, \text{mmol g}^{-1}$, while Polyhipe loadings range from 0.3 to $1.8 \, \text{mmol g}^{-1}$. The aforementioned materials have been largely supplanted by polyethylene glycol-polystyrene graft supports which swell in a range of solvents and have excellent physical and mechanical properties for both batchwise and continuous-flow SPPS (Hellermann et al., 1983; Zalipsky et al., 1985, 1994; Bayer and Rapp, 1986; Bayer et al., 1990; Barany and Albericio 1991; Barany et al., 1997; Kates et al., 1998; Gooding et al., 1999). These latter materials are marketed as PEG-PS, TentaGel, or ArgoGel, and the loadings are usually $0.1–0.4 \, \text{mmol g}^{-1}$, but higher loadings can be achieved by branching strategies. Also suitable for both batchwise and continuous-flow SPPS are poly(N-[2-(4-hydroxyphenyl)ethyl]acrylamide) (core Q), polyethylene glycol dimethylacrylamide (PEGA) resins, and cross-linked ethoxylate acrylate resin (CLEAR). Core Q is available at high loading capacities ($5 \, \text{mmol g}^{-1}$) (Epton et al., 1987; Baker et al., 1990). PEGA resins, which consist of poly(N,N-dimethylacrylamide-co-bisacrylamido polyethylene glycol-co-monoacrylamido polyethylene glycol) (Meldal, 1992; Renil et al. 1998), are freely permeable to macromolecules. CLEAR is suitable for both peptide and organic solid-phase synthesis, due to its excellent swelling behavior in polar and nonpolar solvents ranging from water to DCM (Kempe and Barany, 1996).

Attachment to Support

Almost all syntheses by the solid-phase method are carried out in the $C \rightarrow N$ direction, and therefore generally start with the intended C-terminal residue of the desired peptide being linked to the support either directly or via a suitable handle. Anchoring linkages have been designed so that eventual cleavage provides either a free acid or amide at the C-terminus, although, in specialized cases, other useful end groups can be obtained. The discussion that follows focuses on linkers that are either commercially available, readily prepared, and/or of special interest (table 3-2); more complete listings are available (Barany et al., 1987; Fields and Noble, 1990; Barany and Kempe, 1997; Songster and Barany, 1997; Blackburn, 2000).

Table 3-2 Resin linkers and handles[a]

Linker/Handle/Resin	Cleavage conditions	Resulting C-terminus	Reference(s)
4-Chloromethyl resin $Cl-CH_2$—[ ]—Resin	strong acid	acid	Gutte & Merrifield, 1971 Stewart & Young, 1984
4-Hydroxymethylphenylacetic acid (PAM) $HO-CH_2$—[ ]—CH_2CO_2H	strong acid	acid	Mitchell et al., 1978 Tam et al., 1979
3-Nitro-4-(2-hydroxyethyl)benzoic acid (NPE) $HO-(CH_2)_2$—[ ]—CO_2H, O_2N	piperidine DBU	acid	Eritja et al., 1991 Albericio et al., 1991b
9-(Hydroxymethyl)-2-fluoreneacetic acid (HMFA) CH_2CO_2H, CH_2OH, H	piperidine	acid	Mutter & Bellof, 1984 Liu et al., 1990 Sakakibara, 1999
2-Hydroxypropyl-dithio-2'-isobutyric acid (HPDI) CH_3 $HO-CH-CH_2-S-S-C-CO_2H$ CH_3 CH_3	CN^- or $P(CH_2CH_2CO_2H)_3$	acid	Brugidou & Mery, 1994
4-Nitrobenzophenone oxime resin $HO-N=$[ ]—Resin, NO_2	HOPip AA^- $^+N(nBu)_4$ AA-NH_2	acid[b] amide	DeGrado & Kaiser, 1982[c] Findeis & Kaiser, 1989 Scarr & Findeis, 1990

(continued)

Table 3-2 (*Continued*)

Linker/Handle/Resin	Cleavage conditions	Resulting C-terminus	Reference(s)
α-Bromophenacyl 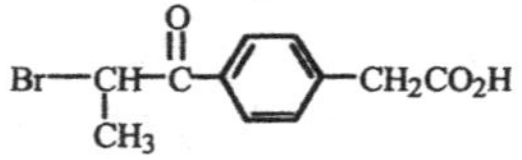 $$Br-CH-C-\!\!\!\!<\!\!\bigcirc\!\!>\!\!-CH_2CO_2H$$ (O above C, CH₃ below)	hv (350 nm)	acid	Wang, 1976
4-Alkoxybenzyl alcohol resin $HO-CH_2-\bigcirc-OCH_2-\bigcirc-Resin$ 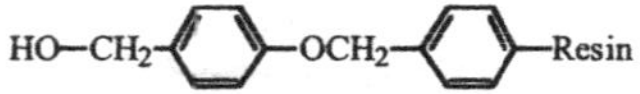	TFA	acid	Wang, 1973[c] Lu *et al.*, 1981
4-Hydroxymethylphenoxyacetic acid (HMPA/PAC) 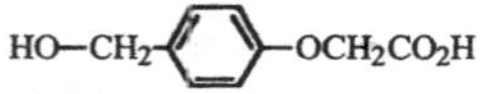 $HO-CH_2-\bigcirc-OCH_2CO_2H$	TFA	acid	Sheppard & Williams, 1982
3-(4-Hydroxymethylphenoxy)propionic acid (PAB) 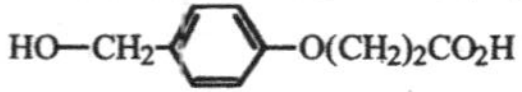 $HO-CH_2-\bigcirc-O(CH_2)_2CO_2H$	TFA	acid	Albericio & Barany, 1985
3-Methoxy-4-hydroxymethylphenoxyacetic acid 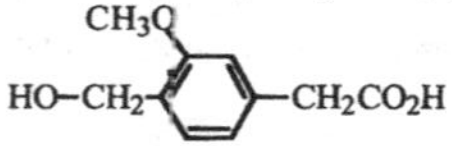 $HO-CH_2-\bigcirc(CH_3O)-CH_2CO_2H$	dilute TFA	acid	Sheppard & Williams, 1982
4-(2',4'-Dimethoxyphenylhydroxymethyl) phenoxymethyl resin $HO-CH(CH_3O, OCH_3)-\bigcirc-O-Resin$ 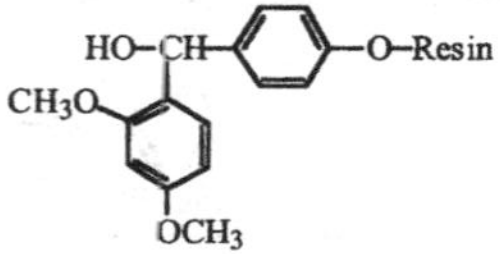	dilute TFA	acid	Rink, 1987
2-Methoxy-4-alkoxybenzyl alcohol resin (SASRIN) 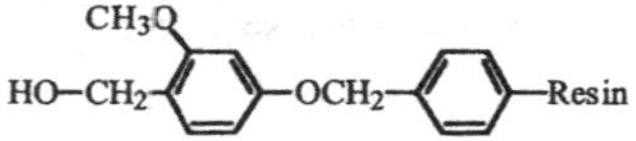 $HO-CH_2-\bigcirc(CH_3O)-OCH_2-\bigcirc-Resin$	dilute TFA	acid	Mergler *et al.*, 1988a

124

Linker	Cleavage reagent	Bond	Reference
2-Chlorotrityl chloride resin 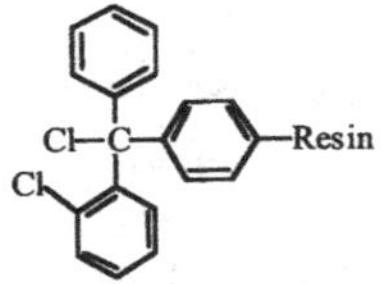	dilute TFA or HOAc	acid	Barlos et al., 1989 Barlos et al., 1991a
4-(4-Hydroxymethyl-3-methoxyphenoxy)butyric acid (HMPB) 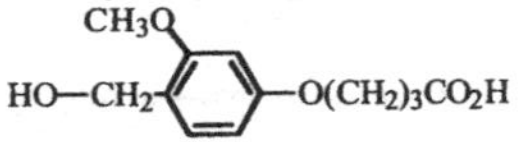 $HO-CH_2$ CH_3O $O(CH_2)_3CO_2H$	dilute TFA	acid	Flörsheimer & Riniker, 1991
5-(4-Hydroxymethyl-3,5-dimethoxyphenoxy)valeric acid (HAL) 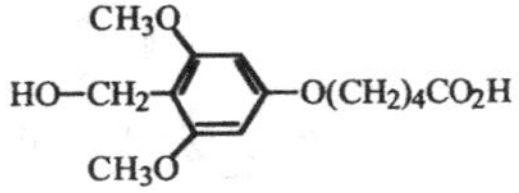 $HO-CH_2$ CH_3O CH_3O $O(CH_2)_4CO_2H$	dilute TFA	acid	Albericio & Barany, 1991
Hydroxy-crotonyl-aminomethyl resin (HYCRAM) 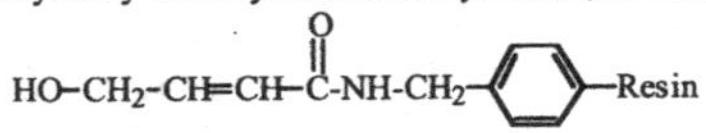 $HO-CH_2-CH=CH-C-NH-CH_2$—Resin	Pd(0) NMM or dimedone	acid	Kunz & Dombo, 1988 Guibé et al., 1989 Lloyd-Williams et al., 1991b
3-Nitro-4-hydroxymethylbenzoic acid (ONb) 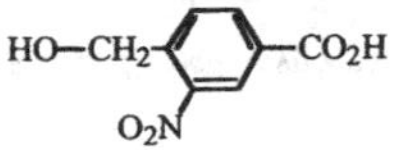 $HO-CH_2$ CO_2H O_2N	hv (350 nm)	acid	Rich & Gurwara, 1975c Giralt et al., 1982 Barany & Albericio, 1985 Kneib-Cordonier et al., 1990
4-[1-Amino-2-(trimethylsilyl)ethyl]-phenoxyacetic acid (SAL) 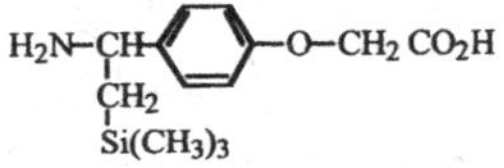 H_2N-CH $O-CH_2 CO_2H$ CH_2 $Si(CH_3)_3$	moderate acid	amide	Chao et al., 1993

(continued)

125

Table 3-2 (*Continued*)

Linker/Handle/Resin	Cleavage conditions	Resulting C-terminus	Reference(s)
4-Methylbenzhydrylamine resin (MBHA) 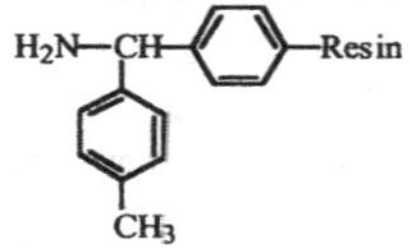	strong acid	amide	Matsueda & Stewart, 1981; Gaehde & Matsueda, 1981
4-(2',4'-Dimethoxyphenylaminomethyl)phenoxymethyl resin 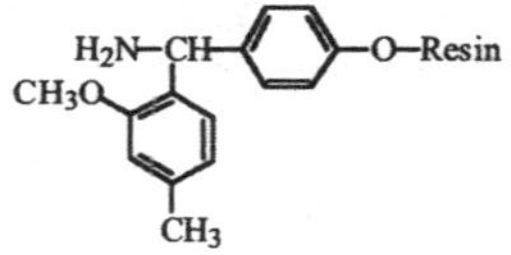	dilute TFA	amide	Rink, 1987
4-(4'-Methoxybenzhydryl)phenoxyacetic acid (Dod) 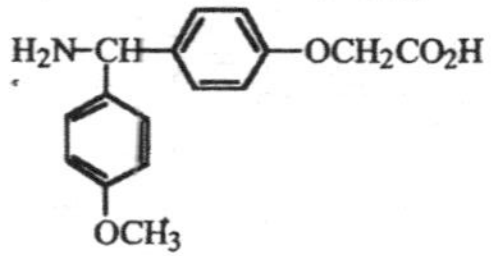	TFA	amide	Stüber et al., 1989
3-(Amino-4-methoxybenzyl)-4,6-dimethoxyphenylpropionic acid 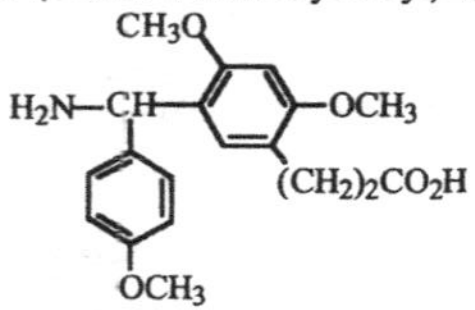	dilute TFA	amide	Breipohl et al., 1989
5-(4-Aminomethyl-3,5-dimethoxyphenoxy)valeric acid (PAL) 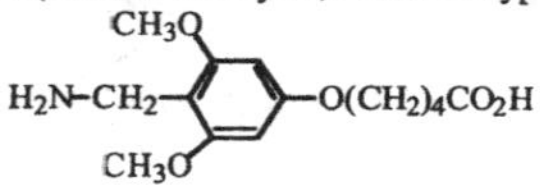	TFA	amide	Albericio & Barany, 1987b; Albericio et al., 1990a

126

5-(9-Aminoxanthen-3-oxy)valeric acid (XAL)

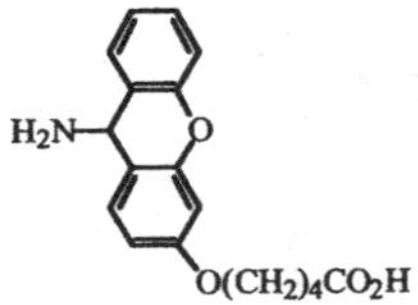

dilute TFA

amide

Sieber, 1987c;[c]
Barany & Albericio,1991
Han et al., 1996b

3-Nitro-4-aminomethylbenzoic acid (Nonb)

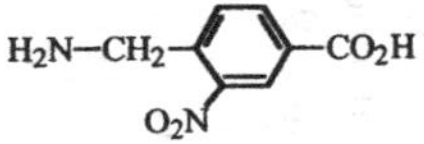

hv (350 nm)

amide

Hammer et al., 1990

4-(2'-Aminoethyl)-2-methoxy-5-nitro-phenoxypropionic acid

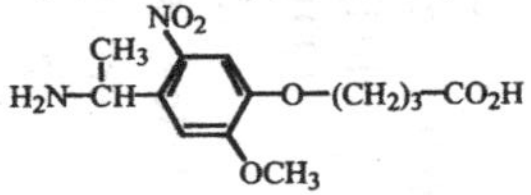

hv (356 nm)

amide

Holmes & Jones, 1995

---

[a] Structural diagrams are oriented so that the resin or point of attachment to support is on the far right, and the site for anchoring the C-terminal amino acid residue is on the far left. Benzyl ester linkages may also be cleaved by a range of nucleophiles to create acids, esters, or other derivatives. See text discussion under "Cleavage," and consult Barany and Merrifield (1979) and Barany et al. (1987) for further examples.

[b] Cleavage by aminolysis results in 0.6–2% racemization (DeGrado and Kaiser, 1980).

[c] Reference for historical reasons; preparation of linker/resin has been improved in later references.

Note that several handles (table 3-2) have a free or activated carboxyl group that is intended for attachment to the polymeric support. Such handles are most frequently coupled onto supports that have been functionalized with amino groups. Aminomethyl-polystyrene resin is optimally prepared essentially as described by Mitchell et al., (1978), except that methanesulfonic acid (0.75 g per 1 g polystyrene) is preferred as the catalyst instead of the originally described TFMSA (S. B. H. Kent and K. M. Otteson, unpublished results). Amino groups are introduced onto a variety of polyamide supports by treatment with ethylenediamine to displace carboxylate derivatives (Atherton and Sheppard, 1989). All else being equal, there are significant advantages to those anchoring methods in which the key step is amide bond formation by reaction of an activated handle carboxyl with an amino support, since such reactions can be readily made to go to completion. This approach allows control of loading levels, and obviates difficulties that may arise due to extraneous or unreacted functionalized groups. As indicated earlier, the best control is achieved by coupling "preformed handles," which are protected amino acid derivatives that have been synthesized and purified in solution prior to the solid-phase anchoring step.

Peptide Acids

For Boc chemistry, the most common approach to peptide acids uses substituted benzyl esters which are cleaved in strong acid at the same time as other benzyl-type protecting groups are removed (figure 3-2). The classical procedures starting with chloromethyl-resin are still favored by many (Gutte and Merrifield, 1971; Gisin, 1973; Stewart and Young, 1984), although preformed handle approaches with 4-(hydroxymethyl)phenylacetic acid (PAM) (Mitchell et al., 1978; Tam et al., 1979; Clark-Lewis and Kent, 1989) are preferable for a number of applications. Other resins and linkers proposed for Boc chemistry are cleaved by orthogonal modes, allowing their use for the preparation of partially protected peptide segments (see below, Auxiliary Issues). The 4-(2-hydroxyethyl)-3-nitrobenzoic acid (NPE) (Eritja et al., 1991; Albericio et al., 1991b) and 9-(hydroxymethyl)-2-fluoreneacetic acid (HMFA) (Liu et al., 1990; Sakakibara, 1999) linkers are cleaved by bases (see below, Cleavage). The HMFA linker is also cleaved by free N^α-amino groups from the peptide-resin; the addition of HOBt during SPPS inhibits premature cleavage from this source (Liu et al., 1990). A 4-nitrobenzophenone oxime resin (DeGrado and Kaiser, 1980, 1982; Findeis and Kaiser, 1989; Scarr and Findeis, 1990) yields a peptide acid upon cleavage by either N-hydroxypiperidine (HOPip) or amino acid tetra-n-butylammonium salts [AA$^-$ $^+$N(nBu)$_4$] (Findeis and Kaiser, 1989; Lansbury et al., 1989;

Sasaki and Kaiser, 1990). Note that in the latter mode of oxime resin cleavage, the penultimate residue of the desired peptide is the one initially attached to the support. Finally, the acid-stable 2-bromopropionyl (α-methylphenacyl ester) linker (Wang, 1976) is of interest because it can be cleaved by photolysis ($\lambda = 350$ nm).

For Fmoc chemistry, peptide acids have been generated traditionally using the 4-alkoxybenzyl alcohol resin/4-hydroxymethylphenoxy (HMP/PAB) linker (Wang, 1973; Lu et al., 1981; Sheppard and Williams, 1982; Colombo et al., 1983; Albericio and Barany, 1985; Bernatowicz et al., 1990), which is cleaved in 1–2 h at 25 °C with 50–100% TFA. The precise lability of the resultant 4-alkoxybenzyl esters depends on the spacer between the phenoxy group and the support. The HMP/PAB moiety can be established directly on the resin, or can be introduced as a handle; preformed handles are best coupled to amino-functionalized resins as their 2,4,5-trichlorophenyl or 2,4-dichlorophenyl activated esters, sometimes in the presence of HOBt (Albericio and Barany, 1985; Bernatowicz et al., 1990; Albericio and Barany, 1991).

A number of supports and linkers are available that can be cleaved in dilute acid; under optimal circumstances these can be used to prepare protected peptide segments retaining side-chain *tert*-butyl protection. These include 3-methoxy-4-hydroxymethylphenoxyacetic acid (Sheppard and Williams, 1982), 4-(2′,4′-dimethoxyphenyl-hydroxymethyl)phenoxy resin (Rink acid) (Rink, 1987; Rink and Ernst, 1991), 2-methoxy-4-alkoxybenzyl alcohol (SASRIN) (Mergler et al., 1988a), 2-chlorotrityl-chloride resin (Barlos et al., 1989b; Barlos et al., 1991a), 4-(4-hydroxymethyl-3-methoxyphenoxy)butyric acid (HMPB) (Flörsheimer and Riniker, 1991), and 5-(4-hydroxymethyl-3,5-dimethoxyphenoxy)valeric acid (HAL) (Albericio and Barany, 1991). Because of its acute acid lability, and in order to prevent premature loss of peptide chains, the Rink acid linker is used in conjunction with N^{α}-protected amino acid preformed symmetrical anhydrides or esters in the presence of excess DIEA (3 equiv.) (Rink and Ernst, 1991). The HMP/PAB and SASRIN linkers are available as the corresponding chlorides or bromides (Colombo et al., 1983; Mergler et al., 1989a; Bernatowicz et al., 1990).

Linkers for preparing peptide acids that are compatible with both Boc and Fmoc chemistries include hydroxy-crotonyl (HYCRAM) (Kunz and Dombo, 1988; Kunz, 1990; Lloyd-Williams et al., 1991), which is cleaved by Pd(0) catalyzed transfer of the allyl linker to a weak nucleophile, and 3-nitro-4-hydroxymethylbenzoic acid (ONb) (Rich and Gurwara, 1975; Giralt et al., 1982; Barany and Albericio, 1985; Kneib-Cordonier et al., 1990), which is cleaved photolytically at $\lambda = 350$ nm.

All of the anchoring linkages that ultimately provide peptide acids are esters; rates and yields of reactions for ester bond formation (table 3-3) are less than those for corresponding methods for amide bond formation (see below, Formation of Peptide Bond). Consequently, compromises are needed in order to achieve reasonable loading reaction times and substitution levels, while ensuring that the extent of racemization remains acceptably low. As a point of departure, esterification of N^α-protected amino acid PSA catalyzed by 1 equiv. of 4-dimethylaminopyridine (DMAP) in DMA results in significant (1.5–20%) racemization (Atherton et al., 1981a). In general, racemization levels can be reduced to acceptable levels (0.2–1.2%) when catalytic (0.06 equiv.) amounts of DMAP are used and loadings are performed with carbodiimides in situ (Mergler et al., 1988a), sometimes in the presence of N-methylmorpholine (NMM) (0.9 equiv.) (D. Hudson, personal communication). Alternatively, in situ carbodiimide loading with HOBt (2 equiv.) and DMAP (1 equiv.) at low temperature (0–3 °C) provides a good compromise of minimized racemization (0.1–0.3%) and reasonable loading times (16 h) (van Nispen et al., 1985). No racemization was detected (< 0.05%) when in situ loadings were carried out at 25 °C with only N,N'-dicyclohexylcarbodiimide (DCC) (4 equiv.) and HOBt (3 equiv.), without DMAP (Grandas et al., 1989). Esterifications of Fmoc-amino acids mediated by N,N-dimethylformamide dineopentyl acetal (Albericio and Barany, 1984, 1985), 2,6-dichlorobenzoyl chloride (DCBC) (Sieber, 1987a), diethyl azodicarboxylate (DEAD) (Sieber, 1987a), or 2,4,6-mesitylene-sulfonyl-3-nitro-1,2,4-triazolide (MSNT) (Blankemeyer-Menge et al., 1990) have all been reported to suppress racemization, as is the case with preformed Fmoc-amino acid 2,5-diphenyl-2,3-dihydro-3-oxo-4-hydroxythiophene dioxide (OTDO) esters (Kirstgen et al., 1987, 1988), Fmoc-amino acid chlorides (Akaji et al., 1990a), or Fmoc-amino acid N-carboxyanhydrides (NCAs) (Fuller et al., 1990). The well-established cesium salt method (Gisin, 1973) also allows loading of N^α-protected amino acids to chloromethyl linkers and resins with low levels of racemization (Colombo et al., 1983; Mergler et al., 1989b) while effectively preventing alkylation of susceptible residues (Cys, His, Met) (Gisin, 1973). Bromomethylated-linkers may be loaded directly by Boc-amino acids in the presence of KF (Tam et al., 1979) or by Fmoc-amino acids in the presence of DIEA (Bernatowicz et al., 1990), in each case with little racemization. While premature removal of the Fmoc group during loading can result in dipeptide formation, the efficient ester bond formation methods described in this paragraph minimize this side reaction. However, care should be taken when preparing cesium salts of Fmoc-amino acids, as Cs_2CO_3 may promote partial removal of the Fmoc group. Of the ester bond formation methods discussed here, the most generally applicable are esterification of N^α-protected amino acids in situ by carbodiimide (DCC or DIPCDI) in the presence of catalytic amounts of DMAP (0.06-0.1 equiv.), or by

Table 3-3 Formation of ester bonds to attach N^α-protected amino acids to linkers[a]

Reagents(s)	Linker[a]	Stoichiometries[b]	Conditions[c,d,e]	Racemization (%)[c]	Loading (%)[c]	Reference
Boc-AA-OH : DCHA	Br-PAM*	2:2:1	DMF (4 h, 50 °C + 14 h, 25 °C)	<0.1	86	Mitchell et al., 1978
Boc-AA-OH : KF	Br-PAM*	2.2:1.1:1	CH$_3$CN (18-48 h)	N.R.	~100	Tam et al., 1979
DMF-dineopentyl acetal	HO-HMP*,#	1.7:1.7:1	DMF (72 h)	<0.05	67-75	Albericio & Barany, 1984; Albericio & Barany, 1985
Fmoc-AA-OH : DMAP : HOBt : DCC	HO-HMP*	2:2:4:1	DMF (18 h, 0 °C)	0.1-0.3	47	van Nispen et al., 1985
Fmoc-AA-OH : DCBC : pyridine	HO-HMP*	2:2:3.3:1	DMF (15-20 h)	0.1-0.7; Arg(Mtr) <1.0; His(Trt) 27.0; His(Bum) 2.2; Cys(Acm) 1.2; Cys(Trt) 4.0	55-66; His(Bum) 16	Sieber, 1987a
Fmoc-AA-OH : DEAD : Ph$_3$P	HO-HMP*	3:3:3:1	THF (16 h, 0 °C)	0.3	53-61	Sieber, 1987a
Fmoc-AA-OH : DCC : DMAP	HO-SASRIN*	1.5:1.2:0.01-0.1	DMF-DCM (1:3) (20 h, 0 °C)	Cys(Acm) 1.6; 0.2-1.0; Ile 1.2; Cys(Acm) 4.0; Cys(Trt) 18.3; His(Trt) 26.0	~80	Mergler et al., 1988a; Mergler et al., 1989b
Fmoc-AA-OH : DCC : HOBt	HO-HMP*	4:4:3:1	DMA (17 h)	<0.05	60-94; Arg 6; Asn 31; Gln 29; Pro 54	Grandas et al., 1989a
Fmoc-AA-OH : MSNT : MeIm	HO-HMP*	2:2:1.5:1	DCM[f] (0.5 h, twice) His(Bum) CHCl$_3$	0-0.6; Cys(StBu) 2.1; Asp(OBu) 1.4	72-100	Blankemeyer-Menge et al., 1990
Fmoc-AA-OH : DIEA	Br-HMP*	1.1:1:1	DMF (2-3 h)	<0.1	61-99	Bernatowicz et al., 1990
Fmoc-AA-OH : DIEA	Cl-Trt#	0.3-1.0:2.5:1.6	DCE (0.5 h)	<0.05	61-99; Trp 51	Barlos et al., 1991a
Fmoc-AA-O$^-$Cs$^+$	Cl-CH$_2$-®#	1:1	DMF (16 h, 50 °C)	N.R.	72-95	Gisin, 1973
Fmoc-AA-O$^-$Cs$^+$	Cl-HMP#	2:1	DMA (15-24 h, 50 °C)	0.01-0.07	89-98	Colombo et al., 1983
Fmoc-AA-O$^-$Cs$^+$: NaI	Cl-SASRIN#	1.5-3.0:1:1	DMA (24 h)	0.1-0.7; Cys(Trt) 2.5	N.R.	Mergler et al., 1989b
Fmoc-AA-OTDO : DIEA	HO-HMP*	40 mM : 1 : 1	DCM[f] (1-2 h) Ile 10 h	Cys(Acm) 1.0	81-97	Kirstgen et al., 1987; Kirstgen & Steglich, 1989
Fmoc-AA-Cl	HO-HMP*	5:1	pyridine-DCM (2:3) (1 h)	<0.5; Met 1.7; Ala 0.7	~100	Akaji et al., 1990a
Fmoc-NCA : NMM	Rink acid#	3:0.02:1	toluene[f] (0.5-1 h)	<0.1	N.R.	Fuller et al., 1990

[a] The column entitled "Linker" distinguishes between ester bond formation *first* to provide a preformed handle (*) (often the preferred route, as discussed further in the text) and ester bond formation *directly* to the linker-resin (#).

[b] Stoichiometries (equivalents) are stated in the same order as the reagents are listed, followed last by the linker or resin.

[c] When publications state overnight reactions, this table indicates 18 h.

[d] Reactions are at room temperature (25 °C) unless stated otherwise.

[e] Values are representative for incorporation of most amino acids. Individual amino acids falling outside of the general range are also listed.

[f] The solvent must be dry, or yields decrease dramatically.

N,N-dimethylformamide dineopentyl acetal. For esterifications in the presence of DMAP, time and temperature should be carefully mediated, and HOBt (1–2 equiv.) may be included.

Fmoc-His and Fmoc-Cys derivatives are particularly difficult to load efficiently while suppressing racemization. Low-racemization loadings have been documented using the cesium salts of Fmoc-His(Trt) (0.4% D-His) and Fmoc-Cys(Acm) (0.5% D-Cys) (Mergler et al., 1989b), and for Fmoc-His(Bum) (0.3% D-His) esterification by MSNT (Blankemeyer-Menge, 1990). Since the last-mentioned result is undoubtedly due to the fact that the Bum group blocks N^{π} of His, it may be noted that a more efficient loading procedure involves Fmoc-His(Bum) (2 equiv.) esterified in situ by N,N'-diisopropylcarbodiimide (DIPCDI) (2 equiv.) and DMAP (0.16 equiv.) in DCM–DMF (1 : 3) for 1 h (Fields and Fields, 1990). The best reported results for loading Fmoc-Cys(Trt) and -Cys(StBu) are 2.1% D-Cys during Fmoc-Cys(StBu) esterification by MSNT (Blankemeyer-Menge, 1990) and 2.0–2.5% D-Cys during Fmoc-Cys(Trt) cesium salt loading (Mergler et al., 1989b) or in situ Fmoc-Cys(Trt) loading with DCC and HOBt (Atherton et al., 1991). Loading of Fmoc-Cys(Acm), Fmoc-Cys(Trt), and Fmoc-His(Boc) to bromomethylated-linkers in the presence of DIEA has been reported to result in < 0.1% D-isomers (Bernatowicz et al., 1990). Racemization may be avoided by anchoring Cys to the resin via its side chain. For this purpose, Fmoc-Cys-OtBu is condensed with 5-(9-hydroxyxanthen-2-oxy)valeric acid (XAL) to provide the Fmoc-Cys(2-XAL$_4$)-OtBu preformed handle, which is subsequently coupled to an amino functionalized PEG-PS resin (Han et al., 1997b; Barany, 1999). Preformed handle coupling was performed using DIPCDI/HOBt (4 equiv. each) or BOP/HOBt/NMM (4, 4, and 8 equiv., respectively) for 2 h.

Peptide Amides

Most anchoring linkages that ultimately provide C-terminal peptide amides in a useful and general manner are benzhydrylamide derivatives. The attachment step is a direct coupling of an N^{α}-protected amino acid via its carboxyl to an appropriate benzhydrylamine-resin, with eventual cleavage at a different locus providing the desired carboxamide. The 4-methylbenzhydrylamine (MBHA) linkage has been "fine-tuned" with an electron-donating 4-methyl group (Matsueda and Stewart, 1981; Gaehde and Matsueda, 1981; Adams et al., 1998b) to cleave in strong acid with good yields, yet is completely stable to the conditions of Boc chemistry. The benzhydrylamide system has also been "fine-tuned" with electron-donating methoxy groups to create the TFA-sensitive 4-(2',4'-dimethoxyphenylaminomethyl)phenoxymethyl (Rink amide) (Rink, 1987),

4-(4′-methoxybenzhydryl)phenoxyacetic acid (Dod) (Stüber et al., 1989), and 3-(amino-4-methoxybenzyl)-4,6-dimethoxyphenylpropionic acid (Breipohl amide) (Breipohl et al., 1989) linkers for use in Fmoc chemistry. Other structural themes are compatible with Fmoc chemistry and provide anchoring linkages which cleave in TFA to give peptides amides. These include the 5-(4-aminomethyl-3,5-dimethoxyphenoxy)valeric acid (PAL) (Albericio et al., 1990a) and 5-(9-amino-xanthen-3-oxy)valeric acid (XAL) (Han et al., 1996b) handles, both of which have highly desirable features by direct comparison to alternative structures. Also, the photolabile 3-nitro-4-aminomethylbenzoic acid (Nonb) handle is an option with both Boc and Fmoc chemistries (Hammer et al., 1990), as is the 4-(2′-aminoethyl)-2-methoxy-5-nitro-phenoxypropionic acid variant (Holmes and Jones, 1995). Dod, Breipohl amide, XAL, PAL, and Nonb handles in their N-protected (usually Fmoc) forms are attached to the appropriate amino-functionalized supports by in situ activation with DIPCDI or BOP/DIEA in the presence of Dhbt-OH or HOBt (Stüber et al., 1989; Breipohl et al., 1989; Albericio et al., 1990a; Hammer et al., 1990; Han et al., 1996b).

Esterification of N^α-protected Asn and Gln can be sluggish (Barany and Merrifield, 1979; Wu et al., 1988; Fields and Fields, 1990). As an alternative, Boc-Glu(OH)-OBzl has been coupled (via an unprotected γ-carboxyl side-chain) to benzhydrylamine resin, with HF cleavage yielding a peptide containing C-terminal Gln (Li et al., 1976). In parallel fashion, Fmoc-Asp(OH)-OtBu or Fmoc-Glu(OH)-OtBu have been coupled (via an unprotected β- or γ-carboxyl side chain) to PAL, Rink amide, or Breipohl amide, with TFA cleavage yielding peptides containing C-terminal Asn or Gln (Albericio et al., 1990b; Breipohl et al., 1990; Fields and Fields, 1990).

The substitution level of Fmoc-amino acid-resins is determined by quantitative spectrophotometric monitoring following piperidine deblocking. Fmoc-amino acid-resin (4–8 mg) is shaken or stirred in piperidine–DMF (3 : 7) (0.5 mL) for 30 min, following which MeOH (6.5 mL) is added, and the resin is allowed to settle. The resultant fulvene-piperidine adduct has UV absorption maxima at $\lambda = 267$ nm ($\varepsilon = 17,500\,\mathrm{M^{-1}\,cm^{-1}}$), 290 nm ($\varepsilon = 5800\,\mathrm{M^{-1}\,cm^{-1}}$), and 301 nm ($\varepsilon = 7800\,\mathrm{M^{-1}\,cm^{-1}}$). For reference, a piperidine–DMF–MeOH solution (0.3 : 0.7 : 39) is prepared. Spectrophotometric analysis is typically carried out at 301 nm, with comparison to a free Fmoc-amino acid (e.g., Fmoc-Ala) of known concentration treated under identical conditions. The substitution level (mmol/g) = $(A_{301} \times 10^6\,\mu\mathrm{mol}/\mathrm{mol} \times 0.007\,\mathrm{L})/(7800\,\mathrm{M^{-1}\,cm^{-1}} \times 1\,\mathrm{cm} \times$ mg of resin) (Meienhofer et al., 1979; D. Hudson, unpublished results).

A logical extension/generalization of side-chain anchoring strategies is the concept of backbone anchoring (backbone amide linker: BAL) (Jensen et al., 1998). As implemented in concert with Fmoc chemistry, the ultimate C-terminal residue (with its α-carboxyl either suitably protected or modified to a different end group, but its α-amino function free) is added to an aldehyde precursor of PAL via a reductive amination reaction (Jensen et al., 1998; Alsina et al., 1999b). A BAL variation compatible with Boc chemistry has also been described (Bourne et al., 1999a, b). The cited papers should be consulted to learn special considerations and the range of applications associated with BAL.

Formation of Peptide Bond

There are currently four major kinds of coupling techniques that serve well for the stepwise introduction of N^{α}-protected amino acids for solid-phase synthesis. In the solid-phase mode, coupling reagents are used in excess to ensure that reactions reach completion. The ensuing discussion will skirt some of the rather complicated mechanistic issues and focus on practical details. Recommendations for coupling methods are included in tables 3-4 and 3-5.

In situ Reagents

See figure 3-5. The classical example of an in situ coupling reagent is N,N'-dicyclohexylcarbodiimide (DCC) (Rich and Singh, 1979; Merrifield et al., 1988). The related N,N'-diisopropylcarbodiimide (DIPCDI) is more convenient to use under some circumstances, as the resultant urea co-product is more soluble in DCM. The generality

Table 3-4 General Boc chemistry SPPS

Description	Solvents/Chemicals	Time
Deprotection	TFA	2 × 1 min
Wash[a]	DMF	1 min flow
Activation and	Boc-amino acid[b] (5 equiv.)	2 min
coupling	HBTU (4.75 equiv.) : DIEA (7.5 equiv.) in DMF	10 min
	or	
	Boc-amino acid[b] (5 equiv.)	2 min
	HATU (4.75 equiv.) : DIEA (7.5 equiv.) in DMF	5–10 min
Wash[a]	DMF	1 min flow

[a] DCM washes may be included to prevent acid catalyzed side reactions of Gln; however coupling efficiency for certain amino acids may suffer (Schnölzer et al., 1992, Alewood et al., 1997).
[b] Boc-Asn and -Gln are side-chain protected.

Table 3-5 General Fmoc chemistry SPPS

Description	Solvents/Chemicals	Time
Wash	NMP	3 × 1 min
Deprotection	piperidine: NMP (1:4)[a]	20 min
Wash	NMP	3 × 1 min
Activation	Fmoc-amino acid (4 equiv.) in NMP	5 min
and	DIPCDI (4 equiv.): HOBt (4 equiv.) in DMF	45 min
coupling	*or*	
	Fmoc-amino acid[b] (4 equiv.) in NMP	5 min
	BOP (3 equiv.): NMM (4.5 equiv.): HOBt (3 equiv.) in DMF	45 min
	or	
	Fmoc-amino acid[b] (4 equiv.) in NMP	5 min
	HBTU (3.6 equiv.): DIEA (7.2 equiv.): HOBt (4 equiv.) in DMF	45 min
	or	
	Fmoc-amino acid[b] (4 equiv.) in NMP	5 min
	HATU (3.6 equiv.): DIEA (7.2 equiv.): HOAt (4 equiv.) in DMF	45 min
	or	
	Fmoc-amino acid preformed ester (4 equiv.) in NMP	60 min
Wash	NMP	3 × 1 min

[a] 0.1 M HOBt may be added to reduce aspartimide formation and resulting side reactions.
[b] Fmoc-Asn and -Gln are side-chain protected.

of carbodiimide-mediated couplings is extended significantly by the use of either 1-hydroxybenzotriazole (HOBt) or 1-hydroxy-7-azaben-zotriazole (HOAt) as an additive, either of which accelerates carbodi-imide-mediated couplings, suppresses racemization, and inhibits dehydration of the carboxamide side chains of Asn and Gln to the corresponding nitriles (König and Geiger, 1970a, 1973; Mojsov et al., 1980; Carpino et al., 1994). Within the past decade, protocols involving benzotriazol-1-yl-oxy-tris(dimethylamino)phosphonium hexafluorophosphate (BOP), 2-(1H-benzotriazol-1-yl)-1,1,3,3-tetra-methyluronium hexafluorophosphate (HBTU), O-(7-azabenzotriazol-1-yl)-1,1,3,3-tetramethyluronium hexafluorophosphate (HATU), 2-(1H-benzotriazol-1-yl)-1,1,3,3-tetramethyluronium tetrafluoroborate (TBTU), and 2-(2-oxo-1(2H)-pyridyl)-1,1,3,3-bispentamethyleneuro-nium tetrafluoroborate (TOPPipU) have achieved popularity. (Interestingly, X-ray crystallographic analysis has shown that the solid-state structures of HBTU and HATU are not tetramethyluro-nium salts, but guanidinium N-oxide isomers [Abdelmoty et al., 1994].) BOP, HBTU, HATU, TBTU, and TOPPipU require a tertiary amine such as NMM or DIEA for optimal efficiency (Dourtoglou et al., 1984; Fournier et al., 1988; Ambrosius et al., 1989; Gausepohl et al., 1989a; Seyer et al., 1990; Fields et al., 1991; Knorr et al., 1991; Reid and Simpson, 1992; Carpino et al., 1994). HOBt has been reported to accelerate further the rates of BOP- and HBTU-mediated couplings (Hudson, 1988; Fields et al., 1991). In situ activation by

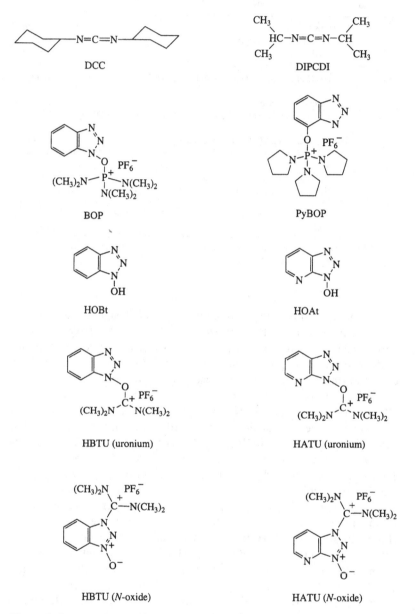

Figure 3-5. In situ coupling reagents and additives for SPPS. Note that HBTU and HATU are shown in both their tetramethyluronium and guanidinium *N*-oxide salt forms.

excess HBTU or TBTU can cap free amino groups (Gausepohl et al., 1992; Story and Aldrich, 1994); it is not known whether HOBt can suppress this side reaction. Acylations using BOP result in the liberation of the carcinogen hexamethylphosphoramide, which might limit its use in large scale work. The modified BOP reagent benzotriazole-1-yl-oxy-tris-pyrrolidinophosphonium hexafluorophosphate (PyBOP) liberates potentially less toxic by-products (Coste et al., 1990). Protocols have been reported for the use of BOP to incorporate side-chain unprotected Thr and Tyr (Fournier et al., 1988, 1989). A caveat to using PyAOP, PyBOP, and PyBroP is that some commercial sources are contaminated with small amounts of pyrrolidine, and this contaminant can be incorporated to form pyrrolide by-products in those cases when reactions of the activated carboxylates are particularly slow, for example, some cyclizations or couplings to relatively weak nucleophilic amines (Alsina et al., 1999a). The side reaction does not occur when the reagents are recrystallized before their use in coupling reactions.

Active Esters

See figure 3-6. A long-known but steadfast coupling method involves the use of active esters. The classical 2- and 4-nitrophenyl esters (ONo and ONp, respectively), used in DMF, allow relatively slow but dehydration-free introduction of Asn and Gln (Mojsov et al., 1980). ONo and ONp esters of Boc- and Fmoc-amino acids are prepared from DCC and either 2- or 4-nitrophenol, and the undesired nitrile contaminant is easily separated (Bodanszky et al., 1973, 1980). N-hydroxysuccinimide (OSu) esters of Fmoc-amino acids have been used successfully in SPPS (Fields et al., 1988), but are not recommended for general use due to formation of succinimidoxycarbonyl-β-alanine-N-hydroxysuccinimide ester (Gross and Bilk, 1968; Weygand et al., 1968b).

Workers later concentrated on pentafluorophenyl (OPfp), benzotriazyl (OBt), 3-hydroxy-2,3-dihydro-4-oxo-benzotriazine (ODhbt), and substituted 1-phenylpyrazolinone enol esters. Boc- and Fmoc-amino acid OPfp esters are prepared from DCC and pentafluorophenol (Kisfaludy et al., 1973; Penke et al., 1974; Kisfaludy and Schön, 1983) or pentafluorophenyl trifluoroacetate (Green and Berman, 1990). Although OPfp esters alone couple slowly, the addition of HOBt (1–2 equiv.) increases the reaction rate (Atherton et al., 1988a; Hudson, 1990b). Fmoc-Asn-OPfp allows for efficient incorporation of Asn with little side-chain dehydration (Gausepohl et al., 1989b). OBt esters of Fmoc-amino acids are formed rapidly, with HOBt plus DIPCDI, and are highly reactive (Harrison et al., 1989; Fields et al., 1989), as are Boc-amino acid OBt esters (Geiser et al.,

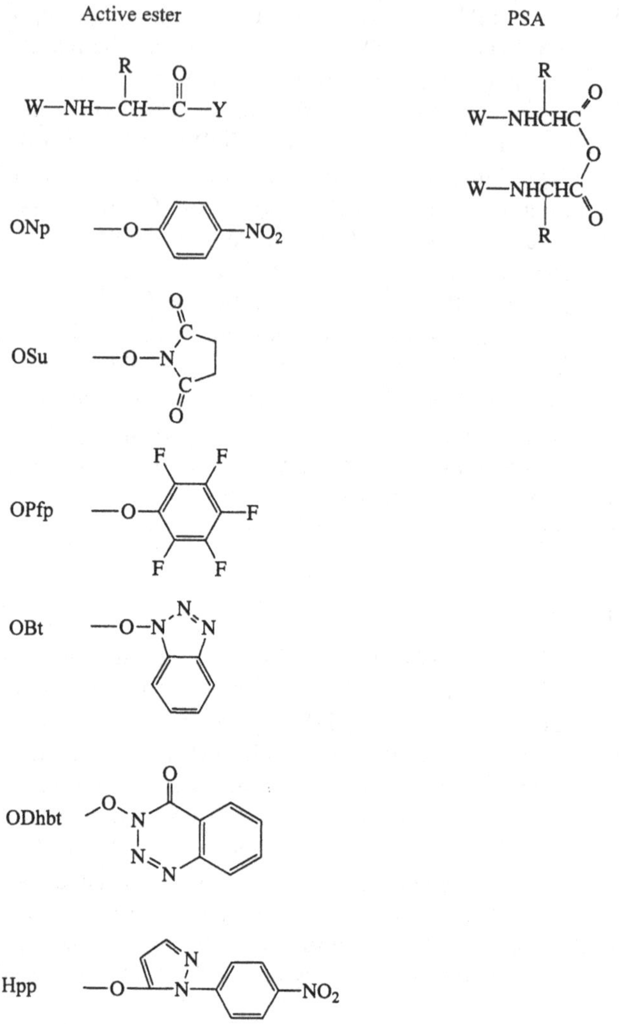

Active ester   PSA

$$W—NH—\underset{\underset{}{\overset{\overset{R}{|}}{CH}}}{}—\overset{\overset{O}{||}}{C}—Y$$

ONp $—O—\bigcirc—NO_2$

OSu

OPfp

OBt

ODhbt

Hpp $—O—\cdots—NO_2$

Figure 3-6. Activated N^α-protected amino acids. W is either Boc or Fmoc, Y is the structure specified next to the abbreviation of the active ester derivative.

1988). N^α-protected amino acid ODhbt esters suppress racemization and are highly reactive, in similar fashion to OBt esters (König and Geiger, 1970b). Preparation of ODhbt esters, with Dhbt-OH and DCC, is accompanied by the formation of the by-product 3-(2-azido-benzoyloxy)-4-oxo-3,4-dihydro-1,2,3-benzotriazine (König and Geiger, 1970c). Fmoc-amino acid ODhbt esters are far more stable than HOBt esters, and can therefore be isolated from the side product before use (Atherton et al., 1988b). Fmoc-amino acid 1-(4-nitrophe-nyl)-2-pyrazolin-5-one (Hpp), 3-phenyl-1-(4-nitrophenyl)-2-pyrazolin-5-one (Pnp), and 3-methyl-1-(4-nitrophenyl)-2-pyrazolin-5-one (Npp) esters also allow for rapid couplings (Hudson, 1990a; Johnson et al., 1992). Competition experiments have shown ester reactivity to be usually Pnp > Hpp ~ Npp > ODhbt > OPfp > OSu > ONp > ONo (Hudson, 1990a, b; Johnson et al., 1992), although Hpp esters were found to be superior to Pnp and Npp esters for "difficult" couplings (Johnson et al., 1992). Both Fmoc-Tyr and Fmoc-Ser have been incorporated successfully as preformed active esters without side-chain protection (Fields et al., 1989; Otvös et al., 1989a).

Preformed Symmetrical Anhydrides

See figure 3-6. Preformed symmetrical anhydrides (PSAs) are favored by some workers because of their high reactivity. They are generated in situ from the corresponding N^α-protected amino acid (2 or 4 equiv.) plus DCC (1 or 2 equiv.) in DCM; following removal of the urea by filtration, the solvent is exchanged to DMF for optimal couplings. Detailed synthetic protocols based on PSAs have been described for Boc (Merrifield et al., 1982; Yamashiro, 1987; Geiser et al., 1988; Kent and Parker, 1988; Wallace et al., 1989) and Fmoc (Chang et al., 1980a; Heimer et al., 1981; Atherton and Sheppard, 1989) chemistries.

The use of the PSA procedure to introduce Boc/Fmoc-Gly or -Ala occasionally results in inadvertent coupling of a diglycyl or dialanyl unit (Merrifield et al., 1974, 1985; Benoiton and Chen, 1987). Also, side-chain unprotected Asn and Gln, all Arg derivatives, and N^τ-protected His should not be used as PSAs due to potential side reactions discussed previously (see above, Protection Schemes). The solubilities of some Fmoc-amino acids make PSAs a less-than-optimum activated species. Not all Fmoc-amino acids are readily soluble in DCM, thus requiring significant DMF for solubilization. Optimum activation conditions, which require neat DCM (Rich and Singh, 1979), cannot be achieved. In addition, the resulting Fmoc-amino acid PSAs are even less soluble than the parent Fmoc-amino acid (Harrison et al., 1989).

Preformed Fmoc-N-carboxyanhydrides (NCAs) are stable derivatives which do not have to be generated in situ. Their reactivity is comparable to PSAs. In addition, NCAs are thermally stable,

---

Efficient couplings using Boc-amino acid PSAs are critically dependent on the concentration of the activated species in solution. It has been recommended that Boc-amino acid PSA couplings proceed at a concentration of 0.15 M, and that double couplings be standard practice for syntheses of >50 residues (Kent and Parker, 1988).

Fmoc-amino acids have variable solubility properties in relatively nonpolar solvents, such as DCM. Fmoc-Asp(O*t*Bu), -Glu(O*t*Bu), -Ile,-Leu, -Lys(Boc), -Ser(*t*Bu), -Thr(*t*Bu), and -Val are soluble in DCM, while Fmoc-Ala, -Gly, -Met, -Trp, and -Tyr(*t*Bu) require the presence of a more polar co-solvent (e.g., DMF) for solubilization. Fmoc-Asn, -Gln, -His(Bum), and -Phe require at least 60% DMF to remain in solution. Conversion of Fmoc-amino acids to the corresponding PSAs results in poorer solubilities in nonpolar solvents. Thus, whether Fmoc-amino acids are coupled in situ or as preactivated species, relatively polar solvents (DMF or NMP) should be used to ensure that all reactants are in solution.

allowing for couplings at elevated temperatures. Such protocols can be advantageous for coupling sterically hindered amino acids. NCAs are sensitive to moisture and should therefore be used in dry solvents (Fuller et al., 1992).

Acid Halides

N^α-protected amino acid chlorides have a long history of use in solution synthesis. Their use in solid-phase synthesis has been limited, as the Boc group is not completely stable to reagents used in the preparation of acid chlorides. The Fmoc group, however, survives acid chloride preparation with thionyl chloride (Carpino et al., 1986), while both the Fmoc and Boc groups are stable to acid fluoride preparation with cyanuric fluoride (Carpino et al., 1990, 1991a; Bertho et al., 1991). For derivatives with *t*Bu side-chain protection, only the acid fluoride procedure can be used (Carpino et al., 1990). Fmoc-amino acid chlorides and fluorides react rapidly under SPPS conditions in the presence of HOBt/DIEA and DIEA, respectively, with very low levels of racemization (Carpino et al., 1990, 1991b, 1992). Fmoc-amino acid fluorides are especially effective for coupling to *N*-alkyl amino acids (Wenschuh et al., 1994, 1995). For convenience, tetramethylfluoroformamidinium hexafluorophosphate (TFFH) can be used for automated preparation of Fmoc-amino acid fluorides (Carpino and El-Faham, 1995).

Monitoring

A crucial issue for stepwise solid-phase peptide synthesis is the repetitive yield per deprotection/coupling cycle. There are a number of ways of monitoring these steps, including some with a possibility for "real-time" feedback based on the kinetics of appearance or disappearance of appropriate soluble chromophores measured in a flow-through system. Most accurate and meaningful are qualitative and quantitative tests for the presence of unreacted amines after an acylation step. Such tests should ideally be negative before proceeding further in the chain assembly. For certain active ester methods, the leaving group has "self-indicating" properties in so far as a colored complex is noted for as long as unreacted amino groups remain on the support. These various techniques reveal that high efficiencies can indeed be achieved in stepwise synthesis.

The best known qualitative monitoring methods are the ninhydrin (Kaiser et al., 1970) and isatin (Kaiser et al., 1980) tests for free N^α-amino and -imino groups, respectively, where a positive colorimetric response to an aliquot of peptide-resin indicates the presence of unreacted N^α-amino/imino groups. These tests are easy, reliable, and require only a few minutes to perform, allowing the chemist to make a quick decision on how to proceed. A highly accurate quantitative modification of the ninhydrin procedure has been developed (see Box).

Other monitoring techniques exist that are generally nondestructive (noninvasive) and can therefore be carried out on the total batch. Resin-bound N^α-amino groups can be titrated with picric acid, 4,4'-dimethoxytrityl chloride, bromophenol blue dye, and quinoline yellow dye. Picric acid is removed from resin-bound amines with 5% DIEA in DCM, and the resulting chromophore quantitated at $\lambda = 362$ nm (Hodges and Merrifield, 1975; Arad and Houghten, 1990). For trityl monitoring, 4,4'-dimethoxytritylchloride and tetra-*n*-butylammonium perchlorate are reacted with the resin, released with 2% dichloroacetic acid in DCM, and quantitated at 498 nm (Horn and Novak, 1987; Reddy and Voelker, 1988). The effect of the dilute acid on Fmoc-amino acid side-chain protecting groups and linkers has not been

---

Free amino groups are quantitated based on their reaction with ninhydrin to produce Ruhemann's purple. Three reagent solutions are required: solution 1 is phenol–ethanol (7 : 3), solution 2 is 0.2 mM KCN in pyridine, and solution 3 is 0.28 M ninhydrin in ethanol. A sample of Boc-peptide-resin (2–5 mg) is incubated with 75 μL of solution 1, 100 μL of solution 2, and 75 μL of solution 3 for 7 min at 100 °C (Sarin et al., 1981; Applied Biosystems, Inc., 1992). For Fmoc-peptide-resins, premature removal of the Fmoc group (by pyridine) is minimized by adding 2–3 drops (20–40 μL) of glacial HOAc to each resin sample and heating the reaction mixture for 5 min instead of 7 min (Applied Biosystems, Inc., 1989b). Immediately following the designated incubation time, 60% aqueous ethanol (4.8 mL) is added to each sample with vigorous mixing. Once the peptide-resin has settled, the absorbance of each sample solution is read at $\lambda = 570$ nm; 60% ethanol is used as a reference. The concentration of free amino groups $(\text{mmol g}^{-1}) = (A_{570} \times 10^6\,\mu\text{mol/mol} \times 0.005\,\text{L})/$ $(15,000\,\text{M}^{-1}\,\text{cm}^{-1} \times 1\,\text{cm} \times \text{mg of resin})$.

reported. For bromophenol blue and quinoline yellow monitoring, the dye is bound to free amino groups following deprotection, then displaced as acylation proceeds. Quantitative monitoring can be carried out at $\lambda = 600$ and 495 nm for bromophenol blue and quinoline yellow, respectively (Krchnák et al., 1988; Flegel and Sheppard, 1990; S. Young et al., 1990). Gel-phase nuclear magnetic resonance (NMR) spectroscopy has been proposed for direct examination of resin-bound reactants (Epton et al., 1980; Giralt et al., 1984; Butwell et al., 1988), but would appear to suffer from problems of sensitivity, expense, and time needed to accumulate data. On-resin monitoring may also be achieved by NMR using the magic angle spinning method with a nanoprobe (Fitch et al., 1994). Magic angle spinning allows for the possible recording of high resolution NMR spectra of the resin-bound growing peptide chain and analysis of the synthesis.

As an alternative to quantify resin-bound species, soluble reactants or co-products can be analyzed. Continuous measurement of electrical conductivity can be used to evaluate coupling and Fmoc deprotection efficiencies (Nielson et al., 1989; Fox et al., 1990; McFerran et al., 1991). The progress of Fmoc chemistry can be evaluated by observing at $\lambda = 300\text{--}312$ nm the decrease of absorbance when Fmoc-amino acids are taken up during coupling, and by the increase in absorbance when the Fmoc group is released with piperidine (Chang et al., 1980a; Atherton and Sheppard, 1985, 1989; Frank and Gausepohl, 1988). Monitoring a decrease in Fmoc-amino acid concentration at $\lambda = 300$ nm can be complicated when OPfp esters are utilized (Atherton et al., 1988b). More straightforward acylation monitoring is possible when Fmoc-amino acid ODhbt, Hpp, Pnp, and Npp esters are used. During the coupling of Fmoc-amino acid ODhbt esters, the liberated Dhbt-OH component binds to free N^α-amino groups, producing a bright yellow color, which diminishes as acylation proceeds (Cameron et al., 1988). Real-time spectrophotometric monitoring proceeds at $\lambda = 440$ nm (Cameron et al., 1988). In a similar way, ionization of the leaving group from Fmoc-amino acid Npp esters by free N^α-amino groups results in a blood red color (Hudson, 1990a). As coupling proceeds the color change (to gold) can be monitored at $\lambda = 488$ nm.

Unfortunately, continuous-flow monitoring is inherently insensitive for direct judgment of reaction endpoints. Assuming an initial two-fold excess of activated incoming amino acid, absorbance will drop from 2.00 units to 1.05 units or 1.01 units respectively with 95% or 99% coupling. It is difficult to distinguish accurately between 1.05 and 1.01. In contrast, if unreacted resin-bound components are titrated for the same two cases, the five-fold difference between 0.05 and 0.01 is easily noted. The sensitivities of techniques monitoring resin-bound components is limited by nonspecific binding or irreversible reactions of the titrant with the protected peptide or resin, which contribute to background readings despite complete reactions.

Invasive monitoring of both synthetic efficiency and amino acid composition of peptide-resins can be achieved by a powerful quantitative variation of the Edman sequential degradation, called "preview analysis" (Tregear et al., 1977; Matsueda et al., 1981; Kent et al., 1982). Crude peptide-resins are sequenced directly; each Edman degradation cycle serves to identify a primary amino acid residue and "preview" the next amino acid in the sequence. Because preview is cumulative, quantitation of peaks after a number of cycles indicates the average level of deletion peptides and thus the overall synthetic efficiency. Since the linkers as well as most side-chain protecting groups used in Boc chemistry are stable to the conditions of Edman degradation, sequencing is a true solid-phase process; moreover, identification of amino acid phenylthiohydantoins requires a set of protected standards (Simmons and Schlesinger, 1980; Steiman et al., 1985). Preview sequence analysis of peptide-resins made by Fmoc chemistry requires initial TFA cleavage, followed by immobilization (covalent or noncovalent) of the crude peptide on a suitable support (Kochersperger et al., 1989; C. G. Fields et al., 1992, 1993c). Most side-chain-protecting groups used in conjunction with Fmoc chemistry are not stable to the conditions of Edman degradation; hence, the usual free side-chain phenylthiohydantoin standards can be used.

Recently, on-resin mass spectrometry (MS) via the MALDI (matrix-assisted laser desorption/ionization) technique has been developed (Egner et al., 1995). This method can be used as a semi-on-line monitoring technique during SPPS (Talbo et al., 1997). An especially elegant approach for on-resin MS is to use the photolabile α-methylphenacyl ester linker, which is cleaved directly upon laser photolysis and MALDI ionization (Egner et al., 1995; Fitzgerald et al., 1996).

The technique of "internal reference amino acids" (IRAA) is often very useful to measure accurately yields of chain assembly and retention of chains on the support during synthesis and after cleavage (Matsueda and Haber, 1980; Atherton and Sheppard, 1989; Albericio et al., 1990a). In addition, amino acid analysis of peptide-resins may be used to monitor synthetic efficiency; the advent of microwave hydrolysis technology may permit rapid analysis (Yu et al., 1988).

Automation of Solid-Phase Synthesis

A significant advantage of solid-phase methods lies in the ready automation of the repetitive steps (see tables 3-4 and 3-5). The first instrument for synthesis of peptides was built by Merrifield, Stewart, and Jernberg (1966). Numerous models for both batchwise and continuous-flow operation at various scales of operation are now commercially available.

Some of these instruments include features to facilitate monitoring (compare to previous section). Supported procedures have also been introduced for generation of large numbers of (usually related) peptide sequences in a reasonably short time, although with some sacrifices in the usual standards for purity and characterization.

Automated Synthesizers, 1–3 Simultaneous Syntheses

The PE Biosystems Model 431A/433A can utilize either Boc or Fmoc chemistries, with reaction mixing by vigorous vortex or gas bubbling. The 431A/433A uses fully automated HBTU + HOBt in situ cycles (Fmoc chemistry in NMP), called *FastMoc* (Fields et al., 1991). The synthesis scale is from 0.1 to 0.25 mmol, with microprocessor-control via an internal computer. Advanced ChemTech Models 90 and 400 use either Boc or Fmoc chemistries, with reaction mixing by nitrogen bubbling or oscillation. Couplings for the Models 90 and 400 can be in situ, preactivation to form symmetrical anhydrides or HOBt esters, or by preformed Pfp esters. Pharmaceutical considerations are fulfilled by the Model 400 (Birr, 1990a, b), as it can utilize 100 g or more of resin. The Rainin/Protein Technologies PS3 features coupling by in situ BOP with Fmoc-amino acids only (all prepackaged), at scales from 0.1 to 0.5 mmol with nitrogen bubbling for reaction mixing. CS Bio offers several synthesizers that accommodate both Boc and Fmoc chemistries. Models 036, 100, 136, and 536 differ primarily in their flexibility (i.e., the 036 adds up to four amino acids in a single run while the 136 adds up to nine) and synthesis scales (0.1–2.5 mmol, 0.05–1 mmol, 0.1–2.5 mmol, and 0.2–25 mmol, respectively). Models 936S and 936 are pilot plant instruments, with scales of up to 600 mmol and 12.5 mol, respectively.

The just-mentioned instruments are designed for batchwise syntheses. Continuous-flow instruments (Fmoc chemistry only) include the PerSeptive Biosystems 9050 Plus and Pioneer and the PE Biosystems 432A (Synergy). The PerSeptive Biosystems 9050 Plus (Kearney and Giles, 1989) is an automated instrument with a synthesis scale of 0.05–5.0 mmol and flow rates typically from 5–15 mL/min. The automated Pioneer Peptide Synthesis System has two columns for synthesis and can operate on a scale of 0.02–2.0 mmol. The PE Biosystems 432A is a single column automated instrument with a synthesis scale of 25 μmol.

Automated and Semi-automated Multiple Peptide Synthesizers

The Advanced ChemTech 348Ω synthesizes up to 48 peptides simultaneously. Either Boc or Fmoc chemistry is used, and the synthesis scale is 5–150 µmol. The CS Bio System 10 synthesizes 10 peptides simultaneously on a 0.1–0.5 mmol scale using either Boc or Fmoc chemistry. The Zinsser Analytic PepSy utilizes four independent pipetting probes to synthesize up to 96 peptides simultaneously via Fmoc chemistry. The ABIMED/Gilson Model AMS 422 (Gausepohl et al., 1990, 1991) uses Fmoc chemistry to synthesize up to 48 peptides on a scale from 5–50 µmol, with activation by in situ PyBOP. A single robotic arm dispenses reagents, while resins are contained in fritted polypropylene tubes. In similar fashion, the Shimadzu PSSM-8 uses a single robotic arm to synthesize up to eight peptides simultaneously. The Rainin/Protein Technologies Symphony/Multiplex uses Fmoc chemistry to synthesize up to 12 peptides simultaneously. Synthesis scale is 5–100 µmol, and peptide-resin cleavage is performed on the instrument. The Spyder Instruments Compas 768 uses tilted plate centrifugation to prepare up to 768 peptides per run by Fmoc chemistry. The PerSeptive Biosystems Pioneer (described above) has a multiple peptide synthesis accessory that allows for the synthesis of 16 or 32 peptides on a 0.005–1.0 mmol scale. Tyler Research Corporation offers the semi-automated SAMPS (Semi-Automatic Multiple Peptide Synthesizer), which synthesizes up to 100 peptides using Fmoc chemistry on a 10–200 µmol scale. The manual Multiple Peptide Synthesis TeaBag method can synthesize up to 120 peptides simultaneously, using either Boc or Fmoc chemistry, and coupling by PSAs or in situ DCC/DIPCDI (Houghten et al., 1986; Beck-Sickinger et al., 1991). The "TeaBag" method has been semi-automated (Beck-Sickinger et al., 1991). Cambridge Research Biochemicals markets the Pepscan/PIN method (Geysen et al., 1984; Hoeprich, et al., 1989; Arendt et al., 1993), which uses either Boc or Fmoc chemistry, and coupling in situ with DCC or BOP/HOBt, to synthesize up to 96 peptides (10–100 nmol) simultaneously. Finally, the Affymax Parallel Chemical Synthesis system (VLSIPS, for "very large-scale immobilized polymer synthesis"), using the photolabile Nvoc N^α-protecting group and preformed HOBt esters, can be used for simultaneous synthesis of an extraordinarily high number of related peptides (e.g., $1024 = 2^{10}$ by 10 stages requiring 2.5 h each) (Fodor et al., 1991).

Automated Solid-Phase Organic Synthesizers

Recent years have seen a dramatic increase in the variety of organic reactions, beyond peptide synthesis, that have utilized solid-phase methodologies. Concurrently, instrumentation to automate solid-phase organic chemistry has been developed. These instruments differ from conventional peptide synthesizers primarily in their durability to a range of caustic reagents. The Zinsser Analytic Sophas M can accommodate a variety of reactors, from 96-well plates to 25 mL reaction vessels.

It can perform up to 864 syntheses in a single run, and allows for heated reactions up to 150 °C. The Sonata/Pilot from Rainin/Protein Technologies performs large scale (0.1–50 mmol) peptide and organic synthesis with heating and cooling capabilities. Advanced ChemTech has several "biomolecular" synthesizers, including the 357FBS, 384HTS, 396MBS, 440MOS, 496MOS, and Labtech Synthesis System. The 357FBS synthesizes up to 36 compounds on a scale of 5–250 μmol. The 384HTS, 440MOS, 496MOS, and LabTech contain four 96-well reactor blocks, one 40-well (8 mL per well) reactor block, one 96-well reactor block, and one 96-well reactor or solution phase block, respectively, and each has a controlled temperature range of −70 to 150°C. The 396MBS accommodates a 96-well reactor block with a synthesis scale of 5–1000 μmol. The PE Biosystems Solaris 530 Organic Synthesis System has 48-position reagent racks, synthesis modules, and product racks, and uses eight liquid-handling tips.

Cleavage

Boc SPPS is designed primarily for simultaneous cleavage of the peptide anchoring linkage and side-chain-protecting groups with strong acid (HF or equivalent), while Fmoc SPPS is designed primarily to accomplish the same cleavages with moderate strength acid (TFA or equivalent). In each case, careful attention to cleavage conditions (reagents, scavengers, temperature, and times) is necessary in order to minimize a variety of side reactions. Considerations for separate removal of acid-stable side-chain-protecting groups has been covered earlier (see above, Protection Schemes). Nonacidolytic methods for cleavage of the anchoring linkage, each with certain advantages as well as limitations, may also be used in conjunction with either Boc or Fmoc chemistries.

Hydrogen Fluoride (HF)

Treatment with HF cleaves PAM or MBHA linkages and simultaneously removes the side-chain protecting groups commonly applied in Boc chemistry (figure 3-2), that is, Bzl (for Asp, Glu, Ser, Thr, and Tyr), 2-BrZ or 2,6-Cl$_2$Bzl (for Tyr), cHex (for Asp), 2-ClZ (for Lys), Bom or Tos (for His), Tos or Mts (for Arg), Xan (for Asn and Gln), and Meb (for Cys) (Tam and Merrifield, 1987; Applied Biosystems, Inc., 1992). HF cleavages are always carried out in the presence of a carbonium ion scavenger, usually anisole (10%, v/v). For cleavages of Cys-containing peptides, further addition of 4-thiocresol (1.8%) is recommended. Additional scavengers, such as dimethyl sulfide and 4-cresol, are used in conjunction with a two-stage "low-high HF" cleavage method that provides extra control and thereby better product

HF cleavage procedures require a special all-fluorocarbon apparatus. The standard method uses HF–anisole (9 : 1) (1 mL per 20 μmol peptide) at 0°C for 1 h, with the addition of 2-mercaptopyridine (10 equiv.) for Met-containing peptides. For Cys- and Trp-containing peptides, HF–anisole–dimethyl sulfide–4-thiocresol (10 : 1 : 1 : 0.2) is recommended. Following cleavage, HF is evaporated carefully under aspirator suction with ice bath cooling, and most of the anisole is removed by vacuum from an oil pump. The vessel is triturated with ether to remove benzylated scavenger adducts; at the same time, the ether facilitates transfer of the cleaved resin (with trapped peptide) to a fritted glass funnel. Next, 30% aqueous HOAc (twice, 1.5 mL per 20 μmol peptide) is used to rinse the cleavage vessel and extract the resin on the fritted glass filter. The combined filtrates are diluted with H_2O to bring the HOAc concentration to < 10%, and the peptide is usually recovered by lyophilization (Stewart and Young, 1984; Tam and Merrifield, 1987; Stewart, 1997). "Low-high" HF cleavages are carried out following the detailed description of the original and later publications (Tam et al., 1983; Tam and Merrifield, 1985, 1987).

purities (Tam and Merrifield, 1987). Both Trp(CHO) and Met(O) can be deprotected under "low HF" conditions [20–25% HF–0 to 5% 4-thiocresol–70 to 80% dimethyl sulfide, 0 °C for 1 h; 4-thiocresol is necessary only for Trp(CHO)] (Tam and Merrifield, 1987). In the presence of the large levels of dimethyl sulfide used in "low HF" conditions, Tyr(Bzl) undergoes little C-alkylation (Tam and Merrifield, 1987).

In strong acid, the γ-carboxyl group of Glu can become protonated and lose water. The resulting acylium ion is then trapped either intra-molecularly by the N^{α}-amide nitrogen to give a pyrrolidone or (more likely) intermolecularly with the commonly used scavenger anisole (Feinberg and Merrifield, 1975). This serious problem can be con-trolled by attenuation of the acid strength (i.e., "low HF" conditions) and caution with regard to cleavage temperature (Tam and Merrifield, 1987). Strong acid can also cause an $N \rightarrow O$ acyl shift in Ser- and Thr-containing peptides resulting in the thermodynamically less stable O-acyl species (Fujino et al., 1978). This process can be reversed for simple cases by treating the cleaved, deprotected peptide with 5% aqueous NH_4HCO_3, pH 7.5, 25 °C for several hours, or in 2% aqu-eous NH_4OH, 0 °C, 0.5 h (Barany and Merrifield, 1979); reversal of the $N \rightarrow O$ acyl shift in multiple Ser- and Thr-containing peptides may be more problematic. HF-liberated Boc groups can modify Met residues (Noble et al., 1976); therefore, the N^{α}-Boc group should be removed prior to HF cleavage. HF deprotection of His(Bom) liberates formal-dehyde, resulting in methylation of susceptible side chains and cycliza-tion of N-terminal Cys residues to a thiazolidine (Mitchell et al., 1990; Gesquiére et al., 1990; Kumagaye et al., 1991). These side reactions may be inhibited by including in the HF cleavage mixture a formaldehyde scavenger, for example, resorcinol (0.27 M), or Cys or CysNH$_2$ (30–90 equiv.), and purifying the peptide immediately after cleavage (Mitchell et al., 1990; Kumagaye et al., 1991). Serious Trp alkylation side reactions has been observed during workup after HF cleavage of peptides containing Cys or Met adjacent to Trp; the problem may be mitigated by adding free Trp (10 equiv.) as a scavenger during cleavage (D. Hudson, unpublished results) or during the initial lyophilization (Ponsati et al., 1990a).

Other Strong Acids

The alternative strong acids listed here, and further examples elsewhere (Barany and Merrifield, 1979; Stewart, 1997), are very likely to promote the same side reactions just described for HF. TFMSA (1 M)–thioanisole (1 M) in TFA cleaves PAM or MBHA linkers (Tam and Merrifield, 1987; Bergot et al., 1987), and simultaneously removes many side-chain-protecting groups used in Boc chemistry, for example, Mts (for Arg), Bzl (for Asp, Glu, Ser, Thr), cHex (for Asp), Meb (for Cys), 2-ClZ (for Lys), 2-BrZ or 2,6-Cl$_2$Bzl (for Tyr), and Bom or Tos (for His) (Tam and Merrifield, 1987), without requiring a special apparatus. A "low-high" method can be used with TFMSA, in similar fashion to HF (Tam and Merrifield, 1987). Tetrafluoroboric acid (HBF$_4$) (1 M)–thioanisole (1 M) in TFA offers a similar range of side-chain deprotection as TFMSA (Kiso et al., 1989; Akaji et al., 1990b).

Trimethylsilyl bromide (TMSBr) and trimethylsilyl trifluoromethanesulfonate (TMSOTf) have also been used for strong acid cleavage and deprotection reactions, which are accelerated by the presence of thioanisole as a "soft" nucleophile (Yajima et al., 1988; Nomizu et al., 1991). TMSBr (1 M)–thioanisole (1 M) removes Mts (for Arg), Bzl (for

---

For TFMSA cleavages, peptide-resin (100 mg) is stirred with thioanisole (187 µL) and 1,2-ethanedithiol (EDT) (64 µL) in an ice bath for 10 min. TFA (1.21 mL) is added, and after equilibration for 10 min, TFMSA (142 µL) is added *slowly*. Cleavage and deprotection proceeds for 1 h (unless the MBHA linker is being cleaved, in which case cleavage proceeds for 1.5–2.5 h). During cleavage and deprotection, the ice bath is removed, but precautions are taken to ensure that the temperature of the reaction does not increase rapidly. Ambient temperature (25 °C) is reached in about 15 min. Subsequently, the cleavage mixture is filtered through a fritted glass funnel directly into methyl *t*Bu ether, to rapidly precipitate the peptide and remove the acid and scavengers. The precipitated peptide should be washed with methyl *t*Bu ether and dried under vacuum overnight (Bergot et al., 1987; Fields and Fields, 1991; C. G. Fields et al., 1993a).

Asp, Glu, Ser, Thr, Tyr), 2,6-Cl$_2$Bzl (for Tyr), and 2-ClZ (for Lys), as well as reducing Met(O) to Met (Yajima et al., 1988). Although not specifically stated, TMSBr–thioanisole probably deprotects His(Tos), Cys(Meb), and Tyr(2-BrZ). TMSOTf (1 M)–thioanisole (1 M) (a.k.a. DEPRO) additionally removes Bom from His (Yajima et al., 1988) and efficiently cleaves PAM or MBHA linkages (Akaji et al., 1989; Nomizu et al., 1991). Stable Cys side-chain protection should be used during TMSBr–thioanisole cleavages.

Trifluoroacetic Acid (TFA)

The combination of side-chain-protecting groups, for example, *t*Bu (for Asp, Glu, Ser, Thr, and Tyr), Boc (for His and Lys), Bum (for His), Tmob (for Asn, Cys, and Gln), Trt (for Asn, Cys, Gln, and His), and/or Xan (for Asn, Cys, and Gln), and anchoring linkages, for example, HMP/PAB or PAL, commonly used in Fmoc chemistry (see figure 3-3) are simultaneously deprotected and cleaved by TFA. Such cleavage of *t*Bu and Boc groups results in *tert*-butyl cations and *tert*-butyl trifluoroacetate formation (Jaeger et al., 1978a, b; Löw et al., 1978a, b; Lundt et al., 1978; Masui et al., 1980). These species are responsible for *tert*-butylation of the indole ring of Trp, the thioether group of Met, and, to a very low degree (0.5–1.0%), the 3′-position of Tyr. Modifications can be minimized during TFA cleavage by utilizing effective *tert*-butyl scavengers. An early comprehensive study showed the advantages of 1,2-ethanedithiol (EDT) (Lundt et al., 1978); this dithiol has the additional virtue of protecting Trp from oxidation that occurs due to acid-catalyzed ozonolysis (Scoffone et al., 1966). To avoid acid-catalyzed oxidation of Met to its sulfoxide, a thioether scavenger such as dimethyl sulfide, ethyl methyl sulfide, or thioanisole should be added (Guttmann and Boissonnas, 1959; King et al., 1990; Guy and Fields, 1997). TFA deprotection of Cys(Trt) is reversible in the absence of a scavenger (Photaki et al., 1970). Realkylation by the Trt carbocation can occur readily following TFA cleavage if solutions of crude peptide in TFA are concentrated by rotary evaporation or lyophilization (D. Hudson, unpublished results). EDT, triethylsilane, or triisopropylsilane are recommended as efficient scavengers to prevent Trt reattachment to Cys (Pearson et al., 1989); this recommendation extends to prevent reattachment of Tmob as well (Munson et al., 1992). TFA deprotection of His(Bum) liberates formaldehyde, in similar fashion to HF deprotection of His(Bom) (Gesquiére et al., 1992). Cyclization of *N*-terminal Cys residues to a thiazolidine is only partially (60%) inhibited by even complex TFA/scavenger mixtures, such as reagent K (see discussion below).

The indole ring of Trp can be alkylated irreversibly by Mtr, Pmc, and Pbf groups from Arg (Sieber, 1987b; Harrison et al., 1989; Riniker and Hartmann, 1990; King et al., 1990; Fields and Fields, 1993), Tmob groups from Asn, Gln, or Cys (Gausepohl et al., 1989b; Sieber and Riniker, 1990), and even by some TFA-labile ester and amide linkers (Atherton et al., 1988a; Riniker and Kamber, 1989; Albericio et al., 1990a; Gesellchen et al., 1990; C. G. Fields et al., 1993c; G. B. Fields et al., 1993). The extent of Pmc modification of Trp is dependent upon the distance between the Arg and Trp residues (Stierandová et al., 1994). Cleavage of the Pmc group may also result in O-sulfation of Ser, Thr, and Tyr (Riniker and Hartmann, 1990; Jaeger et al., 1993). Three efficient cleavage "cocktails" for Mtr/Pmc/Pbf/Tmob quenching, and preservation of Trp, Tyr, Ser, Thr, and Met integrity, are TFA–phenol–thioanisole–EDT–H_2O (82.5:5:5:2.5:5) (reagent K) (King et al., 1990; Van Abel et al., 1995), TFA–thioanisole–EDT–anisole (90:5:3:2) (reagent R) (Albericio et al., 1990a), and TFA–phenol–H_2O–triisopropylsilane (88:5:5:2) (reagent B) (Solé and Barany, 1992; Van Abel et al., 1995). Studies on Trp preservation during amino acid analysis (Bozzini et al., 1991) has led to the development of reagent K′, where EDT is replaced by 1-dodecanethiol (Guy and Fields, unpublished results). Water is an essential component of reagents K and K′, but phenol is necessary only with multiple Trp-containing peptides (King et al., 1990). Thioanisole, a soft nucleophile, accelerates TFA deprotection of both Arg(Mtr) and Arg(Pmc). Triethylsilane (4 equiv.) in MeOH–TFA (1:9) has been reported to efficiently cleave and scavenge Pmc groups (Chan and Bycroft, 1992). Given a choice for Arg protection, Pmc/Pbf is preferred because it is more labile, and gives less Trp alkylation during unscavenged TFA cleavage; the

---

Cleavage and side-chain deprotection of peptide-resins assembled by Fmoc chemistry is carried out in TFA, in the presence of carefully chosen scavengers. The text discussion should be consulted with respect to peptide sequence and potential side reactions. It is recommended that small scale cleavages (< 10 mg peptide-resin) be performed and analyzed before proceeding to large scale cleavages. Standard cleavages of HMP/PAB, Dod, and PAL linkers, and simultaneous side-chain deprotection, proceed by stirring peptide-resin (50–200 mg) in 2 mL of the appropriate, *freshly* prepared cleavage cocktail for 1.5–2.5 h at 25 °C. The resin is filtered over a medium fritted glass filter and rinsed with 1 mL of TFA. The combined filtrate and TFA rinse are either (a) precipitated with methyl *t*Bu ether (~50 mL) or (b) dissolved in ~40 mL H_2O and extracted six times with ~40 mL methyl *t*Bu ether. Following (a), the mixture is centrifuged at 3000 rpm for 2 min and decanted. The peptide pellet is dispersed with a rubber policeman, washed thoroughly with methyl *t*Bu ether, and dried overnight in a lyophilizer (King et al., 1990; Albericio et al., 1990a). Following (b), the H_2O layer is concentrated in vacuo and loaded directly to a semipreparative HPLC column and the peptide is purified.

recommendation for Pmc/Pbf is particularly appropriate for sequences containing multiple Arg residues (Green et al., 1988; Harrison et al., 1989; King et al., 1990). The use of Boc side-chain protection of Trp also significantly reduces alkylation by Pmc/Pbf groups (Choi and Aldrich, 1993; Fields and Fields, 1993). The Trt group, instead of Tmob, is suggested for Asn/Gln side-chain protection in Trp-containing peptides, as Trt cations are easier to scavenge (Sieber and Riniker, 1990).

Nonacidolytic Cleavage Methods

Benzyl ester anchoring linkages can be cleaved usefully under non-acidic conditions. An interesting alternative to standard HF cleavages for Boc chemistry is catalytic transfer hydrogenolysis (CTH), which removes Bzl side-chain protecting groups (from Asp, Glu, Ser, Thr, and Tyr) and cleaves benzyl ester anchors to provide a C-terminal carboxyl group (Anwer and Spatola, 1980, 1983). Peptide-resin (1 g) is treated initially with palladium (II) acetate (1 g) in DMF (13 mL) for 2 h; ammonium formate (1.3 g) is then added, and the reaction proceeds for an additional 2 h (Anwer and Spatola, 1983). CTH can reduce Trp to octahydrotryptophan (Méry and Calas, 1988). Benzyl ester linkages may also be cleaved by 2-dimethylaminoethanol (DMAE)–DMF (1 : 1) for 70 h, with subsequent treatment of the peptide-DMAE ester by DMF–H_2O (1 : 5) for 2 h yielding the peptide acid (Barton et al., 1973). For some applications when peptide amides are required, benzyl ester-type linkages are treated with NH_3 in anhydrous MeOH, 2-propanol, 2,2,2-trifluoroethanol (TFE), or MeOH–DMF for 2–4 days at 25 °C (Atherton et al., 1981c; Stewart and Young, 1984; Story and Aldrich, 1991), although these conditions will also convert Asp(OBzl) and Glu(OBzl) residues to Asn and Gln, respectively. Additionally, ethanolamine in DMF or MeOH cleaves benzyl ester linkages (8–40 h, 45–60 °C) to provide an ethanolamidated peptide C-terminus (Prasad et al., 1982; Fields et al., 1988, 1989; Prosser et al., 1991). Nucleophilic cleavages of benzyl esters can be accompanied by side reactions, including racemization of the C-terminal residue (Barany and Merrifield, 1979). On the other hand, base cleavages of the NPE and HMFA linkers appear to be quite safe and general. Peptide acids are obtained upon treatment with either piperidine–DMF (1 : 4) or DBU (0.1 M)–1,4-dioxane, after 5 min (for HMFA) to 2 h (for NPE) at 25 °C (Liu et al., 1990; Albericio et al., 1991b).

Palladium-catalyzed peptide-resin cleavage is used for the HYCRAM linker (Kunz and Dombo, 1988; Guibé et al., 1989). The fully protected peptide-resin (0.1 mmol of peptide) is shaken for 6–18 h under N_2 or Ar atmosphere in 8 mL of DMSO–THF–0.5 M aqueous HCl (2 : 2 : 1), in the presence of 50–190 equiv. of either NMM (for

Following synthesis, peptide-resins should be well dried, and then stored in a desiccator at 4 °C with the N^α-terminus protected. As a general practice, peptides should never be stored in the solid state after being lyophilized from moderate or strong acid; deamidation, among other side reactions, may proceed rapidly (see below, Auxiliary Issues). If stored in solution following purification, peptides should be used for biological or chemical studies as soon as possible. Met-containing peptides oxidize rapidly upon storage in solution, especially when repeated freeze-thawing occurs (Stewart and Young, 1984), while Asn-containing peptides can hydrolyze spontaneously in solution (see below, Auxiliary Issues).

Boc-peptides) or dimedone (for Fmoc-peptides) as well as 0.015 equiv. of tetrakis(triphenylphosphino)palladium(0) catalyst. The reaction mixture is filtered and washed with DMF, DMF–0.5 M aqueous HCl (1:1), and DMF. The filtrate and washings are combined and evaporated to minimal volume; the peptide is then precipitated with methyl tBu ether (Orpegen, 1990; Lloyd-Williams et al., 1991). The crude peptide acid (0.7 mmol scale) can be converted to an amide by dissolving in dry DMF (50 mL) at 25 °C, adding NMM (80 µL), cooling to −20 °C, adding isobutylchloroformate (90 µL) and, after 8 min, 25% aqueous NH_4OH (0.3 mL), and stirring for 2–12 h. The solution is evaporated and the peptide amide dried over P_2O_5 in vacuo (Orpegen, 1990).

Photolytic cleavage at $\lambda = 350$ nm under inert (N_2, or Ar) atmosphere is used for the ONb, 2-bromopropionyl, and Nonb linkers. The most efficient photolysis of the ONb and Nonb linkers is achieved when peptide-resins are swollen with 20–25% 2,2,2-trifluoroethanol in either DCM or toluene (Giralt et al., 1986; Kneib-Cordonier et al., 1990; Hammer et al., 1990). Photolytic cleavage yields after 9–16 h range from 45–70% for relatively small peptides (five to nine residues). The 2-bromopropionyl linker is cleaved in DMF, with a yield of 70% after 72 h for a tetrapeptide (Wang, 1976). Photolabile linkers combining the o-nitro and α-methyl themes may cleave more rapidly and with higher yields (Holmes and Jones, 1995).

Post-Translational Modifications and Unnatural Structures

Peptides of biological interest often include structural elements beyond the 20 genetically encoded amino acids. This section summarizes the best current methods to duplicate by chemical synthesis the post-translational modifications achieved in nature, including the alignment of half-cysteine residues in disulfide bonds. This section also covers a set of unnatural structures which are of considerable interest for peptide drug design, namely lactams (head-to-tail, side-chain-to-side-chain, etc.), and lastly describes the steps needed to elicit good antibody production from synthetic peptides.

Hydroxylated Residues

Hydroxyproline (Hyp) has been incorporated successfully without side-chain protection in both Boc (Felix et al., 1973; Stewart et al., 1974) and Fmoc (Fields et al., 1987; Fields and Fields, 1992) SPPS. Alternatively, the usual hydroxyl protecting groups Bzl (Cruz et al., 1989) for Boc and tBu (Becker et al., 1989; Grab et al., 1996) for Fmoc can be used. Smaller peptides can be synthesized efficiently without side-chain protection on Hyp (Fields and Noble, 1990), but larger peptides containing multiple Hyp residues should use the appropriate (Bzl or tBu) side-chain protected Hyp derivative (Grab et al., 1996; Henkel et al., 1999).

Hydroxylysine (Hyl) has been incorporated in SPPS as Fmoc-Hyl(Boc, O-$tert$-butyldimethylsilyl [TBDMS]). This derivative was prepared by (a) protecting the N^{ε}-amino group of copper-complexed Hyl by treatment with Boc_2O, (b) disrupting the copper complex, (c) protecting the N^{α}-amino group by treatment with 9-fluorenylmethyl chloroformate, (d) protecting the carboxyl group by treatment with benzyl bromide, (e) blocking the side-chain hydroxyl group with $tert$-butyldimethylsilyl trifluoromethanesulfonate, and (f) hydrogenation to remove the benzyl group (Broddefalk et al., 1998). Hyl cannot be incorporated without side-chain protection of the hydroxyl group, as activation of an unprotected Hyl derivative leads to intramolecular lactone formation (Koeners et al., 1981; Remmer and Fields, 1999).

γ-Carboxyglutamate

Acid-sensitive γ-carboxyglutamate (Gla) residues have been identified in a number of diverse biomolecules, such as prothrombin and the "sleeper" peptide from the venomous fish-hunting cone snail (*Conus geographus*). Fmoc chemistry has been utilized for the efficient SPPS of the 17-residue "sleeper" peptide, with the five Gla residues incorporated as Fmoc-Gla($OtBu$)$_2$ (Rivier et al., 1987). Cleavage and side-chain deprotection of the peptide-resin by TFA–DCM (2 : 3) for 6 h resulted in no apparent conversion of Gla to Glu.

Phosphorylation

Incorporation of side-chain-phosphorylated Ser and Thr by SPPS is especially challenging, as the phosphate group is decomposed by strong acid, and lost with base in a β-elimination process (Perich, 1990). Boc-Ser(PO_3Ph_2) and Boc-Thr(PO_3Ph_2) have been used, where HF or hydrogenolysis cleaves the peptide-resin, and hydrogenolysis removes the phosphate phenyl groups (Perich et al., 1986; Arendt et al., 1989). Fmoc-Ser(PO$_3$Bzl,H) and Fmoc-Thr(PO$_3$Bzl,H) can be used for Fmoc syntheses provided appropriate precautions are taken (Wakamiya et al., 1994; White and Beythien, 1996; Nagata et al., 1997; Perich, 1997; Perich et al., 1999). Alternatively, peptide-resins that were built up by Fmoc chemistry to include unprotected Ser or Thr side chains may be treated with a suitable phosphorylating reagent, for example, N,N-diisopropyl-bis(4-chlorobenzyl)phosphoramidite or dibenzylphosphochloridate, to produce a phosphite triester intermediate. This phosphitylation step is then followed by oxidation with MCPBA, tBuOOH, or aqueous iodine/THF to produce the desired phosphorylated residue(s) (Perich, 1997). The phosphorylated peptide is then obtained in solution, following simultaneous deprotection and cleavage with TFA in the presence of scavengers (Otvös et al., 1989a; de Bont et al., 1990).

Side-chain-phosphorylated Tyr is less susceptible to strong acid decomposition, and is not at all base-labile. Thus, SPPS has been used to incorporate directly Fmoc-Tyr(PO$_3$Me$_2$) (Kitas et al., 1989), Fmoc-Tyr(PO$_3$Bzl$_2$) (Kitas et al., 1991), Fmoc-Tyr(PO$_3$$tBu_2$) (Perich and Reynolds, 1991), Fmoc-Tyr(PO$_3$H$_2$) (Ottinger et al., 1993; Xu et al., 1996; Chan et al., 1997; Lauer et al., 1997), and Boc-Tyr(PO$_3$H$_2$) (Zardeneta et al., 1990). Syntheses incorporating Fmoc-Tyr(PO$_3$Bzl$_2$) use 2% DBU in DMF for N^α-amino deprotection, as piperidine was found to remove the benzyl protecting groups from phosphate (Kitas et al., 1991), whereas the tBu phosphate group is inert to piperidine-mediated dealkylation during Fmoc removal steps (Perich, 1997). BOP/PyBOP and HBTU have been successfully applied for coupling of Fmoc-Tyr(PO$_3$Me$_2$), Fmoc-Tyr(PO$_3$$tBu_2$), and Fmoc-Tyr(PO$_3Bzl_2$). DCC and DIPCDI are not recommended, since interaction with the phosphodiester causes undesired activation (Perich, 1997). TFMSA or TMSBr can be used for peptide-resin cleavage and removal of the methyl phosphate groups without O-dephosphorylation (Kitas et al., 1989; Zardeneta et al., 1990), while TFA is used for removal of the benzyl and tBu phosphate groups (Kitas et al., 1991). For most applications, however, the direct use of side-chain-unprotected Fmoc-Tyr(PO$_3$H$_2$) (Ottinger et al., 1993; Xu et al., 1996; Perich et al., 1999) is particularly convenient and reliable. Coupling is achieved by a BOP/HOBt protocol, and the base used for all subsequent coupling steps needs to be DIEA. An intramolecular side reaction may occur during coupling, leading to pyrophosphate linked byproducts (Ottinger et al., 1996; García-Echeverría, 1995).

Net Tyr phosphorylation can be achieved after chain assembly (Tyr residues incorporated with free phenolic side-chain) by phosphite triester treatment using the dialkyl N,N-diethylphosphoramidates

$(t\text{BuO})_2\text{PNR}_2'$, $(\text{BzlO})_2\text{PNR}_2'$, or $(\text{TmseO})_2\text{PNR}_2'$ ($R' = $ ethyl or iso-propyl) followed by MCPBA, tBuOOH, or aqueous iodine/THF oxidation. If Cys or Met residues are present in the peptide sequence, the use of tBuOOH is preferred to minimize oxidation of these residues (Perich and Johns, 1988). Formation of H-phosphonates during the course of phosphitylation/oxidation routes, together with protocols to minimize occurrence of this side reaction, have been documented (Xu et al., 1997; Perich, 1998).

Sulfation

Gastrin, cholecystokinin, and related hormones contain sulfated Tyr; thus, incorporation of this residue into synthetic peptides is of great interest. Synthesis of Tyr sulfate-containing peptides is difficult, due to the substantial acid lability of the sulfate ester; also, most sulfating agents are more reactive towards the hydroxyls of Ser or Thr with respect to the phenol of Tyr. While there is an elegant history of success in solution chemistry (Beacham et al., 1967; Ondetti et al., 1970; Pluscec et al., 1970; Wünsch et al., 1981), this brief discussion focuses on the best SPPS approaches. Side-chain-unprotected Tyr can be incorporated by Boc or Fmoc chemistry and sulfation while the otherwise fully protected peptide remains anchored to the support is achieved by use of pyridinium acetyl sulfate (Fournier et al., 1989). Base or acid-promoted deprotection/cleavages follows, under conditions carefully optimized to avoid or minimize desulfation. Alternatively, sulfated Tyr can be incorporated directly by use of Fmoc-Tyr($\text{SO}_3^- \cdot \text{Na}^+$)-OPfp, Fmoc-Tyr($\text{SO}_3^- \cdot \text{Na}^+$), or Fmoc-Tyr($\text{SO}_3^- \cdot \text{Ba}_{1/2}^{2+}$) in situ with BOP/HOBt (Penke and Rivier, 1987; Penke and Nyerges, 1989, 1990; Han et al., 1996b). A brief and carefully optimized acidolytic deprotection/cleavage is then used to minimize desulfation. PEG-PS resins functionalized with xanthenylamide (XAL) handles are the solid supports of choice for this application, as cleavage requires only 1–5% TFA. Cholecystokinin containing a sulfonated Tyr (CCK-8) was synthesized successfully using a 15 min treatment with 5% TFA, providing the desired deprotected and cleaved peptide 95% pure (Han et al., 1996b).

O-Sulfated Ser and Thr residues can be obtained by simply incorporating Fmoc-Ser/Thr(SO_3H) into the peptide chain. If an Mtr or Pmc-protected Arg is present in the peptide, the sulfated Ser and/or Thr can alternatively be obtained by cleaving the peptide with 50% TFA in DCM without any scavengers. Under these conditions the sulfate moiety of the Arg-protecting group will be transferred to the hydroxyl groups in a yield of up to 60% (Jaeger et al., 1993). The sulfated and unsulfated peptides can then be separated via HPLC or ion-exchange chromatography (Jaeger et al., 1993).

Glycosylation

Methodology for site-specific incorporation of carbohydrates during chemical synthesis of peptides has developed rapidly. The mild conditions of Fmoc chemistry are more suited for glycopeptide syntheses than Boc chemistry, as repetitive acid treatments as well as final cleavage with a strong acid such as HF can be detrimental to sugar linkages (Kunz, 1987, Meldal, 1994). Fmoc-Ser, -Thr, -Hyl, -Hyp, and -Asn have all been incorporated successfully with glycosylated side chains. Glycosylation is typically performed in the presence of Lewis acids with glycosyl bromides or trichloroacetimidates for Ser, Thr, and Hyp, or by coupling glycosylamines onto Asp (to produce a glycosylated Asn). The side-chain glycosyl is usually hydroxyl-protected by either acetyl (Ac) or benzoyl (Bz) groups (Paulsen et al., 1988, 1990; Jansson et al., 1990; de la Torre et al., 1990; Bardají et al., 1991; Kihlberg et al., 1997; Malkar et al., 2000), although some SPPS have been successful with no protection of glycosyl hydroxyl groups (Otvös et al., 1989b, 1990; Filira et al., 1990). Glycosylated residues are incorporated as preformed Pfp esters or in situ with DCC/HOBt or HATU/HOAt (Paulsen et al., 1988; Filira et al., 1990; Jansson et al., 1990; Meldal and Jensen, 1990; Otvös et al., 1990; Paulsen et al., 1990; Bardají et al., 1991; Jensen et al., 1996; Malkar et al., 2000). These sugars are relatively stable to Fmoc deprotection by piperidine, morpholine, or DBU (Paulsen et al., 1988, 1990; Jansson et al., 1990; Meldal and Jensen, 1990; Otvös et al., 1990; Bardají et al., 1991; Meldal et al., 1994; Jensen et al., 1996; Kihlberg et al., 1997; Malkar et al., 2000), brief treatments (2 h) with TFA for side-chain-deprotection and peptide-resin cleavage (Paulsen et al., 1988, 1990; Filira et al., 1990; Meldal and Jensen, 1990; Jansson et al., 1990; Otvös et al., 1990; Bardají et al., 1991), and palladium treatment for peptide-resin cleavage from HYCRAM (Kunz and Dombo, 1988). Deacetylation and debenzoylation may be performed with hydrazine–MeOH (4:1) prior to glycopeptide-resin cleavage (Kunz, 1987; Bardají et al., 1991). Removal of acetyl and benzoyl groups is usually carried out in solution using catalytic methoxide in methanol. Some side reactions, including β-elimination and Cys-induced degradation of the peptide backbone (Paulsen et al., 1988; Peters et al., 1995; Vuljanic et al., 1996; Elofsson et al., 1997), may occur, but if the apparent pH is controlled carefully to be about 10, these deprotections are generally "safe" (Meldal et al. 1994; Jensen et al., 1996). Acid-labile protecting groups have been developed for the carbohydrate moiety, such as TBDMS, tert-butyldiphenylsilyl (TBDPS), and isopropylidene (Elofsson et al., 1997). The use of DBU in combination with silyl protection should be avoided, due to DBU-induced decomposition of the glycopeptide (Christiansen-Brams et al., 1993).

Disulfide Bond Formation

In the majority of cases, intramolecular disulfides or simple inter-molecular homodimers have been formed from purified linear precursors by nonspecific oxidations in dilute solutions. An even number of Cys residues are brought to the free thiol form by removal of the same S-protecting group, following which disulfide formation is mediated by molecular O_2 (from air), potassium ferricyanide [$K_3Fe(CN)_6$], DMSO, or others from a lengthy catalogue of reagents (Hiskey, 1981; Stewart and Young, 1984; Tam et al., 1991b; Andreu et al., 1994; Annis et al., 1997; Albericio et al., 2000). Accomplishing the same end, but proceeding by a different mechanism, the polythiol can be treated with a mixture of reduced and oxidized glutathione, which catalyze the net oxidation by thiol-disulfide exchange reactions (Ahmed et al., 1975; Lin et al., 1988; Pennington et al., 1991; Andreu et al., 1994; Annis et al., 1997; Albericio et al., 2000). These procedures, which require scrupulous attention to experimental details, have often given the desired disulfide-containing peptide products in acceptable yields. However, even under the best conditions, significant levels of dimeric, oligomeric, or polymeric material are observed. The non-monomeric material has usually proven to be difficult to "recycle" by alternating reduction and reoxidation steps.

A more sophisticated approach, which also requires dilute solutions, involves selective pairwise co-oxidations of two designated free or protected sulfhydryl groups. Such reactions are best carried out in intramolecular fashion because, if the paired groups are on separate chains, there is the problem that homodimers form along with the desired heterodimer. If the thiols have already been deblocked, oxidation follows using procedures mentioned earlier. The prototype oxidative deprotections involve I_2 treatments on Cys(Trt) or Cys(Acm) residues (Kamber et al., 1980). These reactions are carried out in neat TFE, MeOH, 1,1,1,3,3,3-hexafluoroisopropa-nol, HOAc, DCM, or chloroform, or mixtures of the these solvents, and often proceed in modest to high yield; however, side reactions have been observed at Trp residues resulting in Trp-2′-thioethers (Sieber et al., 1980) and β-3-oxindolylalanine (Casaretto and Nyfeler, 1991). Pairwise oxidative removal of appropriate Acm or Tacm Cys protecting groups with Tl(Tfa)$_3$ or methyltrichlorosilane (in the presence of diphenylsulphoxide) also furnishes the disulfide directly (Fujii et al., 1987; Kiso et al., 1990; Akaji et al., 1991; Albericio et al., 2000). However, Trp and Met must be side-chain-protected during such treatments (see, for example, Edwards et al., 1994). Liberation of Acm by Tl(Tfa)$_3$ can also result in a $S \rightarrow O$ Acm shift, a side reaction that can be suppressed by glycerol (Lamthanh et al., 1993). As a final example in this category, Cys(Fm) residues form disulfides directly upon treatment with piperidine (Ponsati et al., 1990b).

Most general, but also most demanding in terms of the range of selectively removable sulfhydryl protecting groups required, are unsymmetrical directed methods of disulfide bridge formation (Barany and Merrifield, 1979; Hiskey, 1981). For example, Cys(Acm) or Cys(Trt) residues in peptides can be reacted with methoxy- or ethoxycarbonylsulfenyl chloride to form Cys(Scm) or Cys(Sce) residues, respectively, which are attacked by the free thiol of a deprotected Cys residue to form a disulfide bond (Kullmann and Gutte, 1978; Mott et al., 1986; Ten Kortenaar and van Nispen, 1988). The S-(N-methyl-N-phenylcarbamoyl)sulfenyl (S-Snm) group is generated and used in the same manner, but is preferred due to its enhanced stability and the specificity of its reactions (Schroll and Barany, 1989; Chen et al., 1997). Disulfide bonds may also be formed by a free thiol attack of Cys(NBoc-NHBoc) residues, which are prepared by treatment of Cys with azodicarboxylic acid di-tBu ester (Romani et al., 1987). Cys(Npys) residues form disulfides upon reaction with deprotected Cys residues (Bernatowicz et al., 1986). Directed methods are particularly suited for linking two separate chains.

As already alluded, directed methods require at least two classes of selectively removable Cys protecting groups; the same holds true for experiments aimed at controlled formation of multiple disulfide bridges by sequential pairwise deprotection/co-oxidations. An overriding concern with all such chemical approaches is to develop conditions that avoid scrambling (disulfide exchange). Oxidation by I_2 in TFE allows for selective disulfide bond formation between Cys(Trt) residues in the presence of Cys(Acm) residues; in DMF, I_2 oxidation is preferred between Cys(Acm) residues in the presence of Cys(Trt) residues (Kamber et al., 1980). Direct I_2 oxidation of Cys(Acm) or Cys(Trt) residues is particularly advantageous in that existing disulfides are not exchanged (Barany and Merrifield, 1979; Kamber et al., 1980; Hiskey, 1981; Gray et al., 1984; Atherton et al., 1985b; Ponsati et al., 1990b). Since the Acm group is essentially stable to HF, a Meb/Acm combination of protecting groups facilitates selective disulfide formation in Boc chemistry (Gray et al., 1984; Tam et al., 1991a). The Tmob/Acm or Xan/Acm combinations work well for the same purpose in Fmoc chemistry (Munson and Barany, 1993; Hargittai and Barany, 1999).

The alternative of carrying out deprotection/oxidation of the Cys residues while the peptide chain remains anchored to a polymeric support is of obvious interest, and has received attention. Such an approach takes advantage of *pseudo-dilution*, which is a kinetic phenomenon expected to favor facile intramolecular processes and thereby minimize dimeric and oligomeric by-products (Barany and

Merrifield, 1979; Andreu et al., 1994). Disulfide bond formation on peptide-resins has been demonstrated by air, $K_3Fe(CN)_6$, dithiobis(2-nitrobenzoic acid), or diiodoethane oxidation of free sulfhydryls, direct deprotection/oxidation of Cys(Acm) residues by Tl(Tfa)$_3$ or I_2, direct conversion of Cys(Fm) residues by piperidine, and nucleophilic attack by a free sulfhydryl on either Cys(Npys) or Cys(Scm) (Gray et al., 1984; Mott et al., 1986; Buchta et al., 1986; Eritja et al., 1987; Ploux et al., 1987; Ten Kortenaar and van Nispen, 1988; García-Echeverría et al., 1989; Albericio et al., 1991a, 2000; Edwards et al., 1994). The most generally applicable and efficient of these methods is direct conversion of Cys(Acm) or Cys(Trt) residues by I_2 (10 equiv. in DMF), Cys(Acm) residues by Tl(Tfa)$_3$ (1.5 equiv. in DMF) (Albericio et al., 1991a; Edwards et al., 1994), and Cys(Fm) residues by (a) piperidine–DMF (1:1) for 3 h at 25 °C (Ponsati et al., 1990b; Albericio et al., 1991a) or (b) piperidine–DMF–2-mercaptoethanol (10:10:0.7) treatment for 1 h at 25 °C, followed by air oxidation in pH 8.0 DMF for 1 h at 25 °C (Albericio et al., 1991a). The best solid-phase yields were at least as good, and in some cases better than, the results from corresponding solution oxidations.

An ingenious new application of the pseudo-dilution concept to the solid-phase-mediated preparation of intramolecular disulfides was reported recently (Annis et al., 1998). Ellman's reagent, 5,5'-dithiobis(2-nitrobenzoic acid), was bound through two sites to a suitable solid support (PEG-PS, modified Sephadex, or controlled-pore glass). The linear oligo(thiol) precursors are dissolved in acidic aqueous media, and treated with these solid-phase Ellman's reagents. An initial "capture" step ensures that the critical intramolecular displacement that leads to disulfide formation occurs while the intermediate is polymer supported, and thus sequestered from other species in solution that could lead to by-products. Upon completion of the oxidation, the product is readily obtained by a simple filtration, while the solid-phase oxidation reagent is readily recovered and can be recycled.

Lactams

Intrachain lactams are formed between the side chains of Lys or Orn and Asp or Glu to conformationally restrain synthetic peptides, with the goal of increasing biological potency and/or specificity (reviewed by Kates et al., 1994; see Seyfarth et al., 1995; Limal et al., 1998; and Hargittai et al., 2000 for recent examples). Lactams can also be formed via side-chain-to-head, side-chain-to-tail, or head-to-tail cyclization (Kates et al., 1994). The residues used to form intrachain lactams must be selectively side-chain-deprotected, while all side-chain protecting groups of other residues remain intact. Selective deprotection is best achieved by using orthogonal side-chain protection, such as Fmoc and Fm for Lys/Orn and Asp/Glu, respectively, in combination with a Boc/Bzl strategy (Felix et al., 1988a, b; Hruby et al., 1990). A more complicated, but equally efficient approach, is to use side-chain protection based on graduated acid lability (Schiller et al., 1985; Sugg et al., 1988; Hruby et al., 1990). In the Fmoc strategy, orthogonality is achieved with Aloc or Dde protection for Lys and Al or Dmab protection for Asp/Glu (Lyttle and Hudson, 1992; Kates et al., 1993b). Optimized automated conditions for removal of Aloc/Al groups are Pd(Ph$_3$)$_4$ (1 equiv.) in chloroform–acetic acid–NMM (37 : 2 : 1) for 2 h at 25 °C, followed by washing steps with a solution of 0.5% DIEA and 0.5% sodium diethyldithiocarbamate in DMF (Kates et al., 1993a). Cyclization is carried out most efficiently with BOP (3–6 equiv., 2 h, 20 °C) in the presence of DIEA (6–7.5 equiv.) while the peptide is still attached to the resin (Felix et al., 1988a; Plaué, 1990), taking advantage of the pseudo-dilution phenomenon discussed in the previous section.

The three-dimensional orthogonal protection scheme of Fmoc/tBu/ Al protecting groups is the strategy of choice for head-to-tail cyclizations (Trzeciak and Bannwarth, 1992; Kates et al., 1993b, 1994). In the cited examples, an amide linker is used for side-chain attachment of a C-terminal Asp/Glu (which are converted to Asn/Gln) and the α-carboxyl group is protected as an allyl ester (Trzeciak and Bannwarth, 1992; Kates et al., 1993b). A more general way to achieve such on-resin head-to-tail cyclizations involves backbone amide linkage (BAL) (Jensen et al., 1998).

For side-chain-to-head cyclizations, the Lys/Orn and Asp/Glu protecting groups should be either very acid sensitive or base labile. The N-terminal residue (head) can simply be introduced as an N^{α}-Boc or N^{α}-Fmoc derivative while the peptide-resin linkage and the other side-chain protecting groups are stable to dilute acid or carry a third dimension of orthogonality.

Peptide Antigens

For the production of antipeptide antibodies, the peptide must be attached to a carrier (van Regenmortel and Muller, 1999). The simplest, although not necessarily most effective, way to accomplish this is to make direct use of peptide-resins in which the side chains have been freed but where the anchoring linkage is stable to the appropriate deprotection conditions. Polyamide-type and polyethylene glycol-polystyrene resins have been applied according to this approach (Chanh et al., 1986; Kennedy et al., 1987; Goddard et al., 1988; Fischer et al., 1989; Bayer, 1991; Kanda et al., 1991), as have synthesis of the peptide directly on beaded cellulose (Englebretsen and Harding, 1994) or protein carriers (Hansen et al., 1993). Peptides may also be synthesized containing topographic immunogenic determinants

(Kaumaya et al., 1992) or on a scaffold that, following cleavage and deprotection, is used directly for immunization. The scaffold, consisting of branched Lys residues, is referred to as a multiple antigen peptide system (MAP) (Tam, 1988; Tam and Lu, 1990). MAPs may be prepared by either Boc (Tam, 1988) or Fmoc (Pessi et al., 1990; Drijfhout and Bloemhoff, 1991; Biancalana et al., 1991) chemistries. Yet another approach, which is possible when the three-dimensional structure of a discontinuous epitope is known, involves linking the key resides with spacers designed to confer geometric constraints that will favor native-like conformation (Borràs et al., 1999).

The more traditional (and still most common) methods for preparing peptide antigens start with free synthetic peptides that have been cleaved from the support and deprotected. In one variation, the peptide is conjugated to a protein carrier, for example, bovine serum albumin, by coupling mediated by water-soluble carbodiimide (Deen et al., 1990). Alternatively, the peptide can be extended at the C- or N-terminus with a Cys residue, which is subsequently used to form a disulfide bridge with a free sulfhydryl on the carrier. Cystine formation is best achieved by the direct attack of a carrier protein thiol onto a peptide Cys(Npys) residue. Thiol groups are introduced on the carrier protein either by reduction to form free Cys (Albericio et al., 1989b; Ponsati et al., 1989) or by functionalization of Lys using S-acetylmercaptosuccinic anhydride (Drijfhout et al., 1988). Peptide-carrier conjugation is quantitated by monitoring the liberation of the Npys group at $\lambda = 329\,nm$ (Drijfhout et al., 1988). For peptides synthesized by Fmoc chemistry, Boc-Cys(Npys) is incorporated as the N-terminal residue, thus avoiding additional piperidine treatments which would remove the Npys group (Albericio et al., 1989b).

Auxiliary Issues

This final section of the chapter covers a variety of practical considerations that researchers in SPPS should be aware of. Included are some potential side reactions that do not fit neatly with categories covered earlier, and ways to mitigate the extent of their occurrence. A logical culmination of expertise in SPPS is the successful preparation of long sequences, and we outline current achievements and possible ways to improve on them in the future.

Diketopiperazine Formation

The free N^α-amino group of an anchored dipeptide is poised for an acid- or base-catalyzed intramolecular attack onto the C-terminal carbonyl (Gisin and Merrifield, 1972; Barany and Merrifield, 1979; Pedroso et al., 1986). Base deprotection (Fmoc chemistry) or neutralization

(Boc chemistry; at the stage when the third residue is being coupled) can thus release a cyclic diketopiperazine while a hydroxymethyl-handle leaving group remains on the resin. With residues that can form cis peptide bonds, for example, Gly, Pro, N-methylamino acids, or D-amino acids, in either the first or second position of the $(C \rightarrow N)$ synthesis, diketopiperazine formation can be substantial (Barany and Merrifield, 1979; Pedroso et al., 1986; Gairi et al., 1990). It can also be a problem when the third amino acid residue is added to a BAL-anchored sequence (Jensen et al., 1998), or when an activated handle leaving group such as o-nitrobenzyloxy is used (Barany and Albericio, 1985). For most other sequences and/or handles, the problem can be adequately controlled. In Boc SPPS, the level of diketopiperazine formation can be suppressed either by removing the Boc group with HCl and coupling the NMM salt of the third Boc-amino acid without neutralization (Suzuki et al., 1975), or else by deprotecting the Boc group with TFA and coupling the third Boc-amino acid in situ using BOP, DIEA, and HOBt without neutralization (Gairi et al., 1990). For susceptible sequences being addressed by Fmoc chemistry, the use of piperidine–DMF (1:1) deprotection for 5 min (Pedroso et al., 1986), or deprotection for 2 min with a 0.1 M solution in DMF of tetrabutyl-ammonium fluoride ("quenched" by MeOH) (Ueki and Amemiya, 1987) has been recommended to minimize cyclization. Alternatively, the second and third amino acids may be coupled as a preformed N^α-protected dipeptide, hence circumventing the diketopiperazine-inducing deprotection/neutralization at the second amino acid. The steric hindrance of the 2-chlorotrityl linker may minimize diketo-piperazine formation of susceptible sequences during Fmoc chemistry (Barlos et al., 1989a, b). In the case of BAL strategies, diketo-piperazine formation is suppressed by incorporation of the penul-timate residue as its N^α-trityl (Trt) or N^α-Ddz derivative, followed by selective deprotection with very dilute TFA (negligible premature loss of chains at this stage), and finally incorporation of the third residue as its N^α-Fmoc derivative under in situ neutralization/coupling conditions mediated by PyAOP/DIEA in DMF (Jensen et al., 1998).

For continuous-flow Fmoc SPPS, diketopiperazine formation is suppressed by deprotecting for 1.5 min with piperidine–DMF (1:4) at an increased flow rate (15 mL/min), washing for 3 min with DMF at the same flow rate, and coupling the third Fmoc-amino acid in situ with BOP, NMM, and HOBt in DMF (MilliGen/Biosearch, 1990). For batchwise SPPS, rapid (a maximum of 5 min) treatments by piperidine–DMF (1:1) should be used, followed by DMF washes and then in situ acylations mediated by BOP or HBTU (Pedroso et al., 1986).

Capping

Some workers choose to "cap" unreacted chains, thereby substituting a family of terminated peptides for a family of deletion peptides. In either case, these by-products must ultimately be separated from the desired product. Intentional termination of chains may be carried out when there is an indication of unreacted sites. In the simplest case, capping is carried out with reactive acetylating agents, such as acetic anhydride (Ac_2O) or N-acetylimidazole in the presence of tertiary base (Bayer et al., 1970; Stewart and Young, 1984). An alternative capping reagent is 2-sulfobenzoic acid cyclic anhydride (OSBA). Application of OSBA/tribenzylamine results in a negatively charged amino terminus; the desired product with the positively charged amino terminus may then be isolated after purification by ion-exchange chromatography (Drijfhout and Bloemhoff, 1988). In a reciprocal strategy, chains capped by acetylation are separated by acylating the N-terminus of the desired peptide with suitable Fmoc derivatives. These derivatives include the 9-(2-sulfo)fluorenylmethyloxycarbonyl (Sulfmoc) group (Merrifield and Bach, 1978), tetrabenzo[a,c,g,i]fluorenyl-17-methyloxy-carbonyl (Tbfmoc) group (Ramage and Raphy, 1992; Brown et al., 1993; Ramage et al., 1998), or modified 9-(hydroxymethyl)fluorene-4-carboxylates, where the carboxylate at the 4-position is in turn derivatized with Lys, Glu, or 2-aminodecanoic acid (Ball et al., 1990, 1991). After an ion-exchange or reversed-phase purification which separates the completed peptide from all truncated ones, the modified Fmoc group is removed by base. A further option for the purification of Tbfmoc-modified peptide is to take advantage of their extremely high affinity for porous graphitized carbon (PGC). Another variation of the principle uses either n-decylsulfonylethyloxycarbonyl or n-hexadecylsulfonylethloxycarbonyl groups, homologues of the base-labile Msc group that are retained longer on reversed-phase chromatography by virtue of their hydrophobic arms (García-Echeverría, 1995).

Incidental capping has been found to occur during SPPS. Capping via acetylation has been attributed to ethyl acetate breakdown products in Boc-amino acids (Schnölzer et al., 1992). This side reaction may be avoided by lyophilizing Boc-amino acids just prior to use.

Deletions

The standard explanation for deletion peptides relates to incomplete couplings, which can usually be diagnosed by qualitative or quantitative monitoring tests (see above, Monitoring). In contrast, a chemically plausible side reaction that suggests a different reason for deletion peptides has been elucidated (Kent, 1983). Resin-bound aldehyde groups can form a Schiff's base with deprotected amino groups of the peptide chain. Those amino groups that are involved in a Schiff's base are prevented from acylation by the next incoming protected amino acid. The blockage is not permanent, that is, terminated peptides are not formed by this mechanism. Rather, ready amine exchange of the Schiff's base renders a *different* set of amino groups temporarily inaccessible at a subsequent coupling, and thus deletion peptides result. The side reaction is *not* minimized by capping steps (see above); in fact, amines blocked as Schiff's bases contribute to negative ninhydrin tests. The formation of aldehyde sites on polystyrene-resins can be minimized by using a strong acid Friedel–Crafts catalyst during the original functionalization step. In any event, it is crucial to quantify aldehyde concentrations of prepared or purchased resins (Kent, 1983).

Acid-Sensitive Side Chains and Bonds

Trp is quite sensitive to acid conditions, undergoing reactions with carbonium ions and molecular oxygen (see above, Cleavage). Synthesis of Trp-containing peptides is thus best approached via Fmoc chemistry, where acidolysis is kept at a minimum. Gramicidins A, B, and C (where either three or four of the 15 residues are Trp) have been synthesized efficiently by Fmoc chemistry; acid was avoided entirely throughout the synthesis and final nucleophilic cleavage was achieved with ethanolamine (Fields et al., 1988, 1989), while efficient Fmoc SPPS of indolicidin (which contains five Trp out of 13 residues) used an optimized TFA/scavenger mixture (reagent K) to prevent modification of Trp during acidolytic cleavage and side-chain deprotection (King et al., 1990; van Abel et al., 1995). Fmoc chemistry has also been suggested for incorporation of ^{2}H-labeled amino acids, as the repetitive acidolyses of Boc chemistry can exchange out the ^{2}H label (Fields and Noble, 1990; Prosser et al., 1991). Finally, Fmoc chemistry may be the better choice for the synthesis of peptides containing the acid-labile Asp-Pro bond. SPPS of baboon β-chorionic gonadotropin 109–145, which contains two Asp-Pro bonds, was reported to be successful by Fmoc chemistry only, as Boc chemistry resulted in acid-promoted destruction of the Asp-Pro bonds (Wu et al., 1988).

Imide Formation

Asn residues can cyclize to form a succinimide, which can yield both α- and β-Asp peptides. Imide formation is largely sequence dependent, with Asn-Gly showing the greatest tendency to rearrange (Stephenson and Clarke, 1989). Succinimide formation can be significant for peptides stored in solution (Stephenson and Clarke, 1989;

Patel and Borchardt, 1990a, b). In addition, peptides containing Asn (and even Gln) stored in the solid state with residual acid can undergo deamidation (Ten Kortenaar et al., 1990). Therefore, Asn- and Gln-containing peptides should never be stored in a solid form with residual acid present, and samples stored in solution should be monitored carefully for deamidation and decomposition.

Solvation and Aggregation

Effective solvation of the peptide-resin is perhaps the most crucial condition for efficient chain assembly. Under proper solvent conditions, there is no decrease in synthetic efficiency up to 60 amino acid residues in Boc SPPS (Sarin et al., 1984). The ability of the peptide-resin to swell increases with increasing peptide length due to a net decrease in free energy from solvation of the linear peptide chains (Sarin et al., 1980). Therefore, there is no theoretical upper limit to efficient amino acid couplings, provided that proper solvation conditions exist (Pickup et al., 1990). In practice, obtaining these conditions is not always straightforward. "Difficult couplings" during SPPS have been attributed to poor solvation of the growing chain by DCM. Infrared and NMR spectroscopies have shown that intermolecular β-sheet aggregates are responsible for lowering coupling efficiencies (Live and Kent, 1983; Mutter et al., 1985; Ludwick et al., 1986). A scale of β-sheet structure-stabilizing potential has been developed for Boc-amino acid derivatives (Narita and Kojima, 1989). Enhanced coupling efficiencies are seen upon the addition of polar solvents, such as DMF, TFE, NMP, and DMSO (Yamashiro et al., 1976; Live and Kent, 1983; Geiser et al., 1988; Narita et al., 1989; Fields et al., 1990; Fields and Fields, 1991; Hyde et al., 1992; Miranda and Alewood, 1999). It has been suggested that chaotropic salts may be added to organic solvents in order to disrupt β-sheet aggregates (Stewart and Klis, 1990; Thaler et al., 1991). Also, using a lower substitution level of resin to avoid interchain crowding can improve the synthesis (Tam and Lu, 1995).

Aggregation also occurs in regions of apolar side-chain-protecting groups, sometimes resulting in a collapsed gel structure (Atherton et al., 1980; Atherton and Sheppard, 1985). In cases where aggregation occurs due to apolar side-chain-protecting groups, increased solvent polarity may not be sufficient to disrupt the aggregate. A relatively unstudied problem of Fmoc chemistry is that the lack of polar side-chain-protecting groups could, during the course of an extended peptide synthesis, inhibit proper solvation of the peptide-resin (Atherton et al., 1980; Fields and Fields, 1991; Bedford et al., 1992). To alleviate this problem, the use of solvent mixtures containing both a polar and nonpolar component, such as THF–NMP (7 : 13) or TFE–DCM

(1 : 4), is recommended (Fields and Fields, 1991). The addition of DMSO (Fields and Fields, 1991; Hyde et al., 1992) or a solvent mixture containing detergents (Zhang et al., 1994) can be effective for disrupting such aggregates. The partial substitution or complete replacement of tBu-based side-chain-protecting groups for carboxyl, hydroxyl, and amino side chains by more polar groups would also aid peptide-resin solvation (Atherton et al., 1980; Fields and Fields, 1991; Bedford et al., 1992).

A different approach to circumvent potential aggregation is the use of reversible protection of the amino acid amide nitrogen. For this purpose, N^α-Fmoc, N^α(2-Fmoc-hydroxy-4-methoxybenzyl) [Fmoc (Fmoc-Hmb)] derivatives have been prepared (Johnson et al., 1993). Formation of β-sheet structure can be minimized by using amide nitrogen protection every sixth to eighth amino acid (Johnson et al., 1993; Hyde et al., 1994).

Long Syntheses (>50 Residues)

Many impressive long chain syntheses (>50 residues), including ribonuclease A (124 residues) (Gutte and Merrifield, 1971), human parathyroid hormone (84 residues) (Fairwell et al., 1983), interleukin-3 (140 residues) (Clark-Lewis et al., 1986), HIV-1 aspartyl protease (99 residues) (Schneider and Kent, 1988; Nutt et al., 1988; Wlodawer et al., 1989; de Lisle Milton et al., 1992), HIV-1 vpr protein (95 residues) (Gras-Masse et al., 1990), gonadotropin-releasing hormone precursor protein (92 residues) (Milton et al., 1992), Ala-interleukin-8 (77 residues) (Sueiras-Diaz and Horton, 1992), and insulin-like growth factor (70 residues) (Bagley et al., 1990), have been carried out using Boc methodology. There have also been successful long chain syntheses by Fmoc chemistry, including HIV-1 Tat protein (86 residues) (Frankel et al., 1989; Chun et al., 1990; Calnan et al., 1991), preprocecropin A (64 residues) (Pipkorn and Bernath, 1990), ubiquitin (76 residues) (Ramage et al., 1989; Ogunjobi and Ramage, 1990), yeast actin-binding protein 539-588 (50 residues) (King et al., 1990), pancreastatin (52 residues) (Funakoshi et al., 1988), human β-chorionic gonadotropin 1–74 (Wu et al., 1989), HIV-1 nucleocapsid protein Ncp7 (72 residues) (de Rocquigny et al., 1991), the "minibody" (61 residues) (Bianchi et al., 1993), bovine pancreatic trypsin inhibitor and analogs (58 residues) (Ferrer et al., 1992; Ferrer et al., 1995; Pan et al., 1995), and "mini-collagens" (76–121 residues) (C. G. Fields et al., 1993a, b; Grab et al., 1996). Both chemistries appear susceptible to the same difficult couplings (Meister and Kent, 1983; van Woerkom and van Nispen, 1991; J. Young et al., 1990), and side-by-side syntheses for moderate length chains (~30 residues) are comparable (Atherton et al., 1983; Wade et al., 1986). However, there are two additional considerations when using Fmoc, rather than Boc, chemistry for long chain syntheses. First, the efficient solvation of hydrophobic side-chain-protecting groups used in conjunction with Fmoc chemistry, which was discussed previously, can become more critical for extended syntheses (Fields and Fields, 1991). Second, deprotection of the Fmoc group can proceed slowly in certain sequences (Atherton and Sheppard, 1985; Larsen et al., 1991). By monitoring deprotection as the synthesis proceeds, one can extend base deprotection times and/or alter solvation conditions as necessary (Ogunjobi and Ramage, 1990). The use of the DBU–piperidine mixture described earlier can also alleviate slow deprotection problems.

Segment Condensation

The advantages of segment condensation procedures for the synthesis of large peptides have been well described (Barany and Merrifield, 1979; Kaiser et al., 1989; Kneib-Cordonier et al., 1990), but to date there are relatively few examples for polymer-supported procedures. A significant aspect of the problem involves ready access to pure partially protected peptide segments which are needed as building blocks. The application of solid-phase synthesis to prepare the requisite intermediates depends on several levels of selectively cleavable protecting groups and anchoring linkages. Combination of the Boc/Bzl strategy with the 4-nitrobenzophenone oxime resin (DeGrado and Kaiser, 1982; Kaiser et al., 1989; Lansbury et al., 1989; Sasaki and Kaiser, 1990), base-labile linkers (Liu et al., 1990; Albericio et al., 1991b), palladium-labile linkers (Kunz and Dombo, 1988; Guibé et al., 1989; Lloyd-Williams et al., 1991), or photolabile linkers (Rich and Gurwara, 1975; Albericio et al., 1987b), and of the Fmoc/tBu strategy with dilute acid-labile linkers (Mergler et al., 1988b; Barlos et al., 1989b; Atherton et al., 1990; Albericio and Barany, 1991; Barlos et al., 1991c) or photolabile linkers (Kneib-Cordonier et al., 1990) has proven successful for the generation of N^α-amino and side-chain-protected segments with free C^α-carboxyl groups. Methods for subsequent solubilization and purification of the protected segments are nontrivial (Atherton et al., 1990; Lloyd-Williams et al., 1991; Albericio et al., 1997; Sakakibara, 1999), and beyond the scope of this review.

Solid-phase assembly of protected segments has proven successful for a 44-residue model of apolipoprotein A-1 (Nakagawa et al., 1985), human cardiodilatin 99-126 (Nokihara et al., 1989), human gastrin-I (17 residues) (Kneib-Cordonier et al., 1990), *Androctonus australis* Hector toxin II (64 residues), λ-Cro DNA binding protein (66 residues) (Atherton et al., 1990), prothymosin α (109 residues) (Barlos et al., 1991c), β-amyloid protein (42 residues) (Hendrix and Lansbury, 1992; Hendrix et al., 1992), the N-terminal domain of γ-zein protein (48 residues) (Dalcol et al., 1995), a variety of triple-helical "mini-collagens" (C. G. Fields et al., 1993a, b; Feng et al., 1996a, b, 1997), and a template assembled synthetic protein containing three peptide loops (Peluso et al., 1997). Polystyrene crosslinked with 1% 1,3-divinylbenzene and polyamide resins have been shown to be suitable supports for solid-phase segment condensations (Albericio et al., 1989c). Individual rates for coupling segments are substantially lower than for activated amino acid species by stepwise synthesis, and there is always a risk of racemization at the C-terminus of each segment. Careful attention to synthetic design and execution may minimize these problems.

Chemoselective Ligation

As an alternative to the segment condensation approach, methods have been developed by which unprotected peptide fragments may be linked. Known as "chemoselective ligation," the initial strategies resulted in the formation of thioester or oxime bonds between peptide fragments. For example, reaction of a peptide bearing a C-terminal thioacid with a peptide containing an N-terminal bromoacetyl group results in a synthetic protein product containing a thioester bond (Schnölzer and Kent, 1992; Muir et al., 1997). This approach has been used successfully to construct HIV-1 protease (Schnölzer and Kent, 1992), a four α-helix TASP molecule (Dawson and Kent, 1993), a folded β-sandwich fibronectin domain model (Williams et al., 1994), and a tethered dimer of HIV-1 protease (Baca et al., 1995). A convenient linker that produces C-terminal thioacids in conjunction with Boc chemistry has been described (Canne et al., 1995). A variation of the thioester approach, in which a peptide C-terminal alkyl thiol is reacted with an N-terminal bromoacetyl or maleimido peptide to form a thioether bond, has been used to construct linked cytoplasmic domains from the $\alpha_{IIb}\beta_3$ integrin (Muir et al., 1994), a β-meander TASP molecule (Nefzi et al., 1995), and a 129-residue tripod protein (McCafferty et al., 1995). Linkers that produce the required modified C-terminal thiols have been described for both Boc (Englebretsen et al., 1995) and Fmoc (Ramage et al., 1996) chemistries. The other approach, in which peptides can be ligated to form oxime bonds (Tuchscherer, 1993; Rose, 1994), require either an aldehyde or N-terminal aminooxyacetyl groups. A peptide aldehyde can additionally be ligated to another peptide containing a weak nucleophilic base, such as hydrazide, N-terminal Cys, or N-terminal Thr, to form hydrazone, thiazolidine, and oxazolidine linkages, respectively (Rao and Tam, 1994; Spetzler and Tam, 1995; Shao and Tam, 1995; Tam and Spetzler, 1997).

   Chemoselective ligation has been made further attractive by "native chemical ligation," in which an amide bond, rather than a thioester, thioether, or oxime bond, is generated between fragments (Muir et al., 1997); see figure 3-7. The original work employed a peptide bearing a

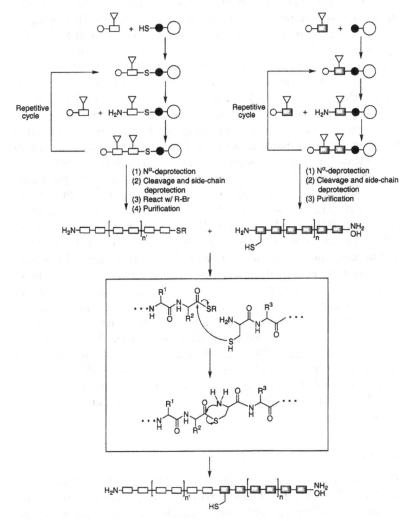

Figure 3-7. Native chemoselective ligation. Two fragments are assembled initially in the solid phase; in this specific example, one has a *C*-terminal thioacid and the other has an *N*-terminal Cys residue. Both fragments are cleaved, side-chain deprotected, and purified. The thioacid is converted to a thioester, allowing for reaction of the two fragments in aqueous solution to form a thioester bond between them. Spontaneous intramolecular rearrangement of the thioester bond results in a native amide bond between the fragments (suggested mechanism in box). Modifications to the process making it possible to link additional peptide fragments have been described.

C-terminal thioacid, which was converted to a 5-thio-2-nitrobenzoic acid ester and subsequently reacted with a peptide bearing an N-terminal Cys residue (Dawson et al., 1994). An intermolecular thioester ensues which undergoes a spontaneous $S \rightarrow N$ acyl shift, leading to the naturally occurring amide bond and regeneration of the free sulfhydryl on Cys. The method has been further refined so that a relatively unreactive thioester can be used in the ligation reaction, which is then carried out in the presence of thiophenol and benzylmercaptan as thiol exchange reagents (Dawson et al., 1997), or in the presence of 2-mercaptoethanesulfonic acid, which has a less noxious odor (Ayers, et al., 1999). Thioesters are readily prepared by Boc chemistry using a thiopropionic acid spacer attached to BHA-resin (Vlattas et al., 1997) or the water-compatible support PEGA (Camarero et al., 1998).

Since thioester linkages decompose in the presence of piperidine used in Fmoc chemistry, alternative deprotection conditions are needed. A solution containing 1-methylpyrrolidine, hexamethyleneimine, and HOBt in NMP–DMSO (1 : 1) has been suggested (Li et al., 1998; Aimoto, 1999). The problem can also be circumvented completely by introducing the thioester functionality after chain elongation. Very recent developments in this rapidly evolving field include a Fmoc/BAL strategy to prepare peptide S-benzyl, S-phenyl, and S-alkyl thioesters (Alsina et al., 1999b), and a novel safety-catch method (Ingenito et al., 1999). Analogues of the glycoprotein diptericin have been assembled (Shin et al., 1999), as well as solid-phase ligation demonstrated (Canne et al., 1999) using the safety-catch approach. It should be noted that amide bonds may also be generated by chemoselective ligation methods which result in thiazolidine linkages via an $O \rightarrow N$ acyl shift to form a hydroxymethyl-thiazolidine (Liu and Tam, 1994, 1996).

Summary

The solid-phase method has made the synthesis of peptides widely accessible. The promise of sophisticated commercial automated instrumentation has been fulfilled. The vast majority of solid-phase peptide syntheses of targets containing up to 50 or so residues can be performed with high efficiencies by either Boc or Fmoc stepwise chemistry. For Boc chemistry, the protection strategy applies Bzl-type and compatible derivatives on the side chains, and for Fmoc chemistry, the corresponding is true with tBu and related derivatives. It is important, however, not to lose sight of the fact that each synthetic procedure has limitations, and that even in the hands of highly experienced workers, certain sequences defy facile preparation. Residue-specific side reactions that may lead to failed syntheses include (a) dehydration of Asn and Gln to the respective nitriles (see Protection Schemes and

Formation of Peptide Bond), (b) racemization of His (see Protection Schemes), (c) aspartimide formation from Asp (see Protection Schemes), (d) alkylation of Trp and Glu (see Cleavage), and (e) acidolysis of Asp-Pro bonds (see Auxiliary Issues). Diketopiperazine-forming sequences (see Auxiliary Issues) and difficult couplings between amino acids with branched side chains (see Auxiliary Issues) are additional, common sources of synthetic problems. In all of the aforementioned examples, deleterious side reactions or other difficulties can be somewhat minimized by careful examination of a peptide sequence *prior* to synthesis. Appropriate precautions as outlined in this chapter (alternative side-chain-protecting groups, use of additional reagents during coupling or cleavage, etc.) can then be taken.

The maturation of high performance liquid chromatography (HPLC) during the 1980s was a major boon to modern peptide synthesis, because the resolving power of this technique facilitates removal of many of the systematic low-level by-products that accrue during chain assembly and upon cleavage. Nowadays, the homogeneity of synthetic materials should be checked by at least two chromatographic or electrophoretic techniques, such as, reversed-phase and ion-exchange HPLC, and capillary zone electrophoresis. Also, determination of a molecular ion by mass spectrometry coupled with a mild ionization method is almost de rigeur for proof of structure. Synthetic peptides must be checked routinely for the proper amino acid composition, and in some cases sequencing data are helpful. Spectroscopic measurements, particularly through use of one and two-dimensional nuclear magnetic resonance (NMR), at the least provide insights on structure and purity, and can often give conformational information as well.

Improvements in the chemistry of SPPS and allied methods have continued since the first edition of this chapter. The current updated version has touched on most of the key issues and discussed the recent status for each of them. Our enthusiasm and optimism for the importance of peptide synthesis as a tool to elucidate an ever-broadening array of biological processes remains undiminished.

Abbreviations

Abbreviations used for amino acids and the designations of peptides follow the rules of the IUPAC-IUB Commission of Biochemical Nomenclature in J. Biol. Chem. 247:977–983 (1972). The following additional abbreviations are used: AA, amino acid; Ac$_4\beta$Gal, 2,3,4,6-tetra-O-acetyl-β-D-galactopyranosyl; Acm, acetamidomethyl; Ada, adamantyl; Al, allyl; Alloc, allyloxycarbonyl; BAL, backbone amide linker; Boc, *tert*-butyloxycarbonyl; Boc-ON, 2-*tert*-butyloxycarbonyloximino-2-phenylacetonitrile; Bom, benzyloxymethyl; BOP, benzotriazolyl *N*-oxytrisdimethylaminophosphonium hexafluorophosphate; 2-BrZ, 2-bromobenzyloxycarbonyl; Bum, *tert*-butoxymethyl; Bzl, benzyl; cHex, cyclohexyl; CLEAR, cross-linked ethoxylate acrylate resin; Cs, cesium salt; 2,6-Cl$_2$Bzl, 2,6-dichlorobenzyl; DBU, 1,8-diazabicyclo[5.4.0]undec-7-ene; DCBC, 2,6-dichlorobenzoyl chloride; DCC, *N,N'*-dicyclohexylcarbodiimide; DCE, 1,2-dichloroethane; DCM, dichloromethane (methylene chloride); Dde, 1-(4,4-dimethyl-2,6-dioxocyclohex-1-ylidene)ethyl; DIEA, *N,N*-diisopropylethylamine; DIPCDI, *N,N'*-diisopropylcarbodiimide; DMA, *N,N*-dimethylacetamide; Dmab, *N*-[1-(4,4,-dimethyl-2,6-dioxocyclohexylidene)-3-methylbutyl]aminobenzyl; DMAP, 4-dimethylaminopyridine; DMF, *N,N*-dimethylformamide; DMSO, dimethyl sulfoxide; Dnp, 2,4-dinitrophenyl; Dod, 4-(4'-methoxybenzhydryl)phenoxyacetic acid; EDT, 1,2-ethanedithiol; Et$_3$N, triethylamine; Fm, 9-fluorenylmethyl; Fmoc, 9-fluorenylmethyloxycarbonyl; Fmoc-OSu, fluorenylmethyl succinimidyl carbonate; HAL, 5-(4-hydroxymethyl-3,5-dimethoxyphenoxy)-valeric acid; HATU, *O*-(7-azabenzotriazol-1-yl)-1,1,3,3-tetramethyluronium hexafluorophosphate or *N*-7-azabenzotriazol-1-yl) (dimethylamino)methylene]-*N*-methylmethanaminium hexafluorophosphate *N*-oxide; HBTU, 2-(1H-benzotriazol-1-yl)-1,1,3,3-tetramethyluronium hexafluorophosphate or *N*-[(1H-benzotriazol-1-yl) (dimethylamino)methylene]-*N*-methylmethanaminium hexafluorophosphate *N*-oxide; HF, hydrogen fluoride; Hmb, 2-hydroxy-4-methoxybenzyl; HMFA, 9-(hydroxymethyl)-2-fluoreneacetic acid; HMP, 4-hydroxymethylphenoxy; HOAc, acetic acid; HOAt, 1-hydroxy-7-azabenzotriazole; HOBt, 1-hydroxybenzotriazole; Hoc, cyclohexyloxycarbonyl; HOPip, *N*-hydroxypiperidine; HPLC, high performance liquid chromatography; Hpp, 1-(4-nitrophenyl)-2-pyrazolin-5-one; HYCRAM, hydroxycrotonylaminomethyl; Hyp, 4-hydroxy-L-proline; MALDI, matrix-assisted laser desorption/ionization; MBHA, 4-methylbenzhydrylamine (resin); MCPBA, *m*-chloroperbenzoic acid; Meb, 4-methylbenzyl; MeIm, 1-methylimidazole; MeOH, methanol; MMA, *N*-methylmercaptoacetamide; Mmt, 4-methoxytrityl; Mob, 4-methoxybenzyl; MS, mass spectrometry; MSNT, 2,4,6-mesitylenesulfonyl-3-nitro-1,2,4-triazolide; Mtr, 4-methoxy-2,3,6-trimethylbenzenesulfonyl; Mts, mesitylene-2-sulfonyl; Mtt; 4-methyltrityl; NCA, *N*-carboxyanhydrides; NMM, *N*-methylmorpholine; NMP, *N*-methylpyrrolidone; Nonb, 3-nitro-4-aminomethylbenzoic acid; NPE, 4-(2-hydroxyethyl)-3-nitrobenzoic acid; Npp, 3-methyl-1-(4-nitrophenyl)-2-pyrazolin-5-one; N.R., not reported; Nvoc, 6-nitroveratryloxycarbonyl   (4,5-dimethoxy-2-nitrobenzyloxycarbonyl); ODhbt, 1-oxo-2-hydroxydihydrobenzotriazine; ONb, 2-nitrobenzyl ester; ONo, 2-nitrophenyl; ONp, 4-nitrophenyl; OPfp, pentafluorophenyl; Orn, ornithine; OSu, *N*-hydroxysuccinimidyl; OTDO, 2,5-diphenyl-2,3-dihydro-3-oxo-4-hydroxythiophene dioxide; PAB,

4-alkoxybenzyl; PAL, 5-4-aminomethyl-3,5-dimethoxyphenoxy)vale-
ric acid; PAM, phenylacetamidomethyl; Pbf, 2,2,4,6,7-pentamethyl-
dihydro-benzofuran-5-sulfonyl; PEG, polyethylene glycol; PEGA,
polyethylene glycol dimethyl acrylamide; PGC, porous graphitized
carbon; Pmc, 2,2,5,7,8-pentamethylchroman-6-sulfonyl; Pnp, 3-phe-
nyl-1-(4-nitrophenyl)-2-pyrazolin-5-one; PSA, preformed symmetrical
anhydride; PyBOP, benzotriazole-1-yl-oxy-tris-pyrrolidinophospho-
nium hexafluorophosphate; SASRIN, 2-methoxy-4-alkoxybenzyl
alcohol; Scm, S-carboxymethylsulfenyl; Snm, S-(N-methyl-N-phenyl-
carbamoyl)sulfenyl; SPPS, solid-phase peptide synthesis; StBu, $tert$-
butylsulfenyl; Tacm, trimethylacetamidomethyl; TASP, template
assembled synthetic protein; TBDMS, $tert$-butyldimethylsilyl; tBu,
$tert$-butyl; TES, triethylsilane; TFA, trifluoroacetic acid; TFE, 2,2,2-
trifluoroethanol; TFMSA, trifluoromethanesulfonic acid; TIS, triisopropylsilane; Tl(Tfa)$_3$, thallium (III) trifluoroacetate; Tmob,
2,4,6-trimethoxybenzyl; TMSBr, trimethylsilyl bromide; TMSOTf, trimethylsilyl trifluoromethanesulfonate; Tos, 4-toluenesulfonyl; Trt, triphenylmethyl; UNCA, urethane-protected N-carboxyanhydride;
XAL, 5-(9-aminoxanthen-2-oxy)valeric acid; Xan, 9-xanthenyl; Z,
benzyloxycarbonyl. Amino acid symbols denote the L-configuration
where applicable, unless indicated otherwise.

ACKNOWLEDGMENTS   This revised chapter is re-dedicated to Professors Miklos
Bodanzsky and Bruce Merrifield in recognition of the high standards they have
set in the development of peptide chemistry. We wish to thank Drs. Fernando
Albericio, Robert Hammer, Derek Hudson, Steven Kates, Stephen Kent, and
Ken Otteson for helpful discussions and unpublished results, and Dr.
Zhenping Tian for his contributions to the earlier edition. Work in the
authors' laboratories has been supported over the years by NIH (CA 77402,
HL 62427, and AR 01929 to G.B.F. and GM 28934, 42722, 43522, and 51628
to G.B.).

References

Abdelmoty, I., Albericio, F., Carpino, L. A., Foxman, B. M., and Kates, S. A.
    (1994). Structural studies of reagents for peptide bond formation:Crystal
    and molecular structures of HBTU and HATU. Lett. Peptide Sci. 1:57–
    67.
Adams, J. H., Cook, R. M., Hudson, D., Jammalamadaka, V., Lyttle, M. H.,
    and Songster, M. F. (1998a). Improved supports and methods for solid
    phase mediated transformation. In *Innovations and Perspectives*, R.
    Epton, ed., Birmingham, UK, Mayflower Scientific Ltd., pp. 23–30.
Adams, J. H., Cook, R. M., Hudson, D. Jammalamadaka, V., Lyttle, M. H.,
    and Songster, M. F. (1998b). A reinvestigation of the preparation, prop-
    erties and applications of aminomethyl and 4-methylbenzhydrylamine
    polystyrene resins. J. Org. Chem. 63:3706–3716.

Adamson, J. G., Blaskovich, M. A., Groenevelt, H., and Lajoie, G. A. (1991). Simple and convenient synthesis of *tert*-butyl ethers of Fmoc-serine, Fmoc-threonine, and Fmoc-tyrosine. J. Org. Chem. 56:3447–3449.

Ahmed, A. K., Schaffer, S. W., and Wetlaufer, D. B. (1975). Nonenzymic reactivation of reduced bovine pancreatic ribonuclease by air oxidation and by glutathione oxidoreduction buffers. J. Biol. Chem. 250:8477–8482.

Aimoto, S. (1999). Polypeptide synthesis by the thioester method. Biopolymers (Pept. Sci.) 51:247–265.

Akaji, K., Fujii, N., Tokunaga, F., Miyata, T., Iwanaga, S., and Yajima, H. (1989). Studies on peptides CLXVIII: Syntheses of three peptides isolated from horseshoe crab hemocytes, tachyplesin I, tachyplesin II, and poly-phemusin I. Chem. Pharm. Bull. 37:2661–2664.

Akaji, K., Tanaka, H., Itoh, H., Imai, J., Fujiwara, Y., Kimura, T., and Kiso, Y. (1990a). Fluoren-9-ylmethyloxycarbonyl (Fmoc) amino acid chloride as an efficient reagent for anchoring Fmoc amino acid to 4-alkoxybenzyl alcohol resin. Chem. Pharm. Bull. 38:3471–3472.

Akaji, K., Yoshida, M., Tatsumi, T., Kimura, T., Fujiwara, Y., and Kiso, Y. (1990b). Tetrafluoroboric acid as a useful deprotecting reagent in Fmoc-based solid-phase peptide synthesis. J. Chem. Soc., Chem. Commun. 288–290.

Akaji, K., Tatsumi, T., Yoshida, M., Kimura, T., Fujiwara, Y., and Kiso, Y. (1991). Synthesis of cystine-peptide by a new disulphide bond-forming reaction using the silyl chloride-sulphoxide system. J. Chem. Soc., Chem. Commun. 167–168.

Albericio, F., and Barany, G. (1984). Application of *N,N*-dimethylformamide dineopentyl acetal for efficient anchoring N^α-9-fluorenylmethyloxycarbo-nylamino acids as *p*-alkoxybenzyl esters in solid-phase peptide synthesis. Int. J. Peptide Protein Res. 23:342–349.

Albericio, F., and Barany, G. (1985). Improved approach for anchoring N^α-9-fluorenylmethyloxycarbonylamino acids as *p*-alkoxybenzyl esters in solid-phase peptide synthesis. Int. J. Peptide Protein Res. 26:92–97.

Albericio, F., and Barany, G. (1987). Mild, orthogonal solid-phase peptide synthesis: Use of N^α-dithiasuccinoyl (Dts) amino acids and *N*-(*iso*-pro-pyldithio)carbonylproline, together with *p*-alkoxybenzyl ester anchoring linkages. Int. J. Peptide Protein Res. 30:177–205.

Albericio, F., and Barany, G. (1991). Hypersensitive acid-labile (HAL) tris-(alkoxy)benzyl ester anchoring for solid-phase synthesis of protected peptide segments. Tetrahedron Lett. 32:1015–1018.

Albericio, F., Grandas, A., Porta, A., Pedroso, E., and Giralt, E. (1987a). One-pot synthesis of *S*-acetamidomethyl-*N*-fluorenylmethoxycarbonyl-L-cysteine (Fmoc-Cys(Acm)-OH). Synthesis 271–272.

Albericio, F., Nicolás, E., Josa, J., Grandas, A., Pedroso, E., Giralt, E., Granier, C., and van Rietschoten, J. (1987b). Convergent solid phase peptide synthesis V: Synthesis of the 1–4, 32–34, and 53–59 protected segments of the toxin II of *Androctonus australis* Hector. Tetrahedron 43:5961–5971.

Albericio, F., Ruiz-Gayo, M., Pedroso, E., and Giralt, E. (1989a). Use of polystyrene-1% divinylbenzene and Kel-F-g-styrene for the simultaneous synthesis of peptides. Reactive Polymers 10:259–268.

Albericio, F., Andreu, D., Giralt, E., Navalpotro, C., Pedroso, E., Ponsati, B., and Ruiz-Gayo, M. (1989b). Use of the Npys thiol protection in solid phase peptide synthesis. Int. J. Peptide Protein Res. 34:124–128.

Albericio, F., Pons, M., Pedroso, E., and Giralt, E. (1989c). Comparative study of supports for solid-phase coupling of protected-peptide segments. J. Org. Chem. 54:360–366.

Albericio, F., Kneib-Cordonier, N., Biancalana, S., Gera, L., Masada, R. I., Hudson, D., and Barany, G. (1990a). Preparation and application of the 5-(4-9-fluorenylmethyloxycarbonyl)aminomethyl-3,5-dimethoxyphenoxy) valeric acid (PAL) handle for the solid-phase synthesis of C-terminal peptide amides under mild conditions. J. Org. Chem. 55:3730–3743.

Albericio, F., Van Abel, R., and Barany, G. (1990b). Solid-phase synthesis of peptides with C-terminal asparagine or glutamine. Int. J. Peptide Protein Res. 35:284–286.

Albericio, F., Nicolás, E., Rizo, J., Ruiz-Gayo, M., Pedroso, E., and Giralt, E. (1990c). Convenient syntheses of fluorenylmethyl-based side chain derivatives of glutamic and aspartic acids, lysine, and cysteine. Synthesis 119–122.

Albericio, F., Hammer, R. P., García-Echeverría, C., Molins, M. A., Chang, J. L., Munson, M. C., Pons, M., Giralt, E., and Barany, G. (1991a). Cyclization of disulfide-containing peptides in solid-phase synthesis. Int. J. Peptide Protein Res. 37:402–413.

Albericio, F., Giralt, E., and Eritja, R. (1991b). NPE-resin, a new approach to the solid-phase synthesis of protected peptides and oligonucleotides II: Synthesis of protected peptides. Tetrahedron Lett. 32:1515–1518.

Albericio, F., Barany, G., Fields, G. B., Hudson, D., Kates, S. A., Lyttle, M. H., and Solé, N. A. (1993). Allyl-based orthogonal solid-phase peptide synthesis. In Peptides 1992, C. H. Schneider and A. N. Eberle, eds., Leiden, Escom, pp. 191–193.

Albericio, F., Lloyd-Williams, P., and Giralt, E. (1997). Convergent solid-phase peptide synthesis. Methods Enzymol. 289:313–336.

Albericio, F., Annis, I., Royo, M., and Barany, G. (2000). Preparation and handling of peptides containing methionine and cysteine. In Fmoc Solid Phase Peptide Synthesis: A Practical Approach, W. C. Chan and P. D. White, eds., Oxford, Oxford University Press, pp. 77–114.

Aletras, A., Barlos, K., Gatos, D., Koutsogianni, S., and Mamos, P. (1995). Preparation of the very acid-sensitive Fmoc-Lys(Mtt)-OH. Int. J. Peptide Protein Res. 45:488–500.

Alewood, P., Alewood, D., Miranda, L., Love, S., Meutermans, W., and Wilson, D. (1997). Rapid in situ neutralization protocols for Boc and Fmoc solid-phase chemistries. Methods Enzymol. 289:14–29.

Al-Obeidi, F., Sanderson, D. G., and Hruby, V. J. (1990). Synthesis of β- and γ-fluorenylmethyl esters of respectively N^α-Boc-L-aspartic acid and N^α-Boc-L-glutamic acid. Int. J. Peptide Protein Res. 35:215–218.

Alsina, J., Barany, G., Albericio, F., and Kates, S. (1999a). Pyrrolidide formation as a side reaction during activation of carboxylic acids by phosphonium salt coupling reagents. Lett. Pept. Sci. 6:243–245.

Alsina, J., Yokum, T. Y., Albericio, F., and Barany, G. (1999b). Backbone amide linker (BAL) strategy for N^α-9-fluorenylmethoxycarbonyl (Fmoc) solid-phase synthesis of unprotected peptide p-nitroanilides and thio-esters. J. Org. Chem. 64:8761–8769.

Ambrosius, D., Casaretto, M., Gerardy-Schahn, R., Saunders, D., Brandenburg, D., and Zahn, H. (1989). Peptide analogues of the anaphylatoxin C3a; synthesis and properties. Biol. Chem. Hoppe-Seyler 370:217–227.

Andreu, D., Albericio, F., Solé, N. A., Munson, M. C., Ferrer, M., and Barany, G. (1994). Formation of disulfide bonds in synthetic peptides and proteins. In *Methods in Molecular Biology*, Vol. 35, M. W. Pennington and B. M. Dunn, eds., Totowa, N.J., Humana Press, pp. 91–169.

Annis, I., Hargittai, B., and Barany, G. (1997). Disulfide bond formation in peptides. Methods Enzymol. 289:198–221.

Annis, I., Chen, L., and Barany, G. (1998). Novel solid-phase reagents for facile formation of intramolecular disulfide bridges in peptides under mild conditions. J. Am. Chem. Soc. 120:7226–7238.

Anwer, M. K., and Spatola, A. F. (1980). An advantageous method for the rapid removal of hydrogenolysable protecting groups under ambient conditions; synthesis of leucine-enkephalin. Synthesis 929–932.

Anwer, M. K., and Spatola, A. F. (1983). Quantitative removal of a pentadecapeptide ACTH fragment analogue from a Merrifield resin using ammonium formate catalytic transfer hydrogenation: Synthesis of $[Asp^{25}, Ala^{26}, Gly^{27}, Gln^{30}]$-ACTH-(25-39)-OH. J. Am. Chem. Soc. 48:3503–3507.

Applied Biosystems, Inc. (1989a). Removal of 2,4-dinitrophenyl (Dnp) protection from peptides synthesized with Boc-His(Dnp). Peptide Synthesizer User Bulletin 28, Foster City, Calif., Applied Biosystems, Inc.

Applied Biosystems, Inc. (1989b). Use of the ninhydrin reaction to monitor Fmoc solid-phase peptide synthesis. Peptide Synthesizer User Bulletin 29, Foster City, Calif., Applied Biosystems, Inc.

Applied Biosystems, Inc. (1992). *Model 431A Peptide Synthesizer User's Manual, Version 2.0*, Foster City, Calif., Applied Biosystems, Inc.

Arad, O., and Houghten, R. A. (1990). An evaluation of the advantages and effectiveness of picric acid monitoring during solid phase peptide synthesis. Peptide Res. 3:42–50.

Arendt, A., Palczewski, K., Moore, W. T., Caprioli, R. M., McDowell, J. H., and Hargrave, P. A. (1989). Synthesis of phosphopeptides containing O-phosphoserine or O-phosphothreonine. Int. J. Peptide Protein Res. 33:468–476.

Arendt, A., McDowell, J., Hugh, J., and Hargrave, P. A. (1993). Optimization of peptide synthesis on polyethylene rods. Peptide Res. 6:346–352.

Arshady, R., Atherton, E., Clive, D. L. J., and Sheppard, R. C. (1981). Peptide synthesis, part 1: Preparation and use of polar supports based on poly(-dimethylacrylamide). J. Chem. Soc. Perkin Trans. I:529–537.

Atherton, E., and Sheppard, R. C. (1985). Detection of problem sequences in solid phase synthesis. In *Peptides: Structure and Function*, C. M. Deber, V. J. Hruby, and K. D. Kopple, eds., Rockford, Ill., Pierce Chemical Co., pp. 415–418.

Atherton, E., and Sheppard, R. C. (1989). *Solid Phase Peptide Synthesis: A Practical Approach*, Oxford, IRL Press.

Atherton, E., Fox, H., Harkiss, D., Logan, C. J., Sheppard, R. C., and Williams, B. J. (1978a). A mild procedure for solid phase peptide synthesis:Use of fluorenylmethoxycarbonylamino-acids. J. Chem. Soc., Chem. Commun. 537–539.

Atherton, E., Fox, H., Harkiss, D., and Sheppard, R. C. (1978b). Application of polyamide resins to polypeptide synthesis: An improved synthesis of β-endorphin using fluorenylmethoxycarbonylamino-acids. J. Chem. Soc., Chem. Commun. 539–540.

Atherton, E., Bury, C., Sheppard, R. C., and Williams, B. J. (1979). Stability of fluorenylmethoxycarbonylamino groups in peptide synthesis: Cleavage by hydrogenolysis and by dipolar aprotic solvents. Tetrahedron Lett. 3041–3042.

Atherton E., Woolley, V., and Sheppard, R. C. (1980). Internal association in solid phase peptide synthesis: Synthesis of cytochrome C residues 66–104 on polyamide supports. J. Chem. Soc., Chem. Commun. 970–971.

Atherton, E., Benoiton, N. L., Brown, E., Sheppard, R. C., and Williams, B. J. (1981a). Racemisation of activated, urethane-protected amino-acids by p-dimethylaminopyridine: Significance in solid-phase peptide synthesis. J. Chem. Soc., Chem. Commun. 336–337.

Atherton, E., Brown, E., Sheppard, R. C., and Rosevear, A. (1981b). A physically supported gel polymer for low pressure, continuous flow solid phase reactions: Application to solid phase peptide synthesis. J. Chem. Soc., Chem. Commun. 1151–1152.

Atherton, E., Logan, C. J., and Sheppard, R. C. (1981c). Peptide synthesis, part 2: Procedures for solid-phase synthesis using N^α-fluorenylmethoxy-carbonylamino-acids on polyamide supports:Synthesis of substance P and of acyl carrier protein 65–74 decapeptide. J. Chem. Soc. Perkin Trans. I:538–546.

Atherton, E., Caviezel, M., Fox, H., Harkiss, D., Over, H., and Sheppard, R. C. (1983). Peptide synthesis, part 3: Comparative solid-phase syntheses of human β-endorphin on polyamide supports using t-butoxycarbonyl and fluorenylmethoxycarbonyl protecting groups. J. Chem. Soc. Perkin Trans. I:65–73.

Atherton, E., Cammish, L. E., Goddard, P., Richards, J. D., and Sheppard, R. C. (1984). The Fmoc-polyamide solid phase method: New procedures for histidine and arginine. In Peptides 1984, U. Ragnarsson, ed., Stockholm, Almqvist & Wiksell, pp. 153–156.

Atherton, E., Pinori, M., and Sheppard, R. C. (1985a). Peptide synthesis, part 6: Protection of the sulphydryl group of cysteine in solid-phase synthesis using N^α-fluorenylmethoxycarbonylamino acids: Linear oxytocin derivatives. J. Chem. Soc. Perkin Trans. I:2057–2064.

Atherton, E., Sheppard, R. C., and Ward, P. (1985b). Peptide synthesis, part 7: Solid-phase synthesis of conotoxin G1. J. Chem. Soc. Perkin Trans. I:2065–2073.

Atherton, E., Cameron, L. R., and Sheppard, R. C. (1988a). Peptide synthesis, part 10: Use of pentafluorophenyl esters of fluorenylmethoxycarbonyla-mino acids in solid phase peptide synthesis. Tetrahedron 44:843–857.

Atherton, E., Holder, J. L., Meldal, M., Sheppard, R. C., and Valerio, R. M. (1988b). Peptide synthesis, part 12:3,4-Dihydro-4-oxo-1,2,3-benzotriazin-

3-yl esters of fluorenylmethoxycarbonyl amino acids as self-indicating reagents for solid phase synthesis. J. Chem. Soc. Perkin Trans. I:2887–2894.

Atherton, E., Cameron, L. R., Cammish, L. E., Dryland, A., Goddard, P., Priestley, G. P., Richards, J. D., Sheppard, R. C., Wade, J. D., and Williams, B. J. (1990). Solid phase fragment condensation—the problems. In *Innovation and Perspectives in Solid Phase Synthesis*, R. Epton, ed., Birmingham, UK, Solid Phase Conference Coordination, Ltd., pp. 11–25.

Atherton, E., Hardy, P. M., Harris, D. E., and Matthews, B. H. (1991). Racemization of *C*-terminal cysteine during peptide assembly. In *Peptides 1990*, E. Giralt and D. Andreu, eds., Leiden, Escom, pp. 243–244.

Ayers, B., Blaschke, U. K., Camarero, J. A., Cotton, G. J., Holford, M., and Muir, T. W. (1999). Introduction of unnatural amino acids into proteins using expressed protein ligation. Biopolymers (Pept. Sci.) 51:343–354.

Baca, M., Muir, T. W., Schnölzer, M., and Kent, S. B. H. (1995). Chemical ligation of cysteine-containing peptides: Synthesis of a 22 kDa tethered dimer of HIV-1 protease. J. Am. Chem. Soc. 117:1881–1887.

Bagley, C. J., Otteson, K. M., May, B. L., McCurdy, S. N., Pierce, L., Ballard, F. J., and Wallace, J. C. (1990). Synthesis of insulin-like growth factor I using *N*-methyl pyrrolidinone as the coupling solvent and trifluromethane sulphonic acid cleavage from the resin. Int. J. Peptide Protein Res. 36:356–361.

Baker, P. A., Coffey, A. F., and Epton, R. (1990). Accelerated continuous flow ultra-high load solid (gel) phase oligopeptide assembly. In *Innovation and Perspectives in Solid Phase Synthesis*, R. Epton, ed., Birmingham, UK, Solid Phase Conference Coordination, Ltd., pp. 435–440.

Ball, H. L., Grecian, C., Kent, S. B. H., and Mascagni, P. (1990). Affinity methods for purifying large synthetic peptides. In *Peptides: Chemistry, Structure and Biology*, J. E. Rivier and G. R. Marshall, eds., Leiden, Escom, pp. 435–436.

Ball, H. L., Kent, S. B. H., and Mascagni, P. (1991). Selective purification of large synthetic peptides using removable chromatographic probes. In *Peptides 1990*, E. Giralt and D. Andreu, eds., Leiden, Escom, pp. 323–325.

Barany, G. (1999). Synthetic approaches to elucidate roles of disulfide bridges in peptides and proteins. In *Peptides: Frontiers of Peptide Science*, J. P. Tam and P. T. P. Kaumaya, eds., Dordrecht, Kluwer Academic Publishers, pp. 226–228.

Barany, G., and Albericio, F. (1985). A three-dimensional orthogonal protection scheme for solid-phase peptide synthesis under mild conditions. J. Am. Chem. Soc. 107:4936–4942.

Barany, G., and Albericio, F. (1991). Peptide synthesis for biotechnology in the 1990's. In *Biotechnology International 1990/1991*, K. North, ed., London, Century Press Ltd., pp. 155–163.

Barany, G., and Kempe, M. (1997). The context of solid-phase synthesis. In *A Practical Guide to Combinatorial Chemistry*, A. W. Czarnik and S. H. DeWitt, eds., Washington, D.C., American Chemical Society, pp. 51–97.

Barany, G., and Merrifield, R. B. (1973). An ATP-binding peptide. Cold Spring Harbor Symp. Quant. Biol. 37:121–125.

Barany, G., and Merrifield, R. B. (1977). A new amino protecting group removable by reduction:Chemistry of the dithiasuccinoyl (Dts) function. J. Am. Chem. Soc. 99:7363–7365.

Barany, G., and Merrifield, R. B. (1979). Solid-phase peptide synthesis. In *The Peptides*, Vol. 2, E. Gross and J. Meienhofer, eds., New York, Academic Press, pp. 1–284.

Barany, G., Kneib-Cordonier, N., and Mullen, D. G. (1987). Solid-phase peptide synthesis: A silver anniversary report. Int. J. Peptide Protein Res. 30:705–739.

Barany, G., Kneib-Cordonier, N., and Mullen, D. G. (1988). Polypeptide synthesis, solid-phase method. In *Encyclopedia of Polymer Science and Engineering*, Vol. 12, 2nd edn., J. I. Kroschwitz, ed., New York, John Wiley and Sons, pp. 811–858.

Barany, G., Albericio, F., Biancalana, S., Bontems, S. L., Chang, J. L., Eritja, R., Ferrer, M., Fields, C. G., Fields, G. B., Lyttle, M. H., Solé, N. A., Tian, Z., Van Abel, R. J., Wright, P. B., Zalipsky, S., and Hudson, D. (1992). Biopolymer syntheses on novel polyethylene glycol-polystyrene (PEG-PS) graft supports. In *Peptides: Chemistry and Biology*, J. A. Smith and J. E. Rivier, eds., Leiden, Escom, pp. 603–604.

Barany, G., Albericio, F., Kates, S. A., and Kempe, M. (1997). Poly(ethylene glycol) containing supports for solid phase synthesis of peptides and combinatorial organic libraries. In *Poly(ethylene glycol): Chemistry and Biological Applications*, J. M. Harris and S. Zalipsky, eds., ACS Symposium Series 680, Washington, D.C., American Chemical Society, pp. 239–264.

Bardají, E., Torres, J. L., Clapés, P., Albericio, F., Barany, G., Rodríguez, R. E., Sacristán, M. P., and Valencia, G. (1991). Synthesis and biological activity of *O*-glycosylated morphiceptin analogues. J. Chem. Soc. Perkin Trans. 1:1755–1759.

Barlos, K., Papaioannou, D., and Theodoropoulos, D. (1982). Efficient "one-pot" synthesis of *N*-trityl amino acids. J. Org. Chem. 47:1324–1326.

Barlos, K., Mamos, P., Papaioannou, D., Patrianakou, S., Sanida, C., and Schäfer, W. (1987). Einsatz von Trt- und Fmoc-gruppen zum schutz polyfunktioneller α-aminosäuren. Liebigs Ann. Chem. 1025–1030.

Barlos, K., Gatos, D., Hondrelis, J., Matsoukas, J., Moore, G. J., Schäfer, W., and Sotiriou, P. (1989a). Darstellung neuer säureempfindlicker harze vom sek.-alkohol-typ und ihre anwendung zur synthese von peptiden. Liebigs Ann. Chem. 951–955.

Barlos, K., Gatos, D., Kallitsis, J., Papaphotiu, G., Sotiriu, P., Wenqing, Y., and Schäfer, W. (1989b). Darstellung geschützter peptid-fragmente unter einsatz substituierter triphenylmethyl-harze. Tetrahedron Lett. 30:3943–3946.

Barlos, K., Chatzi, O., Gatos, D., and Stavropoulos, G. (1991a). 2-Chlorotrityl chloride resin: Studies on anchoring of Fmoc-amino acids and peptide cleavage. Int. J. Peptide Protein Res. 37:513–520.

Barlos, K., Gatos, D., Koutsogianni, S., Schäfer, W., Stavropoulos, G., and Wenqing, Y. (1991b). Darstellung und einsatz von N-Fmoc-O-Trt-hydroxyaminosäuren zur "solid phase" synthese von peptiden. Tetrahedron Lett. 32:471–474.

Barlos, K., Gatos, D., and Schäfer, W. (1991c). Synthesis of prothymosin α (ProTα)—a protein consisting of 109 amino acid residues. Angew. Chem. Int. Ed. Engl. 30:590–593.

Barlos, K., Chatzi, O., Gatos, D., Stavropoulos, G., and Tsegenidis, T. (1991d). Fmoc-His(Mmt)-OH und Fmoc-His(Mtt)-OH. Tetrahedron Lett. 32:475–478.

Barlos, K., Gatos, D., Hatzi, O., Koch, N., and Koutsogianni, S. (1996). Synthesis of the very acid-sensitive Fmoc-Cys(Mmt)-OH and its application in solid-phase peptide synthesis. Int. J. Peptide Protein Res. 47:148–153.

Barlos, K., Gatos, D., and Koutsogianni, S. (1998). Fmoc/Trt-amino acids: Comparison to Fmoc/tBu-amino acids in peptide synthesis. J. Peptide Res. 51:194-200.

Barton, M. A., Lemieux, R. U., and Savoie, J. Y. (1973). Solid-phase synthesis of selectively protected peptides for use as building units in the solid-phase synthesis of large molecules. J. Am. Chem. Soc. 95:4501–4506.

Bayer, E. (1991). Towards the chemical synthesis of proteins. Angew. Chem. Int. Ed. Engl. 30:113–129.

Bayer, E., and Rapp, W. (1986). New polymer supports for solid-liquid-phase peptide synthesis. In Chemistry of Peptides and Proteins, Vol. 3, W. Voelter, E. Bayer, Y. A. Ovchinnikov, and V. T. Ivanov, eds., Berlin, Walter de Gruyter & Co., pp. 3–8.

Bayer, E., Albert, K., Willisch, H., Rapp, W., and Hemmasi, B. (1990). ^{13}C NMR relaxation times of a tripeptide methyl ester and its polymer-bound analogues. Macromolecules 23:1937–1940.

Bayer, E., Eckstein, H., Hägele, K., König, W. A., Brüning, W., Hagenmaier, H., and Parr, W. (1970). Failure sequences in the solid phase synthesis of polypeptides. J. Am. Chem. Soc. 92:1735–1738.

Beacham, J., Bentley, P. H., Kenner, G. W., MacLeod, J. K., Mendive, J. J., and Sheppard, R. C. (1967). Peptides part XXV: The structure and synthesis of human gastrin. J. Chem. Soc. (C) 2520–2529.

Beck, W., and Jung, G. (1994). Convenient reduction of S-oxides in synthetic peptides, lipopeptides and peptide libraries. Lett. Pept. Sci. 1:31–37.

Beck-Sickinger, A. G., Dürr, H., and Jung, G. (1991). Semiautomated T-bag peptide synthesis using 9-fluorenylmethoxycarbonyl strategy and benzotriazol-1-yl-tetramethyluronium tetrafluoroborate activation. Peptide Res. 4:88–94.

Becker, S., Atherton, E., and Gordon, R. D. (1989). Synthesis and characterization of μ-conotoxin IIIa. Eur. J. Biochem. 185:79–84.

Bedford, J., Hyde, C., Johnson, T., Jun, W., Owen, D., Quibell, M., and Sheppard, R. C. (1992). Amino acid structure and "difficult sequences" in solid phase peptide synthesis. Int. J. Peptide Protein Res. 40:300–307.

Belshaw, P. J., Mzengeza, S., and Lajoie, G. A. (1990). Chlorotrimethylsilane mediated formation of ω-allyl esters of aspartic and glutamic acids. Synth. Commun. 20:3157–3160.

Benoiton, N. L., and Chen, F. M. F. (1987). Symmetrical anhydride rearrangement leads to three different dipeptide products. In Peptides 1986, D. Theodoropoulos, ed., Berlin, Walter de Gruyter and Co., pp. 127–130.

Berg, R. H., Almdal, K., Pedersen, W. B., Holm, A., Tam, J. P., and Merrifield, R. B. (1989). Long-chain polystyrene-grafted polyethylene film matrix: A new support for solid-phase peptide synthesis. J. Am. Chem. Soc. 111:8024–8026.

Bergot, B. J., Noble, R. L., and Geiser, T. (1987). TFMSA/TFA cleavage and deprotection in SPPS. In *Peptides 1986*, D. Theodoropoulos, ed., Berlin, Walter de Gruyter and Co., pp. 97–101.

Bernatowicz, M. S., Matsueda, R., and Matsueda, G. R. (1986). Preparation of Boc-[S-3-nitro-2-pyridinesulfenyl)]-cysteine and its use for unsymmetrical disulfide bond formation. Int. J. Peptide Protein Res. 28:107–112.

Bernatowicz, M. S., Daniels, S. B., Coull, J. M., Kearney, T., Neves, R. S., Coassin, P. J., and Köster, H. (1990). Recent developments in solid phase peptide synthesis using the 9-fluorenylmethyloxycarbonyl (Fmoc) protecting group strategy. In *Current Research in Protein Chemistry: Techniques, Structure, and Function*, J. J. Villafranca, ed., San Diego, Academic Press, pp. 63–77.

Bertho, J.-N., Loffet, A., Pinel, C., Reuther, F., and Sennyey, G. (1991). Amino acid fluorides: Their preparation and use in peptide synthesis. Tetrahedron Lett. 32:1303–1306.

Biancalana, S., Hayes, T., Toll, L., and Hudson, D. (1991). Immunological and biological activity of octameric peptides prepared by Fmoc methodology. In *Twelfth American Peptide Symposium Program and Abstracts*, Cambridge, Mass., Massachusetts Institute of Technology, P-234.

Bianchi, E., Sollazzo, M., Tramontano, A., and Pessi, A. (1993). Chemical synthesis of a designed β-protein through the flow-polyamide method. Int. J. Peptide Protein Res. 41:385–393.

Birr, C. (1978). *Aspects of the Merrifield Peptide Synthesis*, Heidelberg, Springer-Verlag.

Birr, C. (1990a). The transition to solid-phase production of pharmaceutical peptides. Biochem. Soc. Trans. 18:1313–1316.

Birr, C. (1990b). Scale-up and production potential of current SPPS strategies. In *Innovation and Perspectives in Solid Phase Synthesis*, R. Epton, ed., Birmingham, UK, Solid Phase Conference Coordination, Ltd., pp. 155–181.

Birr, C., Lochinger, W., Stahnke, G., and Lang, P. (1972). Der α,α-dimethyl-3.5-dimethoxybenzyloxycarbonyl (Ddz)-rest, eine photo- und säurelabile stickstoff-schutzgruppe für die peptidchemie. Justus Liebigs Ann. Chem. 763:162–172.

Blackburn, C. (2000). Solid supports for the synthesis of peptides and small molecules. In *Solid-Phase Peptide Synthesis: A Practical Guide*, S. A. Kates and F. Albericio, eds., New York, Marcel Dekker, pp. 197–273.

Blankemeyer-Menge, B., Nimtz, M., and Frank, R. (1990). An efficient method for anchoring Fmoc-amino acids to hydroxyl-functionalised solid supports. Tetrahedron Lett. 31:1701–1704.

Bloomberg, G. B., Askin, D., Gargaro, A. R., and Tanner, M. J. A. (1993). Synthesis of a branched cyclic peptide using a strategy employing Fmoc chemistry and two additional orthogonal protecting groups. Tetrahedron Lett. 34:4709–4712.

Bodanszky, A., Bodanszky, M., Chandramouli, N., Kwei, J. Z., Martinez, J., and Tolle, J. C. (1980). Active esters of 9-fluorenylmethyloxycarbonyl amino acids and their application in the stepwise lengthening of a peptide chain. J. Org. Chem. 45:72–76.

182   Synthetic Peptides

Bodanszky, M., and Bodanszky, A. (1984). *The Practice of Peptide Synthesis*, Berlin, Springer-Verlag.

Bodanszky, M., and Kwei, J. Z. (1978). Side reactions in peptide synthesis VII: Sequence dependence in the formation of aminosuccinyl derivatives from β-benzyl-aspartyl peptides. Int. J. Peptide Protein Res. 12:69–74.

Bodanszky, M., Funk, K. W., and Fink, M. L. (1973). *o*-Nitrophenyl esters of *tert*-butyloxycarbonylamino acids and their application in the stepwise synthesis of peptide chains by a new technique. J. Org. Chem. 38:3565–3570.

Bodanszky, M., Tolle, J. C., Deshmane, S. S., and Bodanszky, A. (1978). Side reactions in peptide synthesis VI: A reexamination of the benzyl group in the protection of the side chains of tyrosine and aspartic acid. Int. J. Peptide Protein Res. 12:57–68.

Bolin, D. R., Wang, C.-T., and Felix, A. M. (1989). Preparation of N-t-butyloxycarbonyl-O^ω-9-fluorenylmethyl esters of asparatic and glutamic acids. Org. Prep. Proc. Int. 21:67–74.

Borràs, E.; Giralt, E., and Andreu, D. (1999). A rationally designed synthetic peptide mimic of a discontinuous viral antigenic site elicits neutralizing antibodies. J. Am. Chem. Soc. 121:11932–11933.

Bourne, G. T., Meutermans, W. D. F., Alewood, P. F., McGeary, R. P., Scanlon, M., Watson, A. A., and Smythe M. L. (1999a). A backbone linker for BOC-based peptide synthesis and on-resin cyclization: Synthesis of stylostatin 1. J. Org. Chem. 64:3095–3101.

Bourne, G. T., Meutermans, W. D. F., and Smythe M. L. (1999b). The development of solid phase protocols for a backbone amide linker and its application to the Boc-based assembly of linear peptides. Tetrahedron Lett. 40:7271–7274.

Bozzini, M., Bello, R., Cagle, N., Yamane, D., and Dupont, D. (1991). Tryptophan recovery from autohydrolyzed samples using dodecanethiol. *Applied Biosystems Research News*, February, Foster City, Calif., Applied Biosystems, Inc.

Brady, S. F., Paleveda, W. J., and Nutt, R. F. (1988). Studies of acetamidomethyl as cysteine protection: Application in synthesis of ANF analogs. In *Peptides: Chemistry and Biology*, G. R. Marshall, ed., Leiden, Escom, pp. 192–194.

Breipohl, G., Knolle, J., and Stüber, W. (1989). Synthesis and application of acid labile anchor groups for the synthesis of peptide amides by Fmoc-solid-phase peptide synthesis. Int. J. Peptide Protein Res. 34:262–267.

Breipohl, G., Knolle, J., and Stüber, W. (1990). Facile SPS of peptides having *C*-terminal Asn and Gln. Int. J. Peptide Protein Res. 35:281–283.

Broddefalk, J., Bäcklund, J., Almqvist, F., Johansson, M., Holmdahl, R., and Kihlberg, J. (1998). T-cells recognize a glycopeptide derived from type II collagen in a model for rheumatoid arthritis. J. Am. Chem. Soc. 120:7676–7683.

Brown, A. R., Irving, S. L., and Ramage, R. (1993). Affinity purification of synthetic peptides and proteins on porous graphitised carbon. Tetrahedron Lett. 34:7129–7132.

Brown, T., Jones, J. H., and Richards, J. D. (1982). Further studies on the protection of histidine side chains in peptide synthesis: The use of the π-benzyloxymethyl group. J. Chem. Soc. Perkin Trans. I:1553–1561.

Brugidou, J., and Méry, J. (1994). 2-Hydroxypropyl-dithio-2'-isobutyric acid (HPDI) as a multipurpose peptide-resin linker for SPPS. Peptide Res. 7:40–47.

Buchta, R., Bondi, E., and Fridkin, M. (1986). Peptides related to the calcium binding domains II and III of calmodulin: Synthesis and calmodulin-like features. Int. J. Peptide Protein Res. 28:289–297.

Büttner, K., Zahn, H., and Fischer, W. H. (1988). Rapid solid phase peptide synthesis on a controlled pore glass support. In Peptides: Chemistry and Biology, G. R. Marshall, ed., Leiden, Escom, pp. 210–211.

Butwell, F. G. W., Haws, E. J., and Epton, R. (1988). Advances in ultra-high load polymer supported peptide synthesis with phenolic supports 1: A selectively-labile C-terminal spacer group for use with a base-mediated N-terminal deprotection strategy and Fmoc amino acids. Makromol. Chem., Macromol. Symp. 19:69–77.

Bycroft, B. W., Chang, W. C., Chhabra, S. R., and Hone, N. D. (1993). A novel lysine protecting procedure for continuous flow solid phase synthesis of branched peptides. J. Chem. Soc., Chem. Commun. 778–779.

Calnan, B. J., Biancalana, S., Hudson, D., and Frankel, A. D. (1991). Analysis of arginine-rich peptides from the HIV; Tat protein reveals unusual features of RNA-protein Recognition. Genes and Development 5:201–210.

Camarero, J. A., Cotton, G. J., Adeva, A., and Muir, T. W. (1998). Chemical ligation of unprotected peptides directly from a solid support. J. Peptide Res. 51:303–316.

Cameron, L. R., Holder, J. L., Meldal, M., and Sheppard, R. C. (1988). Peptide synthesis, part 13: Feedback control in solid phase synthesis: Use of fluorenylmethoxycarbonyl amino acid 3,4-dihydro-4-oxo-1,2,3-benzotriazin-3-yl esters in a fully automated system. J. Chem. Soc. Perkin Trans. I:2895–2901.

Canne, L. E., Walker, S. M., and Kent, S. B. H. (1995). A general method for the synthesis of thioester resin linkers for use in the solid phase synthesis of peptide-α-thioacids. Tetrahedron Lett. 36:1217–1220.

Canne, L. E., Botti, P., Simon, R. J., Chen, Y., Dennis, E. A., and Kent, S. B. H. (1999). Chemical protein synthesis by solid phase ligation of unprotected peptide segments. J. Am. Chem. Soc. 121:8720–8727.

Carpino, L. A., and El-Faham, A. (1995). Tetramethylfluoroformamidinium hexafluorophospate: A rapid-acting peptide coupling reagent for solution and solid phase peptide synthesis. J. Am. Chem. Soc. 117:5401–5402.

Carpino, L. A., and Han, G. Y. (1972). The 9-fluorenylmethoxycarbonyl amino-protecting group. J. Org. Chem. 37:3404–3409.

Carpino, L. A., and Mansour, E.-S. M.E. (1992). Protected β- and γ-aspartic and -glutamic acid fluorides. J. Org. Chem. 57:6371–6373.

Carpino, L. A., Cohen, B. J., Stephens, Jr., K. E., Sadat-Aalaee, S. Y., Tien, J.-H., and Langridge, D. C. (1986). ((9-Fluorenylmethyl)oxy)carbonyl (Fmoc) amino acid chlorides: Synthesis, characterization, and application to the rapid synthesis of short peptide. J. Org. Chem. 51:3732–3734.

Carpino, L. A., Sadat-Aalaee, D., Chao, H. G., and DeSelms, R. H. (1990). ((9-Fluorenylmethyl)oxy)carbonyl (Fmoc) amino acid fluorides: Convenient new peptide coupling reagents applicable to the Fmoc/tert-butyl strategy for solution and solid-phase syntheses. J. Am. Chem. Soc. 112:9651–9652.

Carpino, L. A., Mansour, E.-S. M. E., and Sadat-Aalaee, D. (1991a). *tert*-Butyloxycarbonyl and benzyloxycarbonyl amino acid fluorides: New, stable rapid-acting acylating agents for peptide synthesis. J. Org. Chem. 56:2611–2614.

Carpino, L. A., Chao, H. G., Beyermann, M., and Bienert, M. (1991b). ((9-Fluorenylmethyl)oxy)carbonyl amino acid chlorides in solid-phase peptide synthesis. J. Org. Chem. 56:2635–2642.

Carpino, L. A., Shroff, H., Triolo, S. A., Mansour, E.-S. M.E., Wenschuh, H., and Albericio, F. (1993). The 2,2,4,6,7-pentamethyldihydrobenzofuran-5-sulfonyl group (Pbf) as arginine side chain protectant. Tetrahedron Lett. 34:7829–7832.

Carpino, L. A., El-Faham, A., Minor, C. A., and Albericio, F. (1994). Advantageous applications of azabenzotriazole (triazolopyridine)-based coupling reagents to solid-phase peptide synthesis. J. Chem. Soc., Chem. Commun. 201–203.

Casaretto, R., and Nyfeler, R. (1991). Isolation, structure and activity of a side product from the synthesis of human endothelin. In *Peptides 1990*, E. Giralt and D. Andreu, eds., Leiden, Escom, pp. 181–182.

Chan, A. S. H., Mobley, J. L., Fields, G. B., and Shimizu, Y. (1997). CD7-mediated regulation of integrin adhesiveness on human T cells involves tyrosine phosphorylation-dependent activation of phosphatidylinositol 3-kinase. J. Immunol. 159:934–942.

Chan, W. C., and Bycroft, B. W. (1992). Deprotection of Arg(Pmc) containing peptides using TFA-trialkylsilane-methanol-EMS; application to the synthesis of propeptides of nisin. In *Peptides: Chemistry and Biology*, J. A. Smith and J. E. Rivier, eds., Leiden, Escom, pp. 613–614.

Chan, W. C., and White, P. D., eds. (2000). *Fmoc Solid Phase Peptide Synthesis: A Practical Approach*, Oxford, Oxford University Press.

Chan, W. C., Bycroft, B. W., Evans, D. J., and White, P. D. (1995). A novel 4-aminobenzyl ester-based carboxy-protecting group for synthesis of atypical peptides by Fmoc-Bu^t solid phase chemistry. J. Chem. Soc., Chem. Commun. 2209–2210.

Chang, C. D., Felix, A. M., Jimenez, M. H., and Meienhofer, J. (1980a). Solid-phase peptide synthesis of somatostatin using mild base cleavage of N^α-fluorenylmethyloxycarbonylamino acids. Int. J. Peptide Protein Res. 15:485–494.

Chang, C. D., Waki, M., Ahmad, M., Meienhofer, J., Lundell, E. O., and Haug, J. D. (1980b). Preparation and properties of N^α-9-fluorenylmethyloxycarbonylamino acids bearing *tert*-butyl side chain protection. Int. J. Peptide Protein Res. 15:59–66.

Chanh, T. C., Dreesman, G. R., Kanda, P., Linette, G. P., Sparrow, J. T., Ho, D. D., and Kennedy, R. C. (1986). Induction of anti-HIV neutralizing antibodies by synthetic peptides. EMBO J. 5:3065–3071.

Chao, H. G., Bernatowicz, M. S., and Matsueda, G. R. (1993). Preparation and use of the 4-[1-[N-(9-fluorenylmethyloxycarbonyl)-amino]-2-(trimethylsilyl)ethyl]phenoxyacetic acid linkage agent for solid-phase synthesis of C-terminal peptide amides:improved yields of tryptophan-containing peptides. J. Org. Chem. 58:2640–2644.

Chen, S.-T., Weng, C.-S., and Wang, K.-T. (1987). The synthesis of
p-methoxybenzyloxycarbonyl amino acids. J. Chin. Chem. Soc. 34:117–
123.

Chen, S.-T., Wu, S.-H., and Wang, K.-T. (1989). A new synthesis of O-benzyl-
L-threonine. Synth. Commun. 19:3589–3593.

Chhabra, S. R., Hothi, B., Evans, D. J., White, P., Bycroft, B., and Chan, W.
C. (1998). An appraisal of new variants of Dde amine protecting group
for solid phase peptide synthesis. Tetrahedron Lett. 39:1603–1606.

Chillemi, F., and Merrifield, R. B. (1969). Use of N^{im}-dinitrophenylhistidine in
the solid-phase synthesis of the tricosapeptides 124–146 of human hemo-
globin β chain. Biochemistry 8:4344–4346.

Choi, H., and Aldrich, J. V. (1993). Comparison of methods for the Fmoc
solid-phase synthesis and cleavage of a peptide containing both tryptophan and arginine. Int. J. Peptide Protein Res. 42:58–63.

Christiansen-Brams, I., Meldal, M., and Bock, K. (1993). Silyl protection in
the solid phase synthesis of N-linked neoglycopeptides: One step deprotection of fully protected neoglycopeptides. Tetrahedron Lett. 34:3315–
3318.

Chun, R., Glabe, C. G., and Fan, H. (1990). Chemical synthesis of biologically
active tat trans-activating protein of human immunodeficiency virus type
1. J. Virol. 64:3074–3077.

Clark-Lewis, I., and Kent, S. (1989). Chemical synthesis, purification, and
characterization of peptides and proteins. In Receptor Biochemistry and
Methodology, Vol. 14: The Use of HPLC in Protein Purification and
Characterization, A. R. Kerlavage, J. C. Venter and L. C. Harrison,
eds., New York, Alan R. Liss, pp. 43–75.

Clark-Lewis, I., Aebersold, R., Ziltener, H., Schrader, J. W., Hood, L. E., and
Kent, S. B. (1986). Automated chemical synthesis of a protein growth
factor for hemopoietic cells, interleukin-3. Science 231:134–139.

Colombo, R., Atherton, E., Sheppard, R. C., and Woolley, V. (1983). 4-
Chloromethylphenoxyacetyl polystyrene and polyamide supports for
solid-phase peptide synthesis. Int. J. Peptide Protein Res. 21:118–126.

Colombo, R., Colombo, F., and Jones, J. H. (1984). Acid-labile histidine side-
chain protection: The π-benzyloxymethyl group. J. Chem. Soc., Chem.
Commun. 292–293.

Cook, R. M., and Hudson, D. (1996). Chemically modified polyolefin particles
for biomolecule synthesis, analysis and display. In Peptides: Chemistry,
Structure and Biology, P. T. P. Kaumaya and R. S. Hodges, eds.,
Kingswinford, Mayflower Scientific Ltd., pp. 39–41.

Cook, R. M., and Hudson, D. (1997). Oxidized polyethylene or polypropylene
particulate supports. U.S. Patent 5,679,539, 1997.

Coste, J., Le-Nguyen, D., and Castro, B. (1990). PyBOP: A new peptide
coupling reagent devoid of toxic by-product. Tetrahedron Lett. 31:205–
208.

Cruz, L. J., Kupryszewski, G., LeCheminant, G. W., Gray, W. R., Olivera, B.
M., and Rivier, J. (1989). μ-Conotoxin GIIIA, a peptide ligand for muscle
sodium channels: Chemical synthesis, radiolabeling, and receptor characterization. Biochemistry 28:3437–3442.

Dalcol, I., Rabanal, F., Ludevid, M.-D., Albericio, F., and Giralt, E. (1995). Convergent solid phase peptide synthesis: An efficient approach to the synthesis of highly repetitive protein domains. J. Org. Chem. 60:7575–7581.

Dawson, P. E., and Kent, S. B. H. (1993). Convenient total synthesis of a 4-helix TASP molecule by chemoselective ligation. J. Am. Chem. Soc. 115:7263–7266.

Dawson, P. E., Muir, T. W., Clark-Lewis, I., and Kent, S. B. H. (1994). Synthesis of proteins by native chemical ligation. Science 266:776–779.

Dawson, P. E., Churchill, M. J., Ghadiri, M. R., and Kent, S. B. H. (1997). Modulation of reactivity in native chemical ligation through the use of thiol additives. J. Am. Chem. Soc. 119:4325–4329.

de Bont, H. B.A., van Boom, J. H., and Liskamp, R. M.J. (1990). Automatic synthesis of phophopeptides by phosphorylation on the solid phase. Tetrahedron Lett. 31:2497–2500.

de la Torre, B. G., Torres, J. L., Bardají, E., Clapés, P., Xaus, N., Jorba, X., Calvet, S., Albericio, F., and Valencia, G. (1990). Improved method for the synthesis of O-glycosylated Fmoc amino acids to be used in solid-phase glycopeptide synthesis. J. Chem. Soc., Chem. Commun. 965–967.

de Lisle Milton, R. C., Becker, E., Milton, S. C. F., Baxter, J. E. J., and Elsworth, J. F. (1987). Improved purities for Fmoc-amino acids from Fmoc-ONSu. Int. J. Peptide Protein Res. 30:431–432.

de Lisle Milton, R. C., Milton, S. C. F., and Kent, S. B. H. (1992). Total chemical synthesis of α D-enzyme: The enantiomers of HIV-1 protease show demonstration of reciprocal chiral substrate specificity. Science 256:1445–1448.

de Rocquigny, H., Ficheux, D., Gabus, C., Fournié-Zaluski, M.-C., Darlix, J.-L., and Roques, B. P. (1991). First large scale chemical synthesis of the 72 amino acid HIV-1 nucleocapsid protein Ncp7 in an active form. Biochem. Biophys. Res. Commun. 180:1010–1018.

Deen, C., Claassen, E., Gerritse, K., Zegers, N. D., and Boersma, W. J. A. (1990). A novel carbodiimide coupling method for synthetic peptides: Enhanced anti-peptide antibody responses. J. Immunol. Methods 129:119–125.

DeGrado, W. F., and Kaiser, E. T. (1980). Polymer-bound oxime esters as supports for solid-phase peptide synthesis: Preparation of protected peptide fragments. J. Org. Chem. 45:1295–1300.

DeGrado, W. F., and Kaiser, E. T. (1982). Solid-phase synthesis of protected peptides on a polymer-bound oxime:Preparation of segments comprising the sequence of a cytotoxic 26-peptide analogue. J. Org. Chem. 47:3258–3261.

DiMarchi, R. D., Tam, J. P., Kent, S. B. H., and Merrifield, R. B. (1982). Weak acid-catalyzed pyrrolidone carboxylic acid formation from glutamine during solid phase peptide synthesis. Int. J. Peptide Protein Res. 19:88–93.

Dölling, R., Beyermann, M., Haenel, J., Kernchen, F., Krause, E., Franke, P., Brudel, M., and Bienert, M. (1994). Piperidine-mediated side product formation for Asp(OBu^t)-containing peptides. J. Chem. Soc., Chem. Commun. 853–854.

Dorman, L. C., Nelson, D. A., and Chow, R. C. L. (1972). Solid phase synthesis of glutamine-containing peptides. In *Progress in Peptide Research*, Vol. 2, S. Lande, ed., New York, Gordon and Breach, pp. 65–68.

Dourtoglou, V., Gross, B., Lambropoulou, V., and Zioudrou, C. (1984). *O*-benzotriazolyl-*N*,*N*,*N'*,*N'*-tetramethyluronium hexafluorophosphate as coupling reagent for the synthesis of peptides of biological interest. Synthesis 572–574.

Drijfhout, J. W., and Bloemhoff, W. (1988). Capping with *O*-sulfobenzoic acid cyclic anhydride (OSBA) in solid-phase peptide synthesis enables facile product purification. In *Peptide Chemistry 1987*, T. Shiba and S. Sakakibara, eds., Protein Research Foundation, Osaka, pp. 191–194.

Drijfhout, J. W., and Bloemhoff, W. (1991). A new synthetic functionalized antigen carrier. Int. J. Peptide Protein Res. 37:27–32.

Drijfhout, J. W., Perdijk, E. W., Weijer, W. J., and Bloemhoff, W. (1988). Controlled peptide-protein conjugation by means of 3-nitro-2-pyridine-sulfenyl protection-activation. Int. J. Peptide Protein Res. 32:161–166.

Edwards, W. B., Fields, C. G., Anderson, C. J., Pajeau, T. S., Welch, M. J., and Fields, G. B. (1994). Generally applicable, convenient solid-phase synthesis and receptor affinities of octreotide analogs. J. Med. Chem. 37:3749–3757.

Egner, B. J., Cardno, M., and Bradley, M. (1995). Linkers for combinatorial chemistry and reaction analysis using solid phase in situ mass spectrometry. J. Chem. Soc., Chem. Commun. 2163–2164.

Eichler, J., Beyermann, M., and Bienert, M. (1989). Application of cellulose paper as support material in simultaneous solid phase peptide synthesis. Collect. Czech. Chem. Commun. 54:1746–1752.

Elofsson, M., Salvador, L. A., and Kihlberg, J. (1997). Preparation of TN and sialyl TN building blocks used in Fmoc-solid phase synthesis of glycopeptide fragments from HIV GP120. Tetrahedron 53:369–390.

Englebretsen, D. R., and Harding, D. R. K. (1994). Fmoc SPPS using Perloza™ beaded cellulose. Int. J. Peptide Protein Res. 43:546–554.

Englebretsen, D. R., Garnham, B. G., Bergman, D. A., and Alewood, P. F. (1995). A novel thioether linker: Chemical synthesis of a HIV-1 protease analogue by thioether ligation. Tetrahedron Lett. 36:8871–8874.

Epton, R., Goddard, P., and Ivin, K. J. (1980). Gel phase [13]C n.m.r. spectroscopy as an analytical method in solid (gel) phase peptide synthesis. Polymer 21:1367–1371.

Epton, R., Wellings, D. A., and Williams, A. (1987). Perspectives in ultra-high load solid (gel) phase peptide synthesis. Reactive Polymers 6:143–157.

Erickson, B. W., and Merrifield, R. B. (1973a). Acid stability of several benzylic protecting groups used in solid-phase peptide synthesis: Rearrangement of *O*-benzyltyrosine to 3-benzyltyrosine. J. Am. Chem. Soc. 95:3750–3756.

Erickson, B. W., and Merrifield, R. B. (1973b). Use of chlorinated benzyloxycarbonyl protecting groups to eliminate N^ε-branching at lysine during solid-phase peptide synthesis. J. Am. Chem. Soc. 95:3757–3763.

Erickson, B. W., and Merrifield, R. B. (1976). Solid-phase peptide synthesis. In *The Proteins*, Vol. II, 3rd edn., H. Neurath and R. L. Hill, eds., New York, Academic Press, pp. 255–527.

Eritja, R., Zichler-Martin, J. P., Walker, P. A., Lee, T. D., Legesse, K., Albericio, F., and Kaplan, B. E. (1987). On the use of S-t-butylsulphenyl group for protection of cysteine in solid-phase peptide synthesis using Fmoc-amino acids. Tetrahedron 43:2675–2680.

Eritja, R., Robles, J., Fernandez-Forner, D., Albericio, F., Giralt, E., and Pedroso, E. (1991). NPE-resin, a new approach to the solid-phase synthesis of protected peptides and oligonucleotides I: Synthesis of the supports and their application to oligonucleotide synthesis. Tetrahedron Lett. 32:1511–1514.

Fairwell, T., Hospattankar, A. V., Ronan, R., Brewer, Jr., H. B., Chang, J. K., Shimizu, M., Zitzner, L., and Arnaud, C. D. (1983). Total solid-phase synthesis, purification, and characterization of human parathyroid hormone-(1-84). Biochemistry 22:2691–2697.

Feinberg, R. S., and Merrifield, R. B. (1975). Modification of peptides containing glutamic acid by hydrogen-fluoride-anisole mixtures: γ-Acylation of anisole or the glutamyl nitrogen. J. Am. Chem. Soc. 97: 3485–3496.

Felix, A. M., Jiminez, M. H., Vergona, R., and Cohen, M. R. (1973). Synthesis and biological studies of novel bradykinin analogues. Int. J. Peptide Protein Res. 5:201–206.

Felix, A. M., Wang, C.-T., Heimer, E. P., and Fournier, A. (1988a). Applications of BOP reagent in solid phase synthesis II: Solid phase side-chain cyclization using BOP reagent. Int. J. Peptide Protein Res. 31:231–238.

Felix, A. M., Heimer, E. P., Wang, C.-T., Lambros, T. J., Fournier, A., Mowles, T. F., Maines, S., Campbell, R. M., Wegrzynski, B. B., Toome, V., Fry, D., and Madison, V. S. (1988b). Synthesis, biological activity and conformational analysis of cyclic GRF analogs. Int. J. Peptide Protein Res. 32:441–454.

Feng, Y., Melacini, G., Taulane, J. P., and Goodman, M. (1996a). Acetyl-terminated and template-assembled collagen-based polypeptides composed of Gly-Pro-Hyp sequences 2: Synthesis and conformational analysis by circular dichroism, ultraviolet absorbance, and optical rotation. J. Am. Chem. Soc. 118:10351–10358.

Feng, Y., Melacini, G., Taulane, J. P., and Goodman, M. (1996b). Collagen-based structures containing the peptoid residue N-isobutylglycine (Nleu): Synthesis and biophysical studies of Gly-Pro-Nleu sequences by circular dichroism, ultraviolet absorbance, and optical rotation. Biopolymers 39:859–872.

Feng, Y., Melacini, G., and Goodman, M. (1997). Collagen-based structures containing the peptoid residue N-isobutylglycine (Nleu): Synthesis and biophysical studies of Gly-Nleu-Pro sequences by circular dichroism and optical rotation. Biochemistry 36:8716–8724.

Ferrer, M., Woodward, C., and Barany, G. (1992). Solid-phase synthesis of bovine pancreatic trypsin inhibitor (BPTI) and two analogues. Int. J. Peptide Protein Res. 40:194–207.

Ferrer, M., Barany, G., and Woodward, C. (1995). Partially folded, molten globule and molten coil states of bovine pancreatic trypsin inhibitor. Nature Struct. Biol. 2:211–217.

Ferrer, T., Nicolás, E., and Giralt, E. (1999). Application of the disulfide trapping approach to explain the antiparallel assembly of dimeric rabbit uteroglobin: A preliminary study using short peptide models. Lett. Pept. Sci. 6:165–172.

Fields, C. G., and Fields, G. B. (1990). New approaches to prevention of side reactions in Fmoc solid phase peptide synthesis. In *Peptides: Chemistry, Structure, and Biology*, J. E. Rivier and G. R. Marshall, eds., Leiden, Escom, pp. 928–930.

Fields, C. G., and Fields, G. B. (1993). Minimization of tryptophan alkylation following 9-fluorenylmethoxycarbonyl solid-phase peptide synthesis. Tetrahedron Lett. 34:6661–6664.

Fields, C. G., Fields, G. B., Noble, R. L., and Cross, T. A. (1989). Solid phase peptide synthesis of [^{15}N]-gramicidins A, B, and C and high performance liquid chromatographic purification. Int. J. Peptide Protein Res. 33:298–303.

Fields, C. G., Lloyd, D. H., Macdonald, R. L., Otteson, K. M., and Noble, R. L. (1991). HBTU activation for automated Fmoc solid-phase peptide synthesis. Peptide Res. 4:95-101.

Fields, C. G., Loffet, A., Kates, S. A., and Fields, G. B. (1992). The development of high-performance liquid chromatographic analysis of allyl and allyloxycarbonyl side-chain-protected phenylthiohydantoin-amino acids. Anal. Biochem. 203:245–251.

Fields, C. G., Lovdahl, C. M., Miles, A. J., Matthias Hagen, V. L., and Fields, G. B. (1993a). Solid-phase synthesis and stability of triple-helical peptide incorporating native collagen sequences. Biopolymers 33:1695–1707.

Fields, C. G., Mickelson, D. J., Drake, S. L., McCarthy, J. B., and Fields, G. B. (1993b). Melanoma cell adhesion and spreading activities of a synthetic 124-residue triple-helical "mini-collagen." J. Biol. Chem. 268:14153–14160.

Fields, C. G., VanDrisse, V. L., and Fields, G. B. (1993c). Edman degradation sequence analysis of resin-bound peptides synthesized by 9-fluorenyl-methoxycarbonyl chemistry. Peptide Res. 6:39–46.

Fields, G. B., ed. (1997). Solid-phase peptide synthesis. Methods Enzymol. 289:1–780.

Fields, G. B., and Fields, C. G. (1991). Solvation effects in solid-phase peptide synthesis. J. Am. Chem. Soc. 113:4202–4207.

Fields, G. B., and Fields, C. G. (1992). Optimization strategies for Fmoc solid phase peptide synthesis: Synthesis of triple-helical collagen-model peptides. In *Innovation and Perspectives in Solid Phase Synthesis—Peptides, Polypeptides and Oligonucleotides—1992*, R. Epton, ed., Andover, Intercept, pp. 153–162.

Fields, G. B., and Noble, R. L. (1990). Solid phase peptide synthesis utilizing 9-fluorenylmethoxycarbonyl amino acids. Int. J. Peptide Protein Res. 35:161–214.

Fields, G. B., Van Wart, H. E., and Birkedal-Hansen, H. (1987). Sequence specificity of human skin fibroblast collagenase: Evidence for the role of collagen structure in determining the collagenase cleavage site. J. Biol. Chem. 262:6221–6226.

Fields, G. B., Fields, C. G., Petefish, J., Van Wart, H. E., and Cross, T. A. (1988). Solid phase synthesis and solid state NMR spectroscopy of [15N-Ala3]-Val-gramicidin A. Proc. Natl. Acad. Sci. USA 85:1384–1388.

Fields, G. B., Otteson, K.M, Fields, C. G., and Noble, R. L. (1990). The versatility of solid phase peptide synthesis. In *Innovation and Perspectives in Solid Phase Synthesis*, R. Epton, ed., Birmingham, UK, Solid Phase Conference Coordination, Ltd., pp. 241–260.

Fields, G. B., Tian, Z., and Barany, G. (1992). Principles and practice of solid-phase peptide synthesis. In *Synthetic Peptides: A User's Guide*, G. A. Grant, ed., New York, W. H. Freeman and Co., pp. 77–183.

Fields, G. B., Carr, S. A., Marshak, D. R., Smith, A. J., Stults, J. T., Williams, L. C., Williams, K. R., and Young, J. D. (1993). Evaluation of peptide synthesis as practiced in 53 different laboratories. In *Techniques in Protein Chemistry*, Vol. IV, R. H. Angeletti, ed., San Diego, Calif., Academic Press, pp. 229–238.

Filira, F., Biondi, L., Cavaggion, F., Scolaro, B., and Rocchi, R. (1990). Synthesis of O-glycosylated tuftsins by utilizing threonine derivatives containing an unprotected monosaccharide moiety. Int. J. Peptide Protein Res. 36:86–96.

Findeis, M. A., and Kaiser, E. T. (1989). Nitrobenzophenone oxime based resins for the solid-phase synthesis of protected peptide segments. J. Org. Chem. 54:3478–3482.

Finn, F. M., and Hofmann, K. (1976). The synthesis of peptides by solution methods with emphasis on peptide hormones. In *The Proteins*, Vol. II, 3rd edn., H. Neurath and R. L. Hill, eds., New York, Academic Press, pp. 105–253.

Fischer, P. M., Comis, A., and Howden, M. E. H. (1989). Direct immunization with synthetic peptidyl-polyamide resin: Comparison with antibody production from free peptide and conjugates with carrier proteins. J. Immunol. Methods 118:119–123.

Fitch, W. L., Detre, G., and Holmes, C. P. (1994). High-resolution ^{1}H-NMR in solid phase organic synthesis. J. Org. Chem. 59:7955–7956.

Fitzgerald, M. C., Harris, K., Shevlin, C. G., and Siuzdak, G. (1996). Direct characterization of solid phase resin-bound molecules by mass spectrometry. Bioorg. Med. Chem. Lett. 6:979–982.

Flegel, M., and Sheppard, R. C. (1990). A sensitive, general method for quantitative monitoring of continuous flow solid phase peptide synthesis. J. Chem. Soc., Chem. Commun. 536–538.

Fletcher, A. R., Jones, J. H., Ramage, W. I., and Stachulski, A. V. (1979). The use of the N(π)-phenacyl group for the protection of the histidine side chain in peptide synthesis. J. Chem. Soc. Perkin Trans. I:2261–2267.

Flörsheimer, A., and Riniker, B. (1991). Solid-phase synthesis of peptides with the highly acid-sensitive HMPB linker. In *Peptides 1990*, E. Giralt and D. Andreu, eds., Leiden, Escom, pp. 131–133.

Fodor, S. P. A., Read, J. L., Pirrung, M. C., Stryer, L., Lu, A. T., and Solas, D. (1991). Light-directed, spatially addressable parallel chemical synthesis. Science 251:767–773.

Forest, M., and Fournier, A. (1990). BOP reagent for the coupling of pGlu and Boc-His(Tos) in solid phase peptide synthesis. Int. J. Peptide Protein Res. 35:89–94.

Fournier, A., Wang, C.-T., and Felix, A. M. (1988). Applications of BOP reagent in solid phase peptide synthesis: Advantages of BOP reagent for difficult couplings exemplified by a synthesis of [Ala^{15}]-GRF(1-29)-NH$_2$. Int. J. Peptide Protein Res. 31:86–97.

Fournier, A., Danho, W., and Felix, A. M. (1989). Applications of BOP reagent in solid phase peptide synthesis III: Solid phase peptide synthesis with unprotected aliphatic and aromatic hydroxyamino acids using BOP reagent. Int. J. Peptide Protein Res. 33:133–139.

Fox, J., Newton, R., Heegard, P., and Schafer-Nielsen, C. (1990). A novel method of monitoring the coupling reaction in solid phase synthesis. In Innovation and Perspectives in Solid Phase Synthesis, R. Epton, ed., Birmingham, UK, Solid Phase Conference Coordination, Ltd., pp. 141–153.

Frank, R., and Döring, R. (1988). Simultaneous multiple peptide synthesis under continuous flow conditions on cellulose paper discs as segmental solid supports. Tetrahedron 44:6031–6040.

Frank, R., and Gausepohl, H. (1988). Continuous flow peptide synthesis. In Modern Methods in Protein Chemistry, Vol. 3, H. Tschesche, ed., Berlin, Walter de Gruyter and Co., pp. 41–60.

Frankel, A. D., Biancalana, S., and Hudson, D. (1989). Activity of synthetic peptides from the tat protein of human immunodeficiency virus type 1. Proc. Natl. Acad. Sci. USA 86:7397–7401.

Franzén, H., Grehn, L., and Ragnarsson, U. (1984). Synthesis, properties, and use of N^{in}-Boc-tryptophan derivatives. J. Chem. Soc., Chem. Commun. 1699–1700.

Fujii, N., Otaka, A., Funakoshi, S., Bessho, K., Watanabe, T., Akaji, K., and Yajima, H. (1987). Studies on peptides CLI: Synthesis of cystine-peptides by oxidation of S-protected cysteine-peptides with thallium (III) trifluoroacetate. Chem. Pharm. Bull. 35:2339–2347.

Fujino, M., Wakimasu, M., Shinagawa, S., Kitada, C., and Yajima, H. (1978). Synthesis of the nonacosapeptide corresponding to mammalian glucagon. Chem. Pharm. Bull. 26:539–548.

Fujino, M., Wakimasu, M., and Kitada, C. (1981). Further studies on the use of multi-substituted benzenesulfonyl groups for protection of the guanidino function of arginine. Chem. Pharm. Bull. 29:2825–2831.

Fujiwara, Y., Akaji, K., and Kiso, Y. (1994). Racemization-free synthesis of C-terminal cysteine-peptide using 2-chlorotrityl resin. Chem. Pharm. Bull. 42:724–726.

Fuller, W. D., Cohen, M. P., Shabankareh, M., Blair, R. K., Goodman, M., and Naider, F. R. (1990). Urethane-protected amino acid N-carboxy anhydrides and their use in peptide synthesis. J. Am. Chem. Soc. 112:7414–7416.

Fuller, W. D., Krotzer, N. J., Naider, F. R., Xue, C.-B., and Goodman, M. (1993). Urethane-protected amino acid N-carboxyanhydrides:Stability, reactivity and solubility. In Peptides 1992, C. H. Schneider and A. N. Eberle, eds., Leiden, Escom, pp. 229–230.

Funakoshi, S., Tamamura, H., Fujii, N., Yoshizawa, K., Yajima, H., Miyasaki, K., Funakoshi, A., Ohta, M., Inagaki, Y., and Carpino, L. A. (1988). Combination of a new amide-precursor reagent and trimethylsilyl bromide deprotection for the Fmoc-based solid phase synthesis of human pancreastatin and one of its fragments. J. Chem. Soc., Chem. Commun. 1588–1590.

Futaki, S., Taike, T., Akita, T., and Kitagawa, K. (1990a). A new approach for the synthesis of tyrosine sulphate containing peptides: Use of the p-(methylsulphinyl)benzyl group as a key protecting group of serine. J. Chem. Soc., Chem. Commun. 523–524.

Futaki, S., Yajami, T., Taike, T., Akita, T., and Kitagawa, K. (1990b). Sulphur trioxide/thiol: A novel system for the reduction of methionine sulphoxide. J. Chem. Soc. Perkin Trans. I:653–658.

Gaehde, S. T., and Matsueda, G. R. (1981). Synthesis of N-tert-butoxycarbonyl-(α-phenyl)aminomethylphenoxyacetic acid for use as a handle in solid-phase synthesis of peptide α-carboxamides. Int. J. Peptide Protein Res. 18:451–458.

Gairi, M., Lloyd-Williams, P., Albericio, F., and Giralt, E. (1990). Use of BOP reagent for the suppression of diketopiperazine formation in Boc/Bzl solid-phase peptide synthesis. Tetrahedron Lett. 31:7363–7366.

García-Echeverría, C. (1995). On the use of hydrophobic probes in the chromatographic purification of solid-phase-synthesized peptides. J. Chem. Soc. Chem. Commun. 779–780.

García-Echeverría, C. (1995). Potential pyrophosphate formation upon use of N^{α}-Fmoc-Tyr(PO$_3$H$_2$)-OH in solid-phase peptide synthesis. Lett. Peptide Sci. 2:93–98.

García-Echeverría, C., Albericio, F., Pons, M., Barany, G., and Giralt, E. (1989). Convenient synthesis of a cyclic peptide disulfide: A type II β-turn structural model. Tetrahedron Lett. 30:2441–2444.

Gausepohl, H., Kraft, M., and Frank, R. (1989a). In situ activation of Fmoc-amino acids by BOP in solid phase peptide synthesis. In Peptides 1988, G. Jung and E. Bayer, eds., Berlin, Walter de Gruyter and Co., pp. 241–243.

Gausepohl, H., Kraft, M., and Frank, R. W. (1989b). Asparagine coupling in Fmoc solid phase peptide synthesis. Int. J. Peptide Protein Res. 34:287–294.

Gausepohl, H., Kraft, M., Boulin, Ch., and Frank, R. W. (1990). Automated multiple peptide synthesis with BOP activation. In Peptides: Chemistry, Structure and Biology, J. E. Rivier and G. R. Marshall, eds., Leiden, Escom, pp. 1003–1004.

Gausepohl, H., Kraft, M., Boulin, C., and Frank, R. W. (1991). A multiple reaction system for automated simultaneous peptide synthesis. In Peptides 1990, E. Giralt and D. Andreu, eds., Leiden, Escom, pp. 206–207.

Gausepohl, H., Pieles, U., and Frank, R. W. (1992). Schiffs base analog formation during in situ activation by HBTU and TBTU. In Peptides: Chemistry and Biology, J. A. Smith and J. E. Rivier, eds., Leiden, Escom, pp. 523–524.

Geiser, T., Beilan, H., Bergot, B. J., and Otteson, K. M. (1988). Automation of solid-phase peptide synthesis. In Macromolecular Sequencing and

Synthesis: Selected Methods and Applications, D. H. Schlesinger, ed., New York, Alan R. Liss, pp. 199–218.

Gesellchen, P. D., Rothenberger, R. B., Dorman, D. E., Paschal, J. W., Elzey, T. K., and Campbell, C. S. (1990). A new side reaction in solid-phase peptide synthesis: Solid support-dependent alkylation of tryptophan. In *Peptides: Chemistry, Structure and Biology*, J. E. Rivier and G. R. Marshall, eds., Leiden, Escom, pp. 957–959.

Gesquiére, J.-C., Diesis, E., and Tartar, A. (1990). Conversion of N-terminal cysteine to thiazolidine carboxylic acid during hydrogen fluoride deprotection of peptides containing N-π-Bom protected histidine. J. Chem. Soc., Chem. Commun. 1402–1403.

Gesquiére, J.-C., Najib, J., Diesis, E., Barbry, D., and Tartar, A. (1992). Investigations of side reactions associated with the use of Bom and Bum groups for histidine protection. In *Peptides: Chemistry and Biology*, J. A. Smith and J. E. Rivier, eds., Leiden, Escom, pp. 641–642.

Geysen, H. M., Meloen, R. H., and Barteling, S. J. (1984). Use of peptide synthesis to probe viral antigens for epitopes to a resolution of a single amino acid. Proc. Natl. Acad. Sci. USA 81:3998–4002.

Giralt, E., Rizo, J., and Pedroso, E. (1984). Application of gel-phase ^{13}C-NMR to monitor solid phase peptide synthesis. Tetrahedron 40:4141–4152.

Giralt, E., Albericio, F., Pedroso, E., Granier, C., and van Rietschoten, J. (1982). Convergent solid phase peptide synthesis II: Synthesis of the 1–6 apamin protected segment on a Nbb-resin: synthesis of apamin. Tetrahedron 38:1193–1208.

Giralt, E., Eritja, R., Pedroso, E., Granier, C., and van Rietschoten, J. (1986). Convergent solid phase peptide synthesis III: Synthesis of the 44-52 protected segment of the toxin II of *Androctonus australis* Hector. Tetrahedron 42:691–698.

Gisin, B. F. (1973). The preparation of Merrifield-resins through total esterification with cesium salts. Helv. Chim. Acta 56:1476–1482.

Gisin, B. F., and Merrifield, R. B. (1972). Carboxyl-catalyzed intramolecular aminolysis: A side reaction in solid-phase peptide synthesis. J. Am. Chem. Soc. 94:3102–3106.

Goddard, P., McMurray, J. S., Sheppard, R. C., and Emson, P. (1988). A solubilisable polymer support suitable for solid phase peptide synthesis and for injection into experimental animals. J. Chem. Soc., Chem. Commun. 1025–1027.

Gooding, O. W., Baudart, S., Deegan, D. L., Heisler, K., Labadie, J. W., Newcomb, W. S., Porco, J. A., Jr., and van Eikeren, P. (1999). On the development of new poly(styrene-oxyethylene) graft copolymer resin supports for solid-phase organic synthesis. J. Comb. Chem. 1:113–122.

Goodman, M., Felix, A., Moroder, L., and and Toniolo, C., eds. (2001). *Houben-Weyl Volume E22b: Synthesis of Peptidomimetics*, Georg Thieme Verlag, Stuttgart, Germany.

Grab, B, Miles, A. J., Furcht, L. T., and and Fields, G. B. (1996). Promotion of fibroblast adhesion by triple-helical peptide models of type I collagen-derived sequences. J. Biol. Chem. 271:12234-12240.

Grandas, A., Jorba, X., Giralt, E., and Pedroso, E. (1989). Anchoring of Fmoc-amino acids to hydroxymethyl resins. Int. J. Peptide Protein Res. 33:386–390.

Gras-Masse, H., Ameisen, J. C., Boutillon, C., Gesquiére, J. C., Vian, S., Neyrinck, J. L., Drobecq, H., Capron, A., and Tartar, A. (1990). A synthetic protein corresponding to the entire vpr gene product from the human immunodeficiency virus HIV-1 is recognized by antibodies from HIV-infected patients. Int. J. Peptide Protein Res. 36:219–226.

Gray, W. R., Luque, A., Galyean, R., Atherton, E., Sheppard, R. C., Stone, B. L., Reyes, A., Alford, J., McIntosh, M., Olivera, B. M., Cruz, L. J., and Rivier, J. (1984). Contoxin GI: Disulfide bridges, synthesis, and preparation of iodinated derivatives. Biochemistry 23:2796–2802.

Green, J., Ogunjobi, O. M., Ramage, R., Stewart, A. S. J., McCurdy, S., and Noble, R. (1988). Application of the N^G-(2,2,5,7,8-pentamethylchroman-6-sulphonyl) derivative of Fmoc-arginine to peptide synthesis. Tetrahedron Lett. 29:4341–4344.

Green, M., and Berman, J. (1990). Preparation of pentafluorophenyl esters of Fmoc protected amino acids with pentafluorophenyl trifluoroacetate. Tetrahedron Lett. 31:5851–5852.

Greene, T. W. (1991). *Protective Groups in Organic Synthesis*, New York, John Wiley and Sons.

Gross, H., and Bilk, L. (1968). Zur reaktion von N-hydroxysuccinimid mit dicyclohexylcarbodiimid. Tetrahedron 24:6935–6939.

Guibé, F., Dangles, O., Balavoine, G., and Loffet, A. (1989). Use of an allylic anchor group and of its palladium catalyzed hydrostannolytic cleavage in the solid phase synthesis of protected peptide fragments. Tetrahedron Lett. 30:2641–2644.

Gutte, B., ed. (1995). *Peptides: Synthesis, Structures, and Applications*, Orlando, Fla., Academic Press.

Gutte, B., and Merrifield, R. B. (1971). The synthesis of ribonuclease A. J. Biol. Chem. 246:1922–1941.

Guttmann, S., and Boissonnas, R. A. (1959). Synthése de l'α-mélanotropine (α-MSH) de porc. Helv. Chim. Acta 42:1257–1264.

Guy, C. A., and Fields, G. B. (1997). Trifluoroacetic acid cleavage and deprotection of resin-bound peptides following synthesis by Fmoc chemistry. Methods Enzymol. 289:67–83.

Hahn, K. W., Klis, W. A., and Stewart, J. M. (1990). Design and synthesis of a peptide having chymotrypsin-like esterase activity. Science 248:1544–1547.

Hammer, R. P., Albericio, F., Gera, L., and Barany, G. (1990). Practical approach to solid-phase synthesis of C-terminal peptide amides under mild conditions based on photolysable anchoring linkage. Int. J. Peptide Protein Res. 36:31–45.

Han, Y., Solé, N., Tejbrant, T., and Barany, G. (1996a). Novel N^m-xanthenyl protecting groups for asparagine and glutamine, and applications to N^α-9-Fluorenylmethyloxycarbonyl (Fmoc) solid-phase peptide synthesis. Peptide Res. 9:166–173.

Han, Y., Bontems, S. L., Heygens, P., Munson, M. C., Minor, C. A., Kates, S. A., Albericio, F., and Barany, G. (1996b). Preparation and application of xanthenylamide (XAL) handles for solid phase synthesis of C-terminal peptide amides under particularly mild conditions. J. Org. Chem. 61:6326–6339.

Han, Y., Albericio, F., and Barany, G. (1997a). Occurrence and minimization of cysteine racemization during stepwise solid-phase peptide synthesis. J. Org. Chem. 62:4307–4312.

Han, Y., Vágner, J., and Barany, G. (1997b). A new side-chain anchoring strategy for solid-phase synthesis of peptide acids with C-terminal cysteine. In *Innovation and Perspectives in Solid Phase Synthesis and Combinatorial Chemical Libraries: Biomedical and Biological Applications, 1996*, R. Epton, ed., Kingswinford, Mayflower Scientific Ltd., pp. 385–388.

Hansen, P. R., Holm, A., and Houen, G. (1993). Solid-phase peptide synthesis on proteins. Int. J. Peptide Protein Res. 41:237–245.

Hargittai, B., and Barany, G. (1999). Controlled syntheses of natural and disulfide-mispaired regioisomers of α-conotoxin SI. J. Peptide Res. 54:468–479.

Hargittai, B., Solé, N. A., Groebe, D. R., Abramson, S. N., and Barany, G. (2000). Chemical syntheses and biological activities of lactam analogues of α-conotoxin SI. J. Med. Chem. 43:4787–4792.

Harrison, J. L., Petrie, G. M., Noble, R. L., Beilan, H. S., McCurdy, S. N., and Culwell, A. R. (1989). Fmoc chemistry: Synthesis, kinetics, cleavage, and deprotection of arginine-containing peptides. In *Techniques in Protein Chemistry*, T. E. Hugli, ed., San Diego, Calif., Academic Press, pp. 506–516.

Heimer, E. P., Chang, C.-D., Lambros, T., and Meienhofer, J. (1981). Stable isolated symmetrical anhydrides of N^α-fluorenylmethyloxycarbonylamino acids in solid-phase peptide synthesis. Int. J. Peptide Protein Res. 18:237–241.

Hellermann, H., Lucas, H.-W., Maul, J., Pillai, V. N. R., and Mutter, M. (1983). Poly(ethylene glycol)s grafted onto crosslinked polystyrenes, 2: Multidetachably anchored polymer systems for the synthesis of solubilized peptides. Makromol. Chem. 184:2603–2617.

Hendrix, J. C., and Lansbury, P. T. (1992). Synthesis of a protected peptide corresponding to residues 1–25 of the β-amyloid protein of Alzheimer's disease. J. Org. Chem. 57:3421–3426.

Hendrix, J. C., Jarrett, J. T., Anisfeld, S. T., and Lansbury, P. T. (1992). Studies related to a convergent fragment-coupling approach to peptide synthesis using the Kaiser oxime resin. J. Org. Chem. 57:3414–3420.

Henkel, W., Vogl, T., Echner, H., Voelter, W., Urbanke, C., Schleuder, D., and Rauterberg, J. (1999). Synthesis and folding of native collagen III model peptides. Biochemistry 38:13610-13622.

Hiskey, R. G. (1981). Sulfhydryl Group Protection in Peptide Synthesis. In *The Peptides*, Vol. 3, E. Gross and J. Meienhofer, eds., New York, Academic Press, pp. 137–167.

Hodges, R. S., and Merrifield, R. B. (1975). Monitoring of solid phase peptide synthesis by an automated spectrophotometric picrate method. Anal. Biochem. 65:241–272.

Hoeprich, P. D., Jr., Langton, B. C., Zhang, J.-w., and Tam, J. P. (1989). Identification of immunodominant regions of transforming growth factor α: implications of structure and function. J. Biol. Chem. 264:19086–19091.

Holmes, C. P., and Jones, D. G. (1995). Reagents for combinatorial organic synthesis: Development of a new o-nitrobenzyl photolabile linker for solid phase synthesis. J. Org. Chem. 60:2318–2319.

Horn, M., and Novak, C. (1987). A monitoring and control chemistry for solid-phase peptide synthesis. Am. Biotech. Lab. 5 (September/October):12–21.

Houghten, R. A., and Li, C. H. (1979). Reduction of sulfoxides in peptides and proteins. Anal. Biochem. 98:36–46.

Houghten, R. A., DeGraw, S. T., Bray, M. K., Hoffmann, S. R., and Frizzell, N. D. (1986). Simultaneous multiple peptide synthesis: The rapid preparation of large numbers of discrete peptides for biological, immunological, and methodological studies. BioTechniques 4:522–528.

Hruby, V. J., Al-Obeidi, F., Sanderson, D. G., and Smith, D. D. (1990). Synthesis of cyclic peptides by solid phase methods. In *Innovation and Perspectives in Solid Phase Synthesis*, R. Epton, ed., Birmingham, UK, Solid Phase Conference Coordination, Ltd., pp. 197–203.

Hudson, D. (1988). Methodological implications of simultaneous solid-phase peptide synthesis 1: Comparison of different coupling procedures. J. Org. Chem. 53:617–624.

Hudson, D. (1990a). New logically developed active esters for solid-phase peptide synthesis. In *Peptides: Chemistry, Structure and Biology*, J. E. Rivier and G. R. Marshall, eds., Leiden, Escom, pp. 914–915.

Hudson, D. (1990b). Methodological implications of simultaneous solid-phase peptide synthesis: A comparison of active esters. Peptide Res. 3:51–55.

Hudson, D. (1990c). Protecting groups for asparagine and glutamine in peptide synthesis. US Patent 4,935,536.

Hudson, D. (1999). Matrix supported synthetic transformations. J. Comb. Chem. Part I. 1:333-360; Part II. 1:403–457.

Hudson, D., Kain, D., and Ng, D. (1986). High yielding fully automatic synthesis of cecropin A amide & analogues. In *Peptide Chemistry 1985*, Y. Kiso, ed., Osaka, Protein Research Foundation, pp. 413–418.

Hyde, C., Johnson, T., and Sheppard, R. C. (1992). Internal aggregation during solid phase peptide synthesis: Dimethyl sulfoxide as a powerful dissociating solvent. J. Chem. Soc., Chem. Commun. 1573–1575.

Hyde, C., Johnson, T., Owen, D., Quibell, M., and Sheppard, R. C. (1994). Some "difficult sequences" made easy: A study of interchain association in solid-phase peptide synthesis. Int. J. Peptide Protein Res. 43:431–440.

Ikeda, S., Yokota, T., Watanabe, K., Kan, M., Takahashi, Y., Ichikawa, T., Takahashi, K., and Matsueda, R. (1986). Solid phase peptide synthesis by the use of 3-nitro-2-pyridinesulfenyl (Npys)-amino acids. In *Peptide Chemistry 1985*, Y. Kiso, ed., Osaka, Protein Research Foundation, pp. 115–120.

Ingenito, R., Bianchi, E., Fattori, D., and Pessi, A. (1999). Solid phase synthesis of peptide C-terminal thioesters by Fmoc/t-Bu chemistry. J. Am. Chem. Soc. 121:11369-11374.

Ishiguro, T., and Eguchi, C. (1989). Unexpected chain-terminating side reaction caused by histidine and acetic anhydride in solid-phase peptide synthesis. Chem. Pharm. Bull. 37:506–508.

Jaeger, E., Thamm, P., Knof, S., Wünsch, E., Löw, M., and Kisfaludy, L. (1978a). Nebenreaktionen bei peptidsynthesen III: Synthese und charakterisierung von N^{in}-tert-butylierten tryptophan-derivaten. Hoppe-Seyler's Z. Physiol. Chem. 359:1617–1628.

Jaeger, E., Thamm, P., Knof, S., and Wünsch, E. (1978b). Nebenreaktionen bei peptidsynthesen IV: Charakterisierung von C- und C,N-tert-butylierten tryptophan-derivaten. Hoppe-Seyler's Z. Physiol. Chem. 359:1629–1636.

Jaeger, E., Remmer, H. A., Jung, G., Metzger, J., Oberthuer, W., Rücknagel, K. P., Schaefer, W., Sonnenbichler, J., and Zetl, I. (1993). Side reactions in peptide synthesis V: O-sulfonation of serine and threonine during removal of the Pmc and Mtr protecting groups from arginine residues in Fmoc solid phase synthesis. Biol. Chem. Hoppe-Seyler 347:349–362.

Jansson, A. M., Meldal, M., and Bock, K. (1990). The active ester N-Fmoc-3-O-[Ac$_4$-α-D-Man$p$-(1 → 2)-Ac$_3$-α-D-Man$p$-1-]-threonine-O-Pfp as a building block in solid-phase synthesis of an O-linked dimannosyl glycopeptide. Tetrahedron Lett. 31:6991–6994.

Jensen, K. J. Hansen, P. R., Venugopal, D., and Barany, G. (1996). Synthesis of 2-acetamido-2-deoxy-β-D-glucopyranose O-glycopeptides from N-dithiasuccinoyl-protected derivatives. J. Am. Chem. Soc. 118:3148–3155.

Jensen, K. J., Alsina, J., Songster, M. F., Vágner, J., Albericio, F., and Barany, G. (1998). Backbone amide linker (BAL) strategy for solid-phase synthesis of C-terminal-modified and cyclic peptides. J. Am. Chem. Soc. 120:5441–5452.

Johnson, C. R., Biancalana, S., Hammer, R. P., Wright, P. B., and Hudson, D. (1992). New active esters and coupling reagents based on pyrazolinones. In Peptides: Chemistry and Biology, J. A. Smith and J. E. Rivier, eds., Leiden, Escom, pp. 585–586.

Johnson, T., Quibell, M., Owen, D., and Sheppard, R. C. (1993). A reversible protecting group for the amide bond in peptides: Use in the synthesis of "difficult sequences." J. Chem. Soc., Chem. Commun. 369–372.

Jones, J. H., Ramage, W. I., and Witty, M. J. (1980). Mechanism of racemization of histidine derivatives in peptide synthesis. Int. J. Peptide Protein Res. 15:301–303.

Kaiser, E., Colescott, R. L., Bossinger, C. D., and Cook, P. I. (1970). Color test for detection of free terminal amino groups in the solid-phase synthesis of peptides. Anal. Biochem. 34:595–598.

Kaiser, E., Bossinger, C. D., Colescott, R. L., and Olsen, D. B. (1980). Color test for terminal prolyl residues in the solid-phase synthesis of peptides. Anal. Chim. Acta 118:149–151.

Kaiser, E. T., Mihara, H., Laforet, G. A., Kelly, J. W., Walters, L., Findeis, M. A., and Sasaki, T. (1989). Peptide and protein synthesis by segment synthesis-condensation. Science 243:187–192.

Kamber, B., Hartmann, A., Eisler, K., Riniker, B., Rink, H., Sieber, P., and Rittel, W. (1980). The synthesis of cystine peptides by iodine oxidation of S-trityl-cysteine and S-acetamidomethyl-cysteine peptides. Helv. Chim. Acta 63:899–915.

Kanda, P., Kennedy, R. C., and Sparrow, J. T. (1991). Synthesis of polyamide supports for use in peptide synthesis and as peptide-resin conjugates for antibody production. Int. J. Peptide Protein Res. 38:385–391.

Kates, S. A., and Albericio, F., eds. (2000). *Solid-Phase Peptide Synthesis: A Practical Guide*, Marcel Dekker, New York.

Kates, S. A., Daniels, S. B., and Albericio, F. (1993a). Automated allyl cleavage for continuous flow synthesis of cyclic and branched peptides. Anal. Biochem. 212:303–310.

Kates, S. A., Solé, N. A., Johnson, C. R., Hudson, D., Barany, G., and Albericio, F. (1993b). A novel, convenient, three-dimensional orthogonal strategy for solid-phase synthesis of cyclic peptides. Tetrahedron Lett. 34:1549–1552.

Kates, S. A., Solé, N. A., Albericio, F., and Barany, G. (1994). Solid-phase synthesis of cyclic peptides. In *Peptides: Design, Synthesis and Biological Activity*, C. Basava and G. M. Anantharamaiah, eds., Boston, Birkhaeuser, pp. 39–57.

Kates, S. A., Solé, N. A., Beyermann, M., Barany, G., and Albericio, F. (1996). Optimized preparation of deca(L-alanyl)-L-valinamide by 9-fluorenylmethyloxycarbonyl (Fmoc) solid-phase synthesis on polyethylene glycol-polystyrene (PEG-PS) graft supports, with 1,8-diazobicyclo[5. 4.0]-undec-7-ene (DBU) deprotection. Peptide Res. 9:106–113.

Kates, S. A., McGuinness, B. F., Blackburn, C., Griffin, G. W., Solé, N. A., Barany, G., and Albericio, F. (1998). "High-load" polyethylene glycol-polystyrene (PEG-PS) graft supports for solid-phase synthesis. Biopolymers 47:365–380.

Kaumaya, P. T. P., Van Buskirk, A. M., Goldberg, E., and Pierce, S. K. (1992). Design and immunological properties of topographic immunogenic determinants of a protein antigen (LDH-C$_4$) as vaccines. J. Biol. Chem. 267:6338–6346.

Kawasaki, K., Miyano, M., Murakami, T., and Kakimi, M. (1989). Amino acids and peptides XI: Simple preparation of N^α-protected histidine. Chem. Pharm. Bull. 37:3112–3113.

Kearney, T., and Giles, J. (1989). Fmoc peptide synthesis with a continuous flow synthesizer. Am. Biotech. Lab. 7(9), October:34–44.

Kemp, D. S., Fotouhi, N., Boyd, J. G., Carey, R. I., Ashton, C., and Hoare, J. (1988). Practical preparation and deblocking conditions for N-α-(2-(p-biphenylyl)-2-propyloxycarbonyl)-amino acid (N-α-Bpoc-XXX-OH) derivatives. Int. J. Peptide Protein Res. 31:359–372.

Kempe, M., and Barany, G. (1996). CLEAR: A novel family of highly cross-linked polymeric supports for solid-phase peptide synthesis. J. Am. Chem. Soc. 118:7083–7093.

Kennedy, R. C., Dreesman, G. R., Chanh, T. C., Boswell, R. N., Allan, J. S., Lee, T.-H., Essex, M., Sparrow, J. T., Ho, D. D., and Kanda, P. (1987). Use of a resin-bound synthetic peptide for identifying a neutralizing antigenic determinant associated with the human immunodeficiency virus envelope. J. Biol. Chem. 262:5769–5774.

Kenner, G. W., and Seely, J. H. (1972). Phenyl esters for C-terminal protection in peptide synthesis. J. Am. Chem. Soc. 94:3259–3260.

Kenner, G. W., Galpin, I. J., and Ramage, R. (1979). Synthetic studies directed towards the synthesis of a lysozyme analog. In *Peptides: Structure and Biological Function*, E. Gross and J. Meienhofer, eds., Rockford, Ill., Pierce Chemical Co., pp. 431–438.

Kent, J. J., Alewood, P., and Kent, S. B. H. (1991). Peptide synthesis using Fmoc amino acids stored in DMF solution for prolonged periods. In *Twelfth American Peptide Symposium Program and Abstracts*, Cambridge, Mass., Massachusetts Institute of Technology, P-386.

Kent, S. B. H. (1983). Chronic formation of acylation-resistant deletion peptides in stepwise solid phase peptide synthesis: Chemical mechanism, occurrence, and prevention. In *Peptides: Structure and Function*, V. J. Hruby and D. H. Rich, eds., Rockford, Ill., Pierce Chemical Co., pp. 99–102.

Kent, S. B. H. (1988). Chemical synthesis of peptides and proteins. Ann. Rev. Biochem. 57:957–989.

Kent, S. B. H., and Merrifield, R. B. (1978). Preparation and properties of *tert*-butyloxycarbonylaminoacyl-4-(oxymethyl)phenylacetamidomethyl-(Kel F-g-styrene) resin, an insoluble, noncrosslinked support for solid phase peptide synthesis. Isr. J. Chem. 17:243–247.

Kent, S. B. H., and Parker, K. F. (1988). The chemical synthesis of therapeutic peptides and proteins. In *Banbury Report 29: Therapeutic Peptides and Proteins: Assessing the New Technologies*, D. R. Marshak, and D. T. Liu, eds., New York, Cold Spring Harbor, pp. 3–16.

Kent, S. B. H., Riemen, M., LeDoux, M., and Merrifield, R. B. (1982). A study of the Edman degradation in the assessment of the purity of synthetic peptides. In *Methods in Protein Sequence Analysis*, M. Elzinga, ed., Clifton, N.J., Humana Press, pp. 205–213.

Kihlberg, J., Elofsson, M., and Salvador, L. A. (1997). Direct synthesis of glycosylated amino acids from carbohydrate preacetates and Fmoc-amino acids: Solid-phase synthesis of biomedically interesting glycopeptides. Methods Enzymol. 289:245–266.

King, D. S., Fields, C. G., and Fields, G. B. (1990). A cleavage method for minimizing side reactions following Fmoc solid phase peptide synthesis. Int. J. Peptide Protein Res. 36:255–266.

Kirstgen, R., and Steglich, W. (1989). Fmoc-amino acid-TDO esters as reagents for peptide coupling and anchoring in solid phase synthesis. In *Peptides 1988*, G. Jung and E. Bayer, eds., Berlin, Walter de Gruyter & Co., pp. 148–150.

Kirstgen, R., Sheppard, R. C., and Steglich, W. (1987). Use of esters of 2,5-diphenyl-2,3-dihydro-3-oxo-4-hydroxythiophene dioxide in solid phase peptide synthesis: A new procedure for attachment of the first amino acid. J. Chem. Soc., Chem. Commun. 1870–1871.

Kirstgen, R., Olbrich, A., Rehwinkel, H., and Steglich, W. (1988). Ester von *N*-(9-fluorenylmethyloxycarbonyl)aminosäuren mit 4-hydroxy-3-oxo-2,5-diphenyl-2,3-dihydrothiophen-1,1-dioxid (Fmoc-aminosäure-TDO-ester) und ihre verwendung zur festphasenpeptidsynthese. Liebigs Ann. Chem. 437–440.

Kisfaludy, L., and Schön, I. (1983). Preparation and applications of pentafluorophenyl esters of 9-fluorenylmethyloxycarbonyl amino acids for peptide synthesis. Synthesis 325–327.

Kisfaludy, L., Löw, M., Nyéki, O., Szirtes, T., and Schön, I. (1973). Die verwendung von pentafluorophenylestern bei peptid-synthesen. Justus Liebigs Ann. Chem. 1421–1429.

200 Synthetic Peptides

Kiso, Y., Yoshida, M., Tatsumi, T., Kimura, T., Fujiwara, Y., and Akaji, K. (1989). Tetrafluoroboric acid, a useful deprotecting reagent in peptide synthesis. Chem. Pharm. Bull. 37:3432–3434.

Kiso, Y., Yoshida, M., Fujiwara, Y., Kimura, T., Shimokura, M., and Akaji, K. (1990). Trimethylacetamidomethyl (Tacm) group, a new protecting group for the thiol function of cysteine. Chem. Pharm. Bull. 38:673–675.

Kitas, E. A., Perich, J. W., Wade, J. D., Johns, R. B., and Tregear, G. W. (1989). Fmoc-polyamide solid phase synthesis of an O-phosphotyrosine-containing tridecapeptide. Tetrahedron Lett. 30:6229–6232.

Kitas, E. A., Wade, J. D., Johns, R. B., Perich, J. W., and Tregear, G. W. (1991). Preparation and use of N^α-fluorenylmethoxycarbonyl-O-dibenzyl-phosphono-L-tyrosine in continuous flow solid phase peptide synthesis. J. Chem. Soc., Chem. Commun. 338–339.

Kneib-Cordonier, N., Albericio, F., and Barany, G. (1990). Orthogonal solid-phase synthesis of human gastrin-I under mild conditions. Int. J. Peptide Protein Res. 35:527–538.

Knorr, R., Trzeciak, A., Bannwarth, W., and Gillessen, D. (1991). 1,1,3,3-Tetramethyluronium compounds as coupling reagents in peptide and protein chemistry. In Peptides 1990, E. Giralt and D. Andreu, eds., Leiden, Escom, pp. 62–64.

Kochersperger, M. L., Blacher, R., Kelly, P., Pierce, L., and Hawke, D. H. (1989). Sequencing of peptides on solid phase supports. Am. Biotech. Lab. 7(3), March:26–37.

Koeners, H. J., Schattenkerk, C., Verhoeven, J. J., and van Boom, J. H. (1981). Synthesis of O-(2-O-α-D-glucopyranosyl)-β-D-galactopyranoside of optically pure δ-hydroxy-L-lysylglycine and δ-hydroxy-L-lysylglycyl-L-glutamyl-L-aspartylglycine. Tetrahedron 37:1763–1771.

König, W., and Geiger, R. (1970a). Eine neue methode zur synthese von peptiden: Aktivierung der carboxylgruppe mit dicyclohexylcarbodiimid unter zusatz von 1-hydroxy-benzotriazolen. Chem. Ber. 103:788–798.

König, W., and Geiger, R. (1970b). Racemisierung bei peptidsynthesen. Chem. Ber. 103:2024–2033.

König, W., and Geiger, R. (1970c). Eine neue methode zur synthese von peptiden: Aktivierung der carboxylgruppe mit dicyclohexylcarbodiimid und 3-hydroxy-4-oxo-3,4-dihydro-1.2.3-benzotriazin. Chem. Ber. 103:2034–2040.

König, W., and Geiger, R. (1973). N-hydroxyverbindungen als katalysatoren für die aminolyse aktivierter ester. Chem. Ber. 106:3626–3635.

Krchnák, V., Vágner, J., Safár, P., and Lebl, M. (1988). Noninvasive continuous monitoring of solid-phase peptide synthesis by acid-base indicator. Coll. Czech. Chem. Commun. 53:2542–2548.

Kullmann, W., and Gutte, B. (1978). Synthesis of an open-chain asymmetrical cystine peptide corresponding to the sequence A^{18-21}-B^{19-26} of bovine insulin by solid phase fragment condensation. Int. J. Peptide Protein Res. 12:17–26.

Kumagaye, K. Y., Inui, T., Nakajima, K., Kimura, T., and Sakakibara, S. (1991). Suppression of a side reaction associated with N^{im}-benzyloxy-methyl group during synthesis of peptides containing cysteinyl residue at the N-terminus. Peptide Res. 4:84–87.

Kunz, H. (1987). Synthesis of glycopeptides: Partial structures of biological recognition components. Angew. Chem. Int. Ed. Engl. 26:294–308.

Kunz, H. (1990). Allylic anchoring groups in the solid phase synthesis of peptides and glycopeptides. In *Innovation and Perspectives in Solid Phase Synthesis*, R. Epton, ed., Birmingham, UK, Solid Phase Conference Coordination, Ltd., pp. 371–378.

Kunz, H., and Dombo, B. (1988). Solid phase synthesis of peptides and glycopeptides on polymeric supports with allylic anchor groups. Angew. Chem. Int. Ed. Engl. 27:711–713.

Kusunoki, M., Nakagawa, S., Seo, K., Hamana, T., and Fukuda, T. (1990). A side reaction in solid phase synthesis: Insertion of glycine residues into peptide chains via $N^{im} \rightarrow N^{\alpha}$ transfer. Int. J. Peptide Protein Res. 36:381–386.

Lajoie, G., Crivici, A., and Adamson, J. G. (1990). A simple and convenient synthesis of ω-*tert*-butyl esters of Fmoc-aspartic and Fmoc-glutamic acids. Synthesis 571–572.

Lamthanh, H., Roumestand, C., Deprun, C., and Ménez, A. (1993). Side reaction during the deprotection of (*S*-acetamidomethyl)cysteine in a peptide with a high serine and threonine content. Int. J. Peptide Protein Res. 41:85–95.

Lansbury, P. T., Jr., Hendrix, J. C., and Coffman, A. I. (1989). A practical method for the preparation of protected peptide fragments using the Kaiser oxime resin. Tetrahedron Lett. 30:4915–4918.

Larsen, B. D., Larsen, C., and Holm, A. (1991). Incomplete Fmoc-deprotection in solid phase synthesis. In *Peptides 1990*, E. Giralt and D. Andreu, eds., Leiden, Escom, pp. 183–185.

Lapatsanis, L., Milias, G., Froussios, K., and Kolovos, M. (1983). Synthesis of *N*-2,2,2-(trichloroethoxycarbonyl)-L-amino acids and *N*-(9-fluorenyl-methoxycarbonyl)-L-amino acids involving succinimidoxy anion as a leaving group in amino acid protection. Synthesis 671–673.

Lauer, J. L., Fields, C. G., and Fields, G. B. (1995). Sequence dependence of aspartimide formation during 9-fluorenylmethoxycarbonyl solid-phase peptide synthesis. Lett. Peptide Sci. 1:197–205.

Lauer, J. L., Furcht, L. T., and Fields, G. B. (1997). Inhibition of melanoma cell binding to type IV collagen by analogs of cell adhesion regulator. J. Med. Chem. 40:3077–3084.

Lebl, M., and Eichler, J. (1989). Simulation of continuous solid phase synthesis: Synthesis of methionine enkephalin and its analogs. Peptide Res. 2:297–300.

Li, C. H., Lemaire, S., Yamashiro, D., and Doneen, B. A. (1976). The synthesis and opiate activity of β-endorphin. Biochem. Biophys. Res. Commun. 71:19–25.

Li, X., Kawakami, T., and Aimoto, S. (1998). Direct preparation of peptide thioesters using an Fmoc solid-phase method. Tetrahedron Lett. 39:8669–8672.

Limal, D., Briand, J.-P., Dalbon, P., and Jolivet, M. (1998). Solid-phase synthesis and on-resin cyclization of a disulfide bond peptide and lactam analogues corresponding to the major antigenic site of HIV gp41 protein. J. Peptide Res. 52:121–129.

Lin, Y.-Z., Caporaso, G., Chang, P.-Y., Ke, X.-H., and Tam, J. P. (1988). Synthesis of a biological active tumor growth factor from the predicted DNA sequence of Shope fibroma virus. Biochemistry 27:5640–5645.

Liu, C.-F., and Tam, J. P. (1994). Peptide segment ligation strategy without use of protecting groups. Proc. Natl. Acad. Sci. USA 91:6584–6588.

Liu, C. F., Rao, C., and Tam, J. P. (1996). Orthogonal ligation of unprotected peptide segments through pseudoproline formation for the synthesis of HIV-1 protease. J. Am. Chem. Soc. 118:307–312.

Liu, Y.-Z., Ding, S.-H., Chu, J.-Y., and Felix, A. M. (1990). A novel Fmoc-based anchorage for the synthesis of protected peptides on solid phase. Int. J. Peptide Protein Res. 35:95–98.

Live, D. H., and Kent, S. B. H. (1982). Fundamental aspects of the chemical applications of cross-linked polymers. In *Elastomers and Rubber Elasticity*, J. E. Mark and J. Lal, eds., Washington, D.C., American Chemical Society, pp. 501–515.

Live, D. H., and Kent, S. B. H. (1983). Correlation of coupling rates with physicochemical properties of resin-bound peptides in solid phase synthesis. In *Peptides: Structure and Function*, V. J. Hruby and D. H. Rich, eds., Rockford, Ill., Pierce Chemical Co., pp. 65–68.

Lloyd-Williams, P., Jou, G., Albericio, F., and Giralt, E. (1991). Solid-phase synthesis of peptides using allylic anchoring groups: An investigation of their palladium-catalysed cleavage. Tetrahedron Lett. 32:4207–4210.

Lloyd-Williams, P., Albericio, F., and Giralt, E., eds. (1997). *Chemical Approaches to the Synthesis of Peptides and Proteins*, Boca Raton, Fla., CRC Press.

Loffet, A., Galeotti, N., Jouin, P., and Castro, B. (1989). Tert-butyl esters of *N*-protected amino acids with *tert*-butyl fluorocarbonate (Boc-F). Tetrahedron Lett. 30:6859–6860.

Löw, M., Kisfaludy, L., Jaeger, E., Thamm, P., Knof, S., and Wünsch, E. (1978a). Direkte *tert*-butylierung des tryptophans: Herstellung von 2,5,7-tri-*tert*-butyltryptophan. Hoppe-Seyler's Z. Physiol. Chem. 359:1637–1642.

Löw, M., Kisfaludy, L., and Sohár, P. (1978b). *tert*-Butylierung des tryptophan-indolringes während der abspaltung der *tert*-butyloxycarbonyl-gruppe bei peptidsynthesen. Hoppe-Seyler's Z. Physiol. Chem. 359:1643–1651.

Lu, G.-s., Mojsov, S., Tam, J. P., and Merrifield, R. B. (1981). Improved synthesis of 4-alkoxybenzyl alcohol resin. J. Org. Chem. 46:3433–3436.

Ludwick, A. G., Jelinski, L. W., Live, D., Kintanar, A., and Dumais, J. J. (1986). Association of peptide chains during Merrifield solid-phase peptide synthesis: A deuterium NMR study. J. Am. Chem. Soc. 108:6493–6496.

Lukszo, J., Patterson, D., Albericio, F., and Kates, S. A. (1996). 3-(1-Piperidinyl)alanine formation during the preparation of *C*-terminal cyceine peptides with Fmoc/*t*-Bu strategy. Lett. Peptide Sci. 3:157–166.

Lundt, B. F., Johansen, N. L., Volund, A., and Markussen, J. (1978). Removal of *t*-butyl and *t*-butoxycarbonyl protecting groups with trifluoroacetic acid. Int. J. Peptide Protein Res. 12:258–268.

Lyttle, M. H., and Hudson, D. (1992). Allyl based side-chain protection for SPPS. In *Peptides: Chemistry and Biology*, J. A. Smith and J. E. Rivier, eds., Leiden, Escom, pp. 583–584.

Malkar, N. B., Lauer-Fields, J. L., and Fields, G. B. (2000). Convenient synthesis of glycosylated hydroxylysine derivatives for use in solid-phase synthesis. Tetrahedron Lett. 41:1137–1140.

Masui, Y., Chino, N., and Sakakibara, S. (1980). The modification of tryptophyl residues during the acidolytic cleavage of Boc-groups I: Studies with Boc-tryptophan. Bull. Chem. Soc. Jpn. 53:464–468.

Matsueda, G. R., and Haber, E. (1980). The use of an internal reference amino acid for the evaluation of reactions in solid-phase peptide synthesis. Anal. Biochem. 104:215–227.

Matsueda, G. R., and Stewart, J. M. (1981). A p-methylbenzhydrylamine resin for improved solid-phase synthesis of peptide amides. Peptides 2:45–50.

Matsueda, G. R., Haber, E., and Margolies, M. N. (1981). Quantitative solid-phase Edman degradation for evaluation of extended solid-phase peptide synthesis. Biochemistry 20:2571–2580.

Matsueda, R., and Walter, R. (1980). 3-nitro-2-pyridinesulfenyl (Npys) group: A novel selective protecting group which can be activated for peptide bond formation. Int. J. Peptide Protein Res. 16:392–401.

McCafferty, D. G., Slate, C. A., Nakhle, B. M., Graham, Jr., H. D., Austell, T. L., Vachet, R. W., Mullis, B. H., and Erickson, B. W. (1995). Engineering of a 129-residue tripod protein by chemoselective ligation of proline-II helices. Tetrahedron 51:9859–9872.

McCurdy, S. N. (1989). The investigation of Fmoc-cysteine derivatives in solid phase peptide synthesis. Peptide Res. 2:147–151.

McFerran, N. V., Walker, B., McGurk, C. D., and Scott, F. C. (1991). Conductance measurements in solid phase peptide synthesis I: Monitoring coupling and deprotection in Fmoc chemistry. Int. J. Peptide Protein Res. 37:382–387.

Meienhofer, J. (1985). Protected amino acids in peptide synthesis. In Chemistry and Biochemistry of the Amino Acids, G. C. Barrett, ed., London, Chapman & Hall, pp. 297–337.

Meienhofer, J., Waki, M., Heimer, E. P., Lambros, T. J., Makofske, R. C., and Chang, C.-D. (1979). Solid phase synthesis without repetitive acidolysis. Int. J. Peptide Protein Res. 13:35–42.

Meister, S. M., and Kent, S. B. H. (1983). Sequence-dependent coupling problems in stepwise solid phase peptide synthesis: Occurrence, mechanism, and correction. In Peptides: Structure and Function, V. J. Hruby and D. H. Rich, eds., Rockford, Ill., Pierce Chemical Co., pp. 103–106.

Meldal, M. (1992). A flow stable polyethylene glycol dimethyl acrylamide copolymer for solid phase synthesis. Tetrahedron Lett. 33:3077–3080.

Meldal, M. (1994). Glycopeptide synthesis. In Neoglycoconjugates: Preparation and Applications, Y. C. Lee and R. T. Lee, eds., San Diego, Calif., Academic Press, pp. 145–198.

Meldal, M., and Jensen, K. J. (1990). Pentafluorophenyl esters for the temporary protection of the α-carboxy group in solid phase glycopeptide synthesis. J. Chem. Soc., Chem. Commun. 483–485.

Meldal, M., Bielfeldt, T., Peters, S., Jensen, K. J., Paulsen, H., and Bock, K. (1994). Susceptibility of glycans to β-elimination in Fmoc-based O-glycopeptide synthesis. Int. J. Protein Res. 43:529–536.

Mergler, M., Tanner, R., Gosteli, J., and Grogg, P. (1988a). Peptide synthesis by a combination of solid-phase and solution methods I: A new very acid-labile anchor group for the solid phase synthesis of fully protected fragments. Tetrahedron Lett. 29:4005–4008.

Mergler, M., Nyfeler, R., Tanner, R., Gosteli, J., and Grogg, P. (1988b). Peptide synthesis by a combination of solid-phase and solution methods II: Synthesis of fully protected peptide fragments on 2-methoxy-4-alkoxy-benzyl alcohol resin. Tetrahedron Lett. 29:4009–4012.

Mergler, M., Nyfeler, R., and Gosteli, J. (1989a). Peptide synthesis by a combination of solid-phase and solution methods III: Resin derivatives allowing minimum-racemization coupling of N^α-protected amino acids. Tetrahedron Lett. 30:6741–6744.

Mergler, M., Nyfeler, R., Gosteli, J., and Tanner, R. (1989b). Peptide synthesis by a combination of solid-phase and solution methods IV: Minimum-racemization coupling of N^α-9-fluorenylmethyloxycarbonyl amino acids to alkoxy benzyl alcohol type resins. Tetrahedron Lett. 30:6745–6748.

Merrifield, B. (1986). Solid phase synthesis. Science 232:341–347.

Merrifield, R. B., and Bach, A. E. (1978). 9-(2-Sulfo)fluorenylmethyloxycarbonyl chloride, a new reagent for the purification of synthetic peptides. J. Org. Chem. 43:4808–4816.

Merrifield, R. B., Stewart, J. M., and Jernberg, N. (1966). Instrument for automated synthesis of peptides. Anal. Chem. 38:1905–1914.

Merrifield, R. B., Mitchell, A. R., and Clarke, J. E. (1974). Detection and prevention of urethane acylation during solid-phase peptide synthesis by anhydride methods. J. Org. Chem. 39:660–668.

Merrifield, R. B., Vizioli, L. D., and Boman, H. G. (1982). Synthesis of the antibacterial peptide cecropin A(1–33). Biochemistry 21:5020–5031.

Merrifield, R. B., Singer, J., and Chait, B. T. (1988). Mass spectrometric evaluation of synthetic peptides for deletions and insertions. Anal. Biochem. 174:399–414.

Méry, J., and Calas, B. (1988). Tryptophan reduction and histidine racemization during deprotection by catalytic transfer hydrogenation of an analog of the luteinizing hormone releasing factor. Int. J. Peptide Protein Res. 31:412–419.

MilliGen/Biosearch (1990). Flow rate programming in continuous flow peptide synthesizer solves problematic synthesis. MilliGen/Biosearch Report 7, MilliGen/Biosearch Division of Millipore, Bedford, Mass., pp. 4–5, 11.

Milton, S. C. F., Brandt, W. F., Schnölzer, M., and de Lisle Milton, R. C. (1992). Total solid-phase synthesis and prolactin-inhibiting activity of the gonadotropin-releasing hormone precursor protein and the gonadotropin-releasing hormone associated peptide. Biochemistry 31:8799–8809.

Miranda, L. P., and Alewood, P. F. (1999). Accelerated chemical synthesis of peptides and small proteins. Proc. Natl. Acad. Sci. USA 96:1181–1186.

Mitchell, A. R., Kent, S. B. H., Engelhard, M., and Merrifield, R. B. (1978). A new synthetic route to tert-butyloxycarbonylaminoacyl-4-(oxymethyl)-phenylacetamidomethyl-resin, an improved support for solid-phase peptide synthesis. J. Org. Chem. 43:2845–2852.

Mitchell, M. A., Runge, T. A., Mathews, W. R., Ichhpurani, A. K., Harn, N. K., Dobrowolski, P. J., and Eckenrode, F. M. (1990). Problems associated with use of the benzyloxymethyl protecting group for histidines: Formaldehyde adducts formed during cleavage by hydrogen fluoride. Int. J. Peptide Protein Res. 36:350–355.

Mojsov, S., Mitchell, A. R., and Merrifield, R. B. (1980). A quantitative evaluation of methods for coupling asparagine. J. Org. Chem. 45:555–560.

Mott, A. W., Slomczynska, U., and Barany, G. (1986). Formation of sulfur-sulfur bonds during solid-phase peptide synthesis: Application to the synthesis of oxytocin. In *Forum Peptides Le Cap d'Agde 1984*, B. Castro and J. Martinez, eds., Nancy, Les Impressions Dohr, pp. 321–324.

Muir, T. W., Williams, M. J., Ginsberg, M. H., and Kent, S. B. H. (1994). Design and chemical synthesis of a neoprotein structural model for the cytoplasmic domain of a multisubunit cell-surface receptor: Integrin α_{IIb} β_3 (platelet GPIIb–IIIa). Biochemistry 33:7701–7708.

Muir, T. W., Dawson, P. E., and Kent, S. B. H. (1997). Protein synthesis by chemical ligation of unprotected peptides in aqueous solution. Methods Enzymol. 289:266–298.

Munson, M. C., and Barany, G. (1993). Synthesis of α-conotoxin SI, a bicyclic tridecapeptide amide with two disulfide bridges: Illustration of novel protection schemes and oxidation strategies. J. Am. Chem. Soc. 115:10203–10210.

Munson, M. C., García-Echeverría, C., Albericio, F., and Barany, G. (1992). S-2,4,6-trimethoxybenzyl (Tmob): A novel cysteine protecting group for the N^{α}-9-fluorenylmethyloxycarbonyl (Fmoc) strategy of peptide synthesis. J. Org. Chem. 57:3013–3018.

Mutter, M., and Bellof, D. (1984). A new base-labile anchoring group for polymer-supported peptide synthesis. Helv. Chim. Acta 67:2009–2016.

Mutter, M., Altmann, K.-H., Bellof, D., Flörsheimer, A., Herbert, J., Huber, M., Klein, B., Strauch, L., and Vorherr, T. (1985). The impact of secondary structure formation in peptide synthesis. In *Peptides: Structure and Function*, C. M. Deber, V. J. Hruby, and K. D. Kopple, eds., Rockford, Ill., Pierce Chemical Co., pp. 397–405.

Nagase, H., Fields, C. G., and Fields, G. B. (1994). Design and characterization of a fluorogenic substrate selectively hydrolyzed by stromelysin 1 (matrix metalloproteinase-3). J. Biol. Chem. 269:20952–20957.

Nagata, K., Nakajima, K., Wakamiya, T., Togashi, R., Nishida, T., Saruta, K., Yasuoka, J., Kusumoto, S., Aimoto, S., and Kumagaye, K. Y. (1997). Synthetic study of phosphopeptides related to heat shock protein HSP27. Bioorg. Med. Chem. 5:135–145.

Nakagawa, S. H., Lau, H. S. H., Kézdy, F. J., and Kaiser, E. T. (1985). The use of polymer-bound oximes for the synthesis of large peptides usable in segment condensation: Synthesis of a 44 amino acid amphiphilic peptide model of apolipoprotein A-1. J. Am. Chem. Soc. 107:7087–7092.

Narita, M., and Kojima, Y. (1989). The β-sheet structure-stabilizing potential of twenty kinds of amino acid residues in protected peptides. Bull. Chem. Soc. Jpn. 62:3572–3576.

Narita, M., Umeyama, H., and Yoshida, T. (1989). The easy disruption of the β-sheet structure of resin-bound human proinsulin C-peptide fragments by strong electron-donor solvents. Bull. Chem. Soc. Jpn. 62:3582–3586.

Nefzi, A., Sun, X., and Mutter, M. (1995). Chemoselective ligation of multi-functional peptides to topological templates via thioether formation for TASP synthesis. Tetrahedron Lett. 36:229–230.

Nicolás, E., Pedroso, E., and Giralt, E. (1989). Formation of aspartimide peptides in Asp-Gly sequences. Tetrahedron Lett. 30:497–500.

Nielson, C. S., Hansen, P. H., Lihme, A., and Heegaard, P. M. H. (1989). Real time monitoring of acylations during solid phase peptide synthesis:A method based on electrochemical detection. J. Biochem. Biophys. Methods 20:69–75.

Nishio, H., Kimura, T., and Sakakibara, S. (1994). Side reaction in peptide synthesis:Modification of tryptophan during treatment with mercury(II) acetate/2-mercaptoethanol in aqueous acetic acid. Tetrahedron Lett. 35:1239–1242.

Nishiuchi, Y., Nishio, H., Inui, T., Kimura, T., and Sakakibara, S. (1996). N^{in}-Cyclohexyloxycarbonyl group as a new protecting group for tryptophan. Tetrahedron Lett. 37:7529–7532.

Noble, R. L., Yamashiro, D., and Li, C. H. (1976). Synthesis of a nonadeca-peptide corresponding to residues 37–55 of ovine prolactin: Detection and isolation of the sulfonium form of methionine-containing peptides. J. Am. Chem. Soc. 98:2324–2328.

Nokihara, K., Hellstern, H., and Höfle, G. (1989). Peptide synthesis by fragment assembly on a polymer support. In *Peptides 1988*, G. Jung and E. Bayer, eds., Berlin, Walter de Gruyter & Co., pp. 166–168.

Nomizu, M., Inagaki, Y., Yamashita, T., Ohkubo, A., Otaka, A., Fujii, N., Roller, P. P., and Yajima, H. (1991). Two-step hard acid deprotection/cleavage procedure for solid phase peptide synthesis. Int. J. Peptide Protein Res. 37:145–152.

Nutt, R. F., Brady, S. F., Darke, P. L., Ciccarone, T. M., Colton, C. D., Nutt, E. M., Rodkey, J. A., Bennett, C. D., Waxman, L. H., Sigal, I. S., Anderson, P. S., and Veber, D. F. (1988). Chemical synthesis and enzymatic activity of a 99-residue peptide with a sequence proposed for the human immunodeficiency virus protease. Proc. Natl. Acad. Sci. USA 85:7129–7133.

Ogunjobi, O., and Ramage, R. (1990). Ubiquitin: Preparative chemical synthesis, purification and characterization. Biochem. Soc. Trans. 18:1322–1323.

Okada, Y., and Iguchi, S. (1988). Amino acid and peptides, part 19: Synthesis of β-1- and β-2-adamantyl aspartates and their evaluation for peptide synthesis. J. Chem. Soc. Perkin Trans. I:2129–2136.

Ondetti, M. A., Pluscec, J., Sabo, E. F., Sheehan, J. T., and Williams, N. (1970). Synthesis of cholecystokinin-pancreozymin I: The C-terminal dodecapeptide. J. Am. Chem. Soc. 92:195–199.

Orlowska, A., Witkowska, E., and Izdebski, J. (1987). Sequence dependence in the formation of pyroglutamyl peptides in solid phase peptide synthesis. Int. J. Peptide Protein Res. 30:141–144.

Orpegen (1990). HYCRAM™ support: Loading and cleavage. Technical Note, Orpegen, Heidelberg, Germany.

Otteson, K. M., Harrison, J. L., Ligutom, A., and Ashcroft, P. (1989). Solid phase peptide synthesis with N-methylpyrrolidone as the solvent for both

Fmoc and Boc synthesis. Poster Presentations at the Eleventh American Peptide Symposium, Foster City, Calif., Applied Biosystems, Inc., pp. 34–38.

Ottinger, E. A., Shekels, L. L., Bernlohr, D. A., and Barany, G. (1993). Synthesis of phosphotyrosine-containing peptides and their use as substrates for protein tyrosine phosphatases. Biochemistry 32:4354–4361.

Ottinger, E. A., Qinghong Xu, Q., and Barany, G. (1996). Intramolecular pyrophosphate formation during N^α-9-Fluorenylmethyloxycarbonyl (Fmoc) solid-phase synthesis of peptides containing adjacent phosphotyrosine residues. Peptide Res. 9:223–228.

Otvös, L., Jr., Elekes, I., and Lee, V. M.-Y. (1989a). Solid-phase synthesis of phosphopeptides. Int. J. Peptide Protein Res. 34:129–133.

Otvös, L., Jr., Wroblewski, K., Kollat, E., Perczel, A., Hollosi, M., Fasman, G. D., Ertl, H. C. J., and Thurin, J. (1989b). Coupling strategies in solid-phase synthesis of glycopeptides. Peptide Res. 2:362–366.

Otvös, L., Jr., Urge, L., Hollosi, M., Wroblewski, K., Graczyk, G., Fasman, G. D., and Thurin, J. (1990). Automated solid-phase synthesis of glycopeptides: Incorporation of unprotected mono- and disaccharide units of N-glycoprotein antennae into T cell epitopic peptides. Tetrahedron Lett. 31:5889–5892.

Pacquet, A. (1982). Introduction of 9-fluorenylmethyloxycarbonyl, trichloroethoxycarbonyl, and benzyloxycarbonyl amine protecting groups into O-unprotected hydroxyamino acids using succinimidyl carbonates. Can. J. Chem. 60:976–980.

Pan, H., Barbar, E., Barany, G., and Woodward, C. (1995). Extensive nonrandom structure in reduced and unfolded bovine pancreatic trypsin inhibitor. Biochemistry 34:13974–13981.

Patchornik, A., Amit, B., and Woodward, R. B. (1970). Photosensitive protecting groups. J. Am. Chem. Soc. 92:6333–6335.

Patel, K., and Borchardt, R. T. (1990a). Chemical pathways of peptide degradation II: Kinetics of deamidation of an asparaginyl residue in a model hexapeptide. Pharm. Res. 7:703–711.

Patel, K., and Borchardt, R. T. (1990b). Chemical pathways of peptide degradation III: Effect of primary sequence on the pathways of deamidation of asparaginyl residues in hexapeptides. Pharm. Res. 7:787–793.

Paulsen, H., Merz, G., and Weichert, U. (1988). Solid-phase synthesis of O-glycopeptide sequences. Angew. Chem. Int. Ed. Engl. 27:1365–1367.

Paulsen, H., Merz, G., Peters, S., and Weichert, U. (1990). Festphasensynthese von O-glycopeptiden. Liebigs Ann. Chem. 1165–1173.

Pearson, D. A., Blanchette, M., Baker, M. L., and Guindon, C. A. (1989). Trialkylsilanes as scavengers for the trifluoroacetic acid deblocking of protecting groups in peptide synthesis. Tetrahedron Lett. 30:2739–2742.

Pedroso, E., Grandas, A., de las Heras, X., Eritja, R., and Giralt, E. (1986). Diketopiperazine formation in solid phase peptide synthesis using p-alkoxybenzyl ester resins and Fmoc-amino acids. Tetrahedron Lett. 27:743–746.

Peluso, S., Dumy, P., Eggleston, I. M., Garrouste, P., and Mutter, M. (1997). Protein mimetics (TASP) by sequential condensation of peptide loops to an immobilised topological template. Tetrahedron 53:7231–7236.

Penke, B., and Nyerges, L. (1989). Preparation and application of a new resin for synthesis of peptide amides via Fmoc-strategy. In *Peptides 1988*, G. Jung and E. Bayer, eds., Berlin, Walter de Gruyter & Co., pp. 142–144.

Penke, B., and Nyerges, L. (1990). Solid-phase synthesis of porcine cholecystokinin-33 in a new resin via Fmoc strategy. Peptide Res. 4:289–295.

Penke, B., and Rivier, J. (1987). Solid-phase synthesis of peptide amides on a polystyrene support using fluorenylmethoxycarbonyl protecting groups. J. Org. Chem. 52:1197–1200.

Penke, B., and Tóth, G. K. (1989). An improved method for the preparation of large amounts of ω-cyclohexylesters of aspartic and glutamic acid. In *Peptides 1988*, G. Jung and E. Bayer, eds., Berlin, Walter de Gruyter and Co., pp. 67–69.

Penke, B., Baláspiri, L., Pallai, P., and Kovács, K. (1974). Application of pentafluorophenyl esters of Boc-amino acids in solid phase peptide synthesis. Acta Phys. Chem. 20:471–476.

Penke, B., Zsigo, J., and Spiess, J. (1989). Synthesis of a protected hydroxylysine derivative for application in peptide synthesis. In *The Eleventh American Peptide Symposium Abstracts*, The Salk Institute and University of California at San Diego, La Jolla, Calif., P-335.

Pennington, M. W., Festin, S. M., and Maccecchini, M. L. (1991). Comparison of folding procedures on synthetic ω-conotoxin. In *Peptides 1990*, E. Giralt and D. Andreu, eds., Leiden, Escom, pp. 164–166.

Perich, J. W. (1990). Modern methods of O-phosphoserine- and O-phosphotyrosine-containing peptide synthesis. In *Peptides and Protein Phosphorylation*, B. E. Kemp, ed., Boca Raton, Fla., CRC Press, pp. 289–314.

Perich, J. W. (1997). Synthesis of phosphopeptides using modern chemical approaches. Methods Enzymol. 289:245–266.

Perich, J. W. (1998). Synthesis of phosphopeptides via global phosphorylation on the solid phase: Resolution of H-phosphonate formation. Lett. Peptide Sci. 5:49–55.

Perich, J. W., and Johns, R. B. (1988). Di-*tert*-butyl-N,N-diethylphosphoamidite and dibenzyl-N,N-diethylphosphoamidate: Highly reactive reagents for the "phosphite triester" phosphorylation of serine containing peptides. Tetrahedron Lett. 29:2369–2372.

Perich, J. W., and Reynolds, E. C. (1991). Fmoc/solid-phase synthesis of Tyr(P)-containing peptides through t-butyl phosphate protection. Int. J. Peptide Protein Res. 37:572–575.

Perich, J. W., Valerio, R. M., and Johns, R. B. (1986). Solid-phase synthesis of an O-phosphoseryl-containing peptide using phenyl phosphorotriester protection. Tetrahedron Lett. 27:1377–1380.

Perich, J. W., Ede, N. J., Eagle, S., and Bray, A. M. (1999). Synthesis of phosphopeptides by the Multipin method: Evaluation of coupling methods for the incorporation of Fmoc-Tyr(PO$_3$Bzl,H)-OH, Fmoc-Ser(PO$_3$Bzl,H)-OH, and Fmoc-Thr(PO$_3$Bzl,H)-OH. Lett. Peptide Sci. 6:91–97.

Pessi, A., Bianchi, E., Bonelli, F., and Chiappinelli, L. (1990). Application of the continuous-flow polyamide method to the solid-phase synthesis of a multiple antigen peptide (MAP) based on the sequence of a malaria epitope. J. Chem. Soc., Chem. Commun. 8-9.

Peters, S., Lowary, T. L., Hindsgaul, O., Meldal, M., and Bock, K. (1995). Solid phase synthesis of a fucosylated glycopeptide of human factor IX with a fructose-α-(1 → O)-serine linkage. J. Chem. Soc. Perkin Trans. 1:3017–3022.

Photaki, I., Taylor-Papadimitriou, J., Sakarellos, C., Mazarakis, P., and Zervas, L. (1970). On cysteine and cystine peptides, part V: S-trityl- and S-diphenylmethyl-cysteine and -cysteine peptides. J. Chem. Soc. (C) 2683–2687.

Pickup, S., Blum, F. D., and Ford, W. T. (1990). Self-diffusion coefficients of Boc-amino acid anhydrides under conditions of solid phase peptide synthesis. J. Polym. Sci. Polym. Chem. Ed. 28:931–934.

Pipkorn, R., and Bernath, E. (1990). Solid phase synthesis of preprocecropin A. In *Innovation and Perspectives in Solid Phase Synthesis*, R. Epton, ed., Birmingham, UK, Solid Phase Conference Coordination, Ltd., pp. 537–541.

Planas, M., Bardají, E., Jensen, K. J., and Barany, G. (1999). Use of the dithiasuccinoyl (Dts) amino protecting group for solid-phase synthesis of protected peptide nucleic acid (PNA) oligomers. J. Org. Chem. 64:7281–7289.

Plaué, S. (1990). Synthesis of cyclic peptides on solid support: Application to analogs of hemagglutinin of influenza virus. Int. J. Peptide Protein Res. 35:510–517.

Ploux, O., Chassaing, G., and Marquet, A. (1987). Cyclization of peptides on a solid support: Application to cyclic analogs of substance P. Int. J. Peptide Protein Res. 29:162–169.

Pluscec, J., Sheehan, J. T., Sabo, E. F., Williams, N., Kocy, O., and Ondetti, M. A. (1970). Synthesis of analogs of the C-terminal octapeptide of chole-cystokinin-pancreozymin: Structure-activity relationship. J. Med. Chem. 13:349–352.

Ponsati, B., Giralt, E., and Andreu, D. (1989). A synthetic strategy for simultaneous purification-conjugation of antigenic peptides. Anal. Biochem. 181:389–395.

Ponsati, B., Giralt, E., and Andreu, D. (1990a). Side reactions in post-HF workup of peptides having the unusual Tyr-Trp-Cys sequence: Low-high acidolysis revisited. In *Peptides: Chemistry, Structure and Biology*, J. E. Rivier and G. R. Marshall, eds., Leiden, Escom, pp. 960–962.

Ponsati, B., Giralt, E., and Andreu, D. (1990b). Solid-phase approaches to regiospecific double disulfide formation: Application to a fragment of bovine pituitary peptide. Tetrahedron 46:8255–8266.

Prasad, K. U., Trapane, T. L., Busath, D., Szabo, G., and Urry, D. W. (1982). Synthesis and characterization of 1-^{13}C-D-Leu^{12,14} gramicidin A. Int. J. Peptide Protein Res. 19:162–171.

Prosser, R. S., Davis, J. H., Dahlquist, F. W., and Lindorfer, M. A. (1991). ^{2}H nuclear magnetic resonance of the gramicidin A backbone in a phospholipid bilayer. Biochemistry 30:4687–4696.

Quibell, M., Owen, D., Packman, L. C., and Johnson, T. (1994). Suppression of piperidine-mediated side product formation for Asp(OBu^t)-containing peptides by the use of N-(2-hydroxy-4-methoxybenzyl) (Hmb) backbone amide protection. J. Chem. Soc., Chem. Commun. 2343–2344.

Ramage, R., and Green, J. (1987). N_G-2,2,5,7,8-pentamethylchroman-6-sulphonyl-L-arginine: A new acid labile derivative for peptide synthesis. Tetrahedron Lett. 28:2287–2290.

Ramage, R., and Raphy, G. (1992). Design of an affinity-based N^α-amino protecting group for peptide synthesis: Tetrabenzo[a,c,g,i]fluorenyl-17-methyl urethanes (Tbfmoc). Tetrahedron Lett. 33:385–388.

Ramage, R., Green J., and Ogunjobi, O. M. (1989). Solid phase peptide synthesis of ubiquitin. Tetrahedron Lett. 30:2149–2152.

Ramage, R., Biggin, G. W., Brown, A. R., Comer, A., Davison, A., Draffan, L., Jiang, L., Morton, G., Robertson, N., Shaw, K. T., Tennant, G., Urquhart, K., and Wilken, J. (1996). Methodology for chemical synthesis of proteins. In *Innovation and Perspectives in Solid Phase Synthesis and Combinatorial Libraries 1996*, R. Epton, ed., Kingswinford, Mayflower Scientific, pp. 1–10.

Ramage, R., Swenson, H. R., and Shaw, K. T. (1998). A novel purification protocol for soluble combinatorial peptide libraries. Tetrahedron Lett. 39:8715–8718.

Rao, C., and Tam, J. P. (1994). Synthesis of peptide dendrimer. J. Am. Chem. Soc. 116:6975–6976.

Reddy, M. P., and Voelker, P. J. (1988). Novel method for monitoring the coupling efficiency in solid phase peptide synthesis. Int. J. Peptide Protein Res. 31:345–348.

Reid, G. E., and Simpson, R. J. (1992). Automated solid-phase peptide synthesis: Use of 2-(1H-benzotriazol-1-yl)-1,1,3,3,-tetramethyluronium tetrafluoroborate for coupling of *tert*-butyloxycarbonyl amino acids. Anal. Biochem. 200:301–309.

Remmer, H. A., and Fields, G. B. (1999). Problems encountered with the synthesis of a glycosylated hydroxylysine derivative suitable for Fmoc-solid phase peptide synthesis. In *Peptides: Frontiers of Peptide Science*, J. P. Tam and P. T. P. Kaumaya, eds., Dordrecht, Kluwer Academic Publishers, pp. 287–288.

Renil, M., Ferreras, M., Delaisse, J. M., Foged, N. T., and Meldal, M. (1998). PEGA supports for combinatorial peptide synthesis and solid-phase enzymatic library assays. J. Peptide Sci. 4:195–210.

Rich, D. H., and Gurwara, S. K. (1975). Preparation of a new o-nitrobenzyl resin for solid-phase synthesis of *tert*-butyloxycarbonyl-protected peptide acids. J. Am. Chem. Soc. 97:1575–1579.

Rich, D. H., and Singh, J. (1979). The carbodiimide method. In *The Peptides*, Vol. 1, E. Gross and J. Meienhofer, eds., New York, Academic Press, pp. 241–314.

Riniker, B., and Hartmann, A. (1990). Deprotection of peptides containing Arg(Pmc) and tryptophan or tyrosine: Elucidation of by-products. In *Peptides: Chemistry, Structure and Biology*, J. E. Rivier and G. R. Marshall, eds., Leiden, Escom, pp. 950–952.

Riniker, B., and Kamber, B. (1989). Byproducts of Trp-peptides synthesized on a p-benzyloxybenzyl alcohol polystyrene resin. In *Peptides 1988*, G. Jung and E. Bayer, eds., Berlin, Walter de Gruyter & Co., pp. 115–117.

Riniker, B., and Sieber, P. (1988). Problems and progress in the synthesis of histidine-containing peptides. In *Peptides: Chemistry, Biology,*

Interactions with Proteins, B. Penke and A. Torok, eds., Berlin, Walter de Gruyter & Co., pp. 65–74.

Riniker, B., Flörsheimer, A., Fretz, H., Sieber, P., and Kamber, B. (1993). A general strategy for the synthesis of large peptides: The combined solid-phase and solution approach. Tetrahedron 49:9307–9320.

Rink, H. (1987). Solid-phase synthesis of protected peptide fragments using a trialkoxy-diphenyl-methylester resin. Tetrahedron Lett. 28:3787–3790.

Rink, H., and Ernst, B. (1991). Glycopeptide solid-phase synthesis with an acetic acid-labile trialkoxy-benzhydryl linker. In *Peptides 1990*, E. Giralt and D. Andreu, eds., Leiden, Escom, pp. 418–419.

Rink, H., Sieber, P., and Raschdorf, F. (1984). Conversion of N^G-urethane protected arginine to ornithine in peptide solid phase synthesis. Tetrahedron Lett. 25:621–624.

Rivier, J., Galyean, R., Simon, L., Cruz, L. J., Olivera, B. M., and Gray, W. R. (1987). Total synthesis and further characterization of the γ-carboxy-glutamate-containing "sleeper" peptide from *Conus geographus* venom. Biochemistry 26:8508–8512.

Rohwedder, B., Mutti, Y., Dumy, P., and Mutter, M. (1998). Hydrazinolysis of Dde: Complete orthogonality with Aloc protecting groups. Tetrahedron Lett. 39:1175–1178.

Romani, S., Moroder, L., Göhring, W., Scharf, R., Wünsch, E., Barde, Y. A., and Thoenen, H. (1987). Synthesis of the trypsin fragment 10–25/75–88 of mouse nerve growth factor II: The unsymmetrical double chain cystine peptide. Int. J. Peptide Protein Res. 29:107–117.

Romoff, T. T., and Goodman, M. (1997). Urethane-protected N-carboxy-anhydrides (UNCAs) as unique reactants for the study of intrinsic race-mization tendencies in peptide synthesis. J. Peptide Res. 49:281–292.

Rose, K. (1994). Facile synthesis of homogeneous artificial proteins. J. Am. Chem. Soc. 116:30–33.

Rosen, O., Rubinraut, S., and Fridkin, M. (1990). Thiolysis of the 3-nitro-2-pyridinesulfenyl (Npys) protecting group: An approach towards a general deprotection scheme in peptide synthesis. Int. J. Peptide Protein Res. 35:545–549.

Rzeszotarska, B., and Masiukiewicz, E. (1988). Arginine, histidine and tryptophan in peptide synthesis: The guanidino function of arginine. Org. Prep. Proc. Int. 20:427–464.

Sakakibara, S. (1999). Chemical synthesis of proteins in solution. Biopolymers 51:279–296.

Sarin, V. K., Kent, S. B. H., and Merrifield, R. B. (1980). Properties of swollen polymer networks: Solvation and swelling of peptide-containing resins in solid-phase peptide synthesis. J. Am. Chem. Soc. 102:5463–5470.

Sarin, V. K., Kent, S. B.H., Tam, J. P., and Merrifield, R. B. (1981). Quantitative monitoring of solid-phase peptide synthesis by the ninhydrin reaction. Anal. Biochem. 117:147–157.

Sarin, V. K., Kent, S. B. H., Mitchell, A. R., and Merrifield, R. B. (1984). A general approach to the quantitation of synthetic efficiency in solid-phase peptide synthesis as a function of chain length. J. Am. Chem. Soc. 106:7845–7850.

Sasaki, T., and Kaiser, E. T. (1990). Synthesis and structural stability of helichrome as an artificial hemeproteins. Biopolymers 29:79–88.

Scarr, R. B., and Findeis, M. A. (1990). Improved synthesis and aminoacylation of *p*-nitrobenzophenone oxime polystyrene resin for solid-phase synthesis of protected peptides. Peptide Res. 3:238–241.

Schielen, W. J. G., Adams, H. P. H. M., Nieuwenhuizen, W., and Tesser, G. I. (1991). Use of Mpc-amino acids in solid phase peptide synthesis leads to improved coupling efficiencies. Int. J. Peptide Protein Res. 37:341–346.

Schiller, P. W., Nguyen, T. M.-D., and Miller, J. (1985). Synthesis of side-chain cyclized peptide analogs on solid supports. Int. J. Peptide Protein Res. 25:171–177.

Schneider, J., and Kent, S. B. H. (1988). Enzymatic activity of a synthetic 99 residue protein corresponding to the putative HIV-1 protease. Cell 54:363–368.

Schnölzer, M., and Kent, S. B. H. (1992). Constructing proteins by dovetailing unprotected synthetic peptides: Backbone-engineered HIV protease. Science 256:221–225.

Schnölzer, M., Alewood, P., Jones, A., Alewood, D., and Kent, S. B. H. (1992). *In situ* neutralization in Boc-chemistry solid phase peptide synthesis: Rapid, high yield assembly of difficult sequences. Int. J. Peptide Protein Res. 40:180–193.

Schroll, A. L., and Barany, G. (1989). A new protecting group for the sulfhydryl function of cysteine. J. Org. Chem. 54:244–247.

Scoffone, E., Previero, A., Benassi, C. A., and Pajetta, P. (1966). Oxidative modification of tryptophan residues in peptides. In *Peptides 1963*, L. Zervas, ed., Oxford, Pergamon Press, pp. 183–188.

Seitz, O., and Kunz, H. (1995). A novel allylic anchor for solid-phase synthesis. Synthesis of protected and unprotected *O*-glycosylated mucin-type glycopeptides. Angew. Chem. Int. Ed. Engl. 34:803–805.

Seyer, R., Aumelas, A., Caraty, A., Rivaille, P., and Castro, B. (1990). Repetitive BOP coupling (REBOP) in solid phase peptide synthesis: Luliberin synthesis as model. Int. J. Peptide Protein Res. 35:465–472.

Seyfarth, L., Pineda de Castro, L. F., Liepina, I., Paegelow, I., Liebmann, C., and Reissmann, S. (1995). New cyclic bradykinin antagonists containing disulfide and lactam bridges at the *N*-terminal sequence. Int. J. Peptide Protein Res. 46:155–165.

Shao, J., and Tam, J. P. (1995). Unprotected peptides as building blocks for the synthesis of peptide dendrimers with oxime, hydrazone, and thiazolidine linkages. J. Am. Chem. Soc. 117:3893–3899.

Sheppard, R. C., and Williams, B. J. (1982). Acid-labile resin linkage agents for use in solid phase peptide synthesis. Int. J. Peptide Protein Res. 20:451–454.

Shin, Y., Winans, K. A., Backes, B. J., Kent, S. B. H., Ellman, J. A., and Bertozzi, C. R. (1999). Fmoc-based synthesis of peptide-$^\alpha$thioesters: application to the total chemical synthesis of a glycoprotein by native chemical ligation. J. Am. Chem. Soc. 121:11684–11689.

Sieber, P. (1987a). An improved method for anchoring of 9-fluorenylmethoxycarbonyl-amino acids to 4-alkoxybenzyl alcohol resins. Tetrahedron Lett. 28:6147–6150.

Sieber, P. (1987b). Modification of tryptophan residues during acidolysis of 4-methoxy-2,3,6-trimethylbenzenesulfonyl groups: Effects of scavengers. Tetrahedron Lett. 28:1637–1640.

Sieber, P. (1987c). A new acid-labile anchor group for the solid-phase synthesis of C-terminal peptide amides by the Fmoc method. Tetrahedron Lett. 28:2107–2110.

Sieber, P., and Riniker, B. (1987). Protection of histidine in peptide synthesis: A reassessment of the trityl group. Tetrahedron Lett. 28:6031–6034.

Sieber, P., and Riniker, B. (1990). Side-chain protection of asparagine and glutamine by trityl: Application to solid-phase peptide synthesis. In Innovation and Perspectives in Solid Phase Synthesis, R. Epton, ed., Birmingham, UK, Solid Phase Conference Coordination, Ltd., pp. 577–583.

Sieber, P., and Riniker, B. (1991). Protection of carboxamide functions by the trityl residue: Application to peptide synthesis. Tetrahedron Lett. 32:739–742.

Sieber, P., Kamber, B., Riniker, B., and Rittel, W. (1980). Iodine oxidation of S-trityl- and S-acetamidomethyl-cysteine-peptides containing tryptophan: Conditions leading to the formation of tryptophan-2-thioethers. Helv. Chim. Acta 63:2358–2363.

Sigler, G. F., Fuller, W. D., Chaturvedi, N. C., Goodman, M., and Verlander, M. S. (1983). Formation of oligopeptides during the synthesis of 9-fluorenylmethyloxycarbonyl amino acid derivatives. Biopolymers 22:2157–2162.

Simmons, J., and Schlesinger, D. H. (1980). High-performance liquid chromatography of side-chain-protected amino acid phenylthiohydantoins. Anal. Biochem. 104:254–258.

Small, P. W., and Sherrington, D. C. (1989). Design and application of a new rigid support for high efficiency continuous-flow peptide synthesis. J. Chem. Soc., Chem. Commun. 1589–1591.

Solé, N. A., and Barany, G. (1992). Optimization of solid-phase synthesis of [Ala^8]-dynorphin A. J. Org. Chem. 57:5399–5403.

Songster, M. F., and Barany, G. (1997). Handles for solid-phase peptide synthesis. Methods Enzymol. 289:126–174.

Southard, G. L. Comments. (1971). In Peptides 1969, E. Scoffone, ed., Amsterdam, North-Holland Publishing, pp. 168–170.

Spetzler, J. C., and Tam, J. P. (1995). Unprotected peptides as building blocks for branched peptides and peptide dendrimers. Int. J. Peptide Protein Res. 45:78–85.

Steiman, D. M., Ridge, R. J., and Matsueda, G. R. (1985). Synthesis of side chain-protected amino acid phenylthiohydantoins and their use in quantitative solid-phase Edman degradation. Anal. Biochem. 145:91–95.

Stephenson, R. C., and Clarke, S. (1989). Succinimide formation from aspartyl and asparaginyl peptides as a model for the spontaneous degradation of proteins. J. Biol. Chem. 264:6164–6170.

Stewart, J. M. (1997). Cleavage methods following Boc-based solid-phase peptide synthesis. Methods Enzymol. 289:29–44.

Stewart, J. M., and Klis, W. A. (1990). Polystyrene-based solid phase peptide synthesis: The state of the art. In Innovation and Perspectives in Solid

Phase Synthesis, R. Epton, ed., Birmingham, UK, Solid Phase Conference Coordination, Ltd., pp. 1–9.

Stewart, J. M., and Young, J. D. (1984). *Solid Phase Peptide Synthesis*, 2nd edn., Rockford, Ill., Pierce Chemical Co.

Stewart, J. M., Knight, M., Paiva, A. C.M., and Paiva, T. (1972). Histidine in solid phase peptide synthesis: Thyrotropin releasing hormone and the angiotensins. In *Progress in Peptide Research*, Vol. 2, S. Lande, ed., New York, Gordon and Breach, pp. 59–64.

Stewart, J. M., Ryan, J. W., and Brady, A. H. (1974). Hydroxyproline analogs of bradykinin. J. Med. Chem. 17:537–539.

Stierandová, A., Sepetov, N. F., Nikiforovich, G. V., and Lebl, M. (1994). Sequence-dependent modification of Trp by the Pmc protecting group of Arg during TFA deprotection. Int. J. Peptide Protein Res. 43:31–38.

Story, S. C., and Aldrich, J. V. (1992). A resin for the solid phase synthesis of protected peptide amides using the Fmoc chemical protocol. Int. J. Peptide Protein Res. 39:87–92.

Stüber, W., Knolle, J., and Breipohl, G. (1989). Synthesis of peptide amides by Fmoc-solid-phase peptide synthesis and acid labile anchor groups. Int. J. Peptide Protein Res. 34:215–221.

Sueiras-Diaz, J., and Horton, J. (1992). First solid phase synthesis of endothelial interleukin-8 [Ala-IL8]$_{77}$ using Boc-TBTU chemistry with in situ neutralisation and comparison with synthesis of monocyte interleukin-8 [Ser-IL8]$_{72}$ using the DCC-HOBt method. Tetrahedron Lett. 33:2721–2724.

Sugg, E. E., Castrucci, A. M. de L., Hadley, M. E., van Binst, G., and Hruby, V. J. (1988). Cyclic lactam analogues of Ac-[Nle^4]α-MSH$_{4\text{-}11}$-NH$_2$. Biochemistry 27:8181–8188.

Suzuki, K., Nitta, K., and Endo, N. (1975). Suppression of diketopiperazine formation in solid phase peptide synthesis. Chem. Pharm. Bull. 23:222–224.

Talbo, G., Wade, J. D., Dawson, N., Manoussios, M., and Tregear, G. W. (1997). Rapid semi on-line monitoring of Fmoc solid-phase peptide synthesis by matrix-assisted laser desorption/ionization mass spectrometry. Lett. Peptide Sci. 4:121–127.

Tam, J. P. (1988). Synthetic peptide vaccine design:Synthesis and properties of a high-density multiple antigenic peptide system. Proc. Natl. Acad. Sci. USA 85:5409–5413.

Tam, J. P., and Lu, Y.-A. (1990). Synthetic peptide vaccine engineering: Design and synthesis of unambiguous peptide-based vaccines containing multiple peptide antigens for malaria and hepatitis. In *Innovation and Perspectives in Solid Phase Synthesis*, R. Epton, ed., Birmingham, UK, Solid Phase Conference Coordination, Ltd., pp. 351–370.

Tam, J. P., and Lu, Y.-A. (1995). Coupling difficulty associated with interchain clustering and phase transition in solid phase peptide synthesis. J. Am. Chem. Soc. 117:12058-12063.

Tam, J. P., and Merrifield, R. B. (1985). Solid phase synthesis of gastrin I: Comparison of methods utilizing strong acid for deprotection and cleavage. Int. J. Peptide Protein Res. 26:262–273.

Tam, J. P., and Merrifield, R. B. (1987). Strong acid deprotection of synthetic peptides: Mechanisms and methods. In *The Peptides*, Vol. 9, S.

Udenfriend and J. Meienhofer, eds., New York, Academic Press, pp. 185–248.

Tam, J. P., Kent, S. B. H., Wong, T. W., and Merrifield, R. B. (1979). Improved synthesis of 4-(Boc-aminoacyloxymethyl)-phenylacetic acids for use in solid phase peptide synthesis. Synthesis 955–957.

Tam, J. P., and Spetzler, J. C. (1997). Multiple antigen peptide system. Methods Enzymol. Tam, J. P., Health, W. F., and Merrifield, R. B. (1983). S_N2 deprotection of synthetic peptides with a low concentration of HF in dimethyl sulfide: Evidence and application in peptide synthesis. J. Am. Chem. Soc. 105:6442–6455.

Tam, J. P., Riemen, M. W., and Merrifield, R. B. (1988). Mechanisms of aspartimide formation: The effects of protecting groups, acid, base, temperature and time. Peptide Res. 1:6–18.

Tam, J. P., Liu, W., Zhang, J.-W., Galantino, M., and de Castiglione, R. (1991a). D-Amino acid and alanine scans of endothelin: An approach to study refolding intermediates. In Peptides 1990, E. Giralt and D. Andreu, eds., Leiden, Escom, pp. 160–163.

Tam, J. P., Wu, C.-R., Liu, W., and Zhang, J.-W. (1991b). Disulfide bond formation in peptides by dimethyl sulfoxide:Scope and applications. J. Am. Chem. Soc. 113:6657–6662.

Tamamura, H., Matsumoto, F., Sakano., K., Ibuka, T., and Fujii, N. (1998). Unambiguous synthesis of stromal cell-derived factor-1 by regioselective disulfide bond formation using a DMSO–aqueous HCl system. J. Chem. Soc., Chem. Commun. 151–153.

Ten Kortenaar, P. B. W., and van Nispen, J. W. (1988). Formation of open-chain asymmetrical cystine peptides on a solid support: Synthesis of pGlu-Asn-Cyt-Pro-Arg-Gly-OH. Coll. Czech. Chem. Commun. 53:2537–2541.

Ten Kortenaar, P. B. W., van Dijk, B. G., Peters, J. M., Raaben, B. J., Adams, P. J. H. M., and Tesser, G. I. (1986). Rapid and efficient method for the preparation of Fmoc-amino acids starting from 9-fluorenylmethanol. Int. J. Peptide Protein Res. 27:398–400.

Ten Kortenaar, P. B. W., Hendrix, B. M. M., and van Nispen, J. W. (1990). Acid-catalyzed hydrolysis of peptide-amides in the solid state. Int. J. Peptide Protein Res. 36:231–235.

Tesser, G. I., and Balvert-Geers, I. C. (1975). The methylsulfonylethyloxycarbonyl group, a new and versatile amino protective function. Int. J. Peptide Protein Res. 7:295–305.

Tesser, M., Albericio, F., Pedroso, E., Grandas, A., Eritja, R., Giralt, E., Granier, C., and van Rietschoten, J. (1983). Amino-acids condensations in the preparation of N^α-9-fluorenylmethyloxycarbonylamino-acids with 9-fluorenylmethylchloroformate. Int. J. Peptide Protein Res. 22:125–128.

Thaler, A., Seebach, D., and Cardinaux, F. (1991). Lithium-salt effects in peptide synthesis, part II: Improvement of degree of resin swelling and of efficiency of coupling in solid-phase synthesis. Helv. Chim. Acta 74:628–643.

Tregear, G. W. (1972). Graft copolymers as insoluble supports in peptide synthesis. In Chemistry and Biology of Peptides, J. Meienhofer, ed., Ann Arbor, Mich., Ann Arbor Sci. Publ., pp. 175–178.

Tregear, G. W., van Rietschoten, J., Sauer, R., Niall, H. D., Keutmann, H. T., and Potts, Jr., J. T. (1977). Synthesis, purification, and chemical characterization of the amino-terminal 1–34 fragment of bovine parathyroid hormone synthesized by the solid-phase procedure. Biochemistry 16: 2817–2823.

Trzeciak, A., and Bannwarth, W. (1992). Synthesis of "head to tail" cyclized peptides on solid support by Fmoc chemistry. Tetrahedron Lett. 33:4557–4560.

Tuchscherer, G. (1993). Template assembled synthetic proteins: Condensation of a multifunctional peptide to a topological template via chemoselective ligation. Tetrahedron Lett. 34:8419–8422.

Ueki, M., and Amemiya, M. (1987). Removal of 9-fluorenylmethyloxycarbonyl (Fmoc) group with tetrabutylammonium fluoride. Tetrahedron Lett. 28:6617–6620.

Van Abel, R. J., Tang, Y.-Q., Rao, V. S. V., Dobbs, C. H., Tran, D., Barany, G., and Selsted, M. E. (1995). Synthesis and characterization of indolicidin, a tryptophan-rich antimicrobial peptide from bovine neutrophils. Int. J. Peptide Protein Res. 45:401–409.

van der Eijk, J. M., Nolte, R. J. M., and Zwikker, J. W. (1980). A simple and mild method for the removal of the N^{Im}-tosyl protecting group. J. Org. Chem. 45:547–548.

van Nispen, J. W., Polderdijk, J. P., and Greven, H. M. (1985). Suppression of side-reactions during the attachment of Fmoc-amino acids to hydroxymethyl polymers. Recl. Trav. Chim. Pays-Bas 104:99–100.

van Regenmortel, M. V. H., and Muller, S. (1999). Synthetic Peptides as Antigens: Laboratory Techniques in Biochemistry and Molecular Biology, Vol. 28, Amsterdam, Elsevier.

van Woerkom, W. J., and van Nispen, J. W. (1991). Difficult couplings in stepwise solid phase peptide synthesis: Predictable or just a guess? Int. J. Peptide Protein Res. 38:103–113.

Veber, D. F., Milkowski, J. D., Varga, S. L., Denkewalter, R. G., and Hirschmann, R. (1972). Acetamidomethyl: A novel thiol protecting group for cysteine. J. Am. Chem. Soc. 94:5456–5461.

Vlattas, I., Delluureficio, J., Dunn, R., Sytwwu, I. I., and Stanton, J. (1997). The use of thioesters in solid-phase organic synthesis. Tetrahedron Lett. 38:7321–7324.

Voss, C., and Birr, C. (1981). Synthetic insulin by selective disulfide bridging II: Polymer phase synthesis of the human B chain fragments. Hoppe-Seyler's Z. Physiol. Chem. 362:717–725.

Vuljanic, T., Bergquist, K. E., Clausen, H., Roy, S., and Kihlberg, J. (1996). Piperidine is preferred to morpholine for Fmoc cleavage in solid phase glycopeptide synthesis as exemplified by preparation of glycopeptides related to HIV GP120 and mucins. Tetrahedron 52:7983–8000.

Wade, J. D., Fitzgerald, S. P., McDonald, M. R., McDougall, J. G., and Tregear, G. W. (1986). Solid-phase synthesis of α-human atrial natriuretic factor: Comparison of the Boc-polystyrene and Fmoc-polyamide methods. Biopolymers 25:S21–S37.

Wade, J. D., Bedford, J., Sheppard, R. C., and Tregear, G. W. (1991). DBU as an N^{α}-deprotecting reagent for the fluorenylmethoxycarbonyl group in continuous flow solid-phase peptide synthesis. Peptide Res. 4:194–199.

Wakamiya, T., Saruta, K., Yasuoka, J., and Kusumoto, S. (1994). An efficient procedure for solid-phase synthesis of phosphopeptides by the Fmoc strategy. Chem. Lett. 1099–1102.

Wallace, C. J. A., Mascagni, P., Chait, B. T., Collawn, J. F., Paterson, Y., Proudfoot, A. E. I., and Kent, S. B. H. (1989). Substitutions engineered by chemical synthesis at three conserved sites in mitochondrial cytochrome c. J. Biol. Chem. 264:15199–15209.

Wang, S. S. (1973). p-Alkoxybenzyl alcohol resin and p-alkoxybenzyloxycarbonylhydrazide resin for solid phase synthesis of protected peptide fragments. J. Am. Chem. Soc. 95:1328–1333.

Wang, S. S. (1976). Solid phase synthesis of protected peptides via photolytic cleavage of the α-methylphenacyl ester anchoring linkage. J. Org. Chem. 41:3258–3261.

Wang, S. S., and Merrifield, R. B. (1969). Preparation of some new biphenylisopropyloxycarbonyl amino acids and their application to the solid phase synthesis of a tryptophan-containing heptapeptide of bovine parathyroid hormone. Int. J. Protein Res. 1:235–244.

Wang, S. S., Matsueda, R., and Matsueda, G. R. Automated peptide synthesis under mild conditions. In Peptide Chemistry 1981, T. Shioiri, ed., Osaka, Protein Research Foundation, pp. 37–40.

Wang, S. S., Chen, S. T., Wang, K. T., and Merrifield, R. B. (1987). 4-methoxybenzyloxycarbonyl amino acids in solid phase peptide synthesis. Int. J. Peptide Protein Res. 30:662–667.

Weber, R. W., and Nitschmann, H. (1978). Der einfluss der O-acetylierung auf das konformative verhalten des kollagen-modellpeptides (L-Pro-L-Hyp-Gly)$_{10}$ und von gelatine. Helv. Chim. Acta 61:701–708.

Wenschuh, H., Beyermann, M., Krause, E., Brudel, M., Winter, R., Schümann, M., Carpino, L. A., and Bienert, M. (1994). Fmoc amino acid fluorides: Convenient reagents for the solid-phase assembly of peptides incorporating sterically hindered residues. J. Org. Chem. 59:3275–3280.

Wenschuh, H., Beyermann, M., Haber, H., Seydel, J. K., Krause, E., Bienert, M., Carpino, L. A., El-Faham, A., and Albericio, F. (1995). Stepwise automated solid phase synthesis of naturally occurring peptaibols using Fmoc amino acid fluorides. J. Org. Chem. 60:405–410.

Weygand, F., Steglich, W., and Bjarnason, J. (1968a). Leicht abspaltbare schutzgruppen für die säureamidfunktion 3: Derivate des asparagins und glutamins mit 2,4-dimethoxy-benzyl-und 2. 4.6-trimethoxy-benzylgeschützten amidgruppen. Chem. Ber. 101:3642–3648.

Weygand, F., Steglich, W., and Chytil, N. (1968b). Bildung von N-succinimidoxycarbonyl-β-alanin-amiden bei amidsynthesen mit dicyclohexylcarbodiimid/N-hydroxysuccinimid. Z. Naturforschg. 23b:1391–1392.

White, P. (1992). Fmoc-Trp(Boc)-OH:A new derivative for the synthesis of peptides containing tryptophan. In Peptides: Chemistry and Biology, J. A. Smith and J. E. Rivier, eds., Leiden, Escom, pp. 537–538.

White, P., and Beythien, J. (1996). Preparation of phospho-serine, threonine, and tyrosine containing peptides by the Fmoc methodology using pre-formed phosphoamino acid building blocks. In *Innovation and Perspectives in Solid Phase Synthesis and Combinatorial Libraries 1996*, R. Epton, ed., Kingswinford, Mayflower Scientific, pp. 557–560.

Williams, M. J., Muir, T. W., Ginsberg, M. H., and Kent, S. B. H. (1994). Total chemical synthesis of a folded β-sandwich protein domain: An analog of the tenth fibronectin type 3 module. J. Am. Chem. Soc. 116:10797-10798.

Wlodawer, A., Miller, M., Jaskólski, M., Sathyanarayana, B. K., Baldwin, E., Weber, I. T., Selk, L. M., Clawson, L., Schneider, J., and Kent, S. B. H. (1989). Conserved folding in retroviral proteases: Crystal structure of a synthetic HIV-1 protease. Science 245:616–621.

Wolfe, H. R., and Wilk, R. R. (1989). The RaMPS system: Simplified peptide synthesis for life science researchers. Peptide Res. 2:352–356.

Wu, C.-R., Wade, J. D., and Tregear, G. W. (1988). β-Subunit of baboon chorionic gonadotropin: Continuous flow Fmoc-polyamide synthesis of the C-terminal 37-peptide. Int. J. Peptide Protein Res. 31: 47–57.

Wu, C.-R., Stevens, V. C., Tregear, G. W., and Wade, J. D. (1989). Continuous-flow solid-phase synthesis of a 74-peptide fragment analogue of human β-chorionic gonadotropin. J. Chem. Soc. Perkin Trans. I: 81–87.

Wünsch, E. (1974). Synthese von peptiden. In *Houben-Weyl's Methoden der Organischen Chemie*, Vol. 15, parts 1 and 2, E. Müller, ed., Stuttgart, Thieme.

Wünsch, E., and Spangenberg, R. (1971). Eine neue S-schutzgruppe für cystein. In *Peptides 1969*, E. Scoffone, ed., Amsterdam, North-Holland Publ., pp. 30–34.

Wünsch, E., Moroder, L., Wilschowitz, L., Göhring, W., Scharf, R., and Gardner, J. D. (1981). Zur totalsynthese von cholecystokinin-pankreozymin: Darstellung des verknüpfungsfähigen "schlüsselfragments" der sequenz 24-33. Hoppe-Seyler's Z. Physiol. Chem. 362:143–152.

Xu, Q., Zheng, J., Cowburn, D., and Barany, G. (1996). Synthesis and characterization of branched phosphopeptides: Prototype consolidated ligands for SH(32) domains. Lett. Peptide Sci. 3:31–36.

Xu, Q., Ottinger, E. A., Solé, N. A., and Barany, G. (1997). Detection and minimization of H-phosphonate side reaction during phosphopeptide synthesis by a post assembly global phosphorylation strategy. Lett. Peptide Sci. 3:333–342.

Yajima, H., Takeyama, M., Kanaki, J., and Mitani, K. (1978). The mesitylene-2-sulphonyl group, an acidolytically removable N^G-protecting group for arginine. J. Chem. Soc., Chem. Commun. 482–483.

Yajima, H., Fujii, N., Funakoshi, S., Watanabe, T., Murayama, E., and Otaka, A. (1988). New strategy for the chemical synthesis of proteins. Tetrahedron 44:805–819.

Yamashiro, D. (1987). Preparation and properties of some crystalline symmetrical anhydrides of N^α-*tert*-butyloxycarbonyl-amino acids. Int. J. Peptide Protein Res. 30:9–12.

Yamashiro, D., and Li, C. H. (1973). Protection of tyrosine in solid-phase peptide synthesis. J. Org. Chem. 38:591–592.

Yamashiro, D., Blake, J., and Li, C. H. (1976). The use of trifluoroethanol for improved coupling in solid-phase peptide synthesis. Tetrahedron Lett.:1469–1472.

Yang, Y., Sweeney, W. V., Scheider, K., Thörnqvist, S., Chait, B. T., and Tam, J. P. (1994). Aspartimide formation in base-driven 9-fluorenyl-methoxycarbonyl chemistry. Tetrahedron Lett. 35:9689–9692.

Yoshida, M., Tatsumi, T., Fujiwara, Y., Iinuma, S., Kimura, T., Akaji, K., and Kiso, Y. (1990). Deprotection of the S-trimethylacetamidomethyl (Tacm) group using silver tetrafluoroborate: Application to the synthesis of porcine brain natriuretic peptide-32 (pBNP-32). Chem. Pharm. Bull. 38:1551–1557.

Young, J. D., Huang, A. S., Ariel, N., Bruins, J. B., Ng, D., and Stevens, R. L. (1990). Coupling efficiencies of amino acids in solid phase synthesis of peptides. Peptide Res. 3:194–200.

Young, S. C., White, P. D., Davies, J. W., Owen, D. E.I. A., Salisbury, S. A., and Tremeer, E. J. (1990). Counterion distribution monitoring: A novel method for acylation monitoring in solid-phase synthesis. Biochem. Soc. Trans. 18:1311–1312.

Yu, H.-M., Chen, S.-T., Chiou, S.-H., and Wang, K.-T. (1988). Determination of amino acids on Merrifield resin by microwave hydrolysis. J. Chromatogr. 456:357–362.

Zalipsky, S., Albericio, F., and Barany, G. (1985). Preparation and use of an aminoethyl polyethylene glycol-crosslinked polystyrene graft resin support for solid-phase peptide synthesis. In Peptides: Proceedings of the Ninth American Peptide Symposium, C. M. Deber, V. J. Hruby, and K. D. Kopple, eds., Rockford, Ill., Pierce Chemical Co., pp. 257–260.

Zalipsky, S., Albericio, F., Slomczynska, U., and Barany, G. (1987). A convenient general method for synthesis of N^{α}- or N^{ω}-dithiasuccinoyl (Dts) amino acids and dipeptides: Application of polyethylene glycol as a carrier for funtional purification. Int. J. Peptide Protein Res. 30:740–783.

Zalipsky, S., Chang, J. L., Albericio, F., and Barany, G. (1994). Preparation and applications of polyethylene glycol-polystyrene graft resin supports for solid-phase peptide synthesis. Reactive Polymers 22:243–258.

Zardeneta, G., Chen, D., Weintraub, S. T., and Klebe, R. J. (1990). Synthesis of phosphotyrosyl-containing phosphopeptides by solid-phase peptide synthesis. Anal. Biochem. 190:340–347.

Zhang, L., Goldhammer, C., Henkel, B., Panhaus, G., Zuehl, F., Jung, G., and Bayer, E. (1994). "Magic mixture," a powerful solvent system for solid-phase synthesis of difficult peptides. In Innovation and Perspectives in Solid Phase Synthesis, R. Epton, ed., Birmingham, UK, Mayflower Worldwide, Ltd., pp. 711–716.

4

Evaluation of the Synthetic Product

Gregory A. Grant

In 1987, an article appeared in the *International Journal of Peptide and Protein Research* commemorating the 25th anniversary of the development of solid phase peptide synthesis (Barany et al., 1987). While that article dealt with many aspects of peptide synthesis, one statement in particular stands out as exemplifying the rationale for this chapter. It states: "No synthetic endeavor can be considered complete until the product has been adequately purified and subjected to a battery of analytical tests to verify its structure."

The characterization or evaluation of a synthetic peptide is the one step in its production and experimental utilization that will validate the experimental data obtained. Unfortunately, it is also the one step that many investigators all too often give too little attention. If the synthetic product, upon which the theory and performance of the experimental investigation is based, is not the intended product, the conclusions will be incorrect. Without proper characterization, the investigator will either have to be lucky, or be wrong. Worse yet, he or she will not know which is the case.

Although today the synthesis of a given peptide is often considered routine, the product should never be taken for granted. Peptide synthesis chemistry, although quite sophisticated, is complex and subject to a variety of problems. These problems, which can manifest themselves as unwanted side reactions and decreased reaction efficiency, are subject to a variety of factors such as reagent quality, incompatible chemistries, instrument malfunctions, sequence specific effects, and operator error. Although every effort is made to eliminate their causes and to plan for potential problems in the design and synthesis steps, it is not always successful and the eventual outcome of a synthesis is not always predictable. One must never assume that the final product is the expected one until that has been proven to be the case. To do otherwise may seriously jeopardize the outcome of the research.

Used and performed properly, the evaluation stage is where the fruits of the synthesis are scrutinized and the decision is made to use the peptide as intended, submit it to further purification, or re-synthesize it and possibly change elements of the design or synthesis protocols. The battery of tests referred to in the above quote are the tools that are used to assure the investigator that he or she has obtained the intended peptide and are the main subject of this chapter.

General Considerations

Before embarking on a detailed discussion of the techniques and theory of peptide evaluation, it is perhaps useful to discuss a few general considerations that one should be mindful of within the context of the ensuing sections. First, homogeneity and correct covalent structure are the basic goals of the synthetic endeavor and are the two properties or concepts that pervade any and all considerations of the state of the synthetic product. Second, a familiarity with the properties of the amino acid residues may have direct bearing on the design of the peptide and how it is handled. Finally, the decision to use the synthetic product as is or to undergo further purification depends on what the peptide will be used for and how much purification and characterization is necessary or feasible within the context of that use. That question becomes even more relevant from the viewpoint of synthesis facilities which are a major source of synthetic peptides today, and which produce those peptides for a large group of diverse investigators. Elements of both time and cost must be considered carefully when deciding how much purification and characterization is necessary or feasible.

Strategy for the Evaluation of Synthetic Peptides

The general strategy for the evaluation of synthetic peptides is presented diagrammatically in figure 4-1. This diagram depicts the general course of the evaluation process as it might be carried out and indicates the types of analyses that are available and which

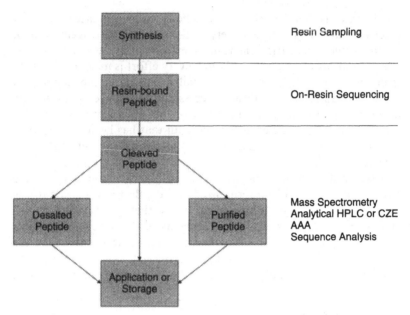

Figure 4-1. Stages in the production of a synthetric peptide for experimental use. The common types of analytical techniques that are applicable to each stage are listed on the right.

might be helpful at any particular stage. The individual components of this diagram are discussed in specific sections of this chapter. When approached in a systematic manner, a clear picture of the quality of the synthesis will emerge and will allow an accurate and clear assessment of the peptide product.

At each stage, starting with the unpurified material and ending with the homogeneous peptide, the product must be evaluated to ascertain whether the desired purity has been reached and whether the covalent nature of the peptide is correct. The latter stages of the evaluation process comprise a purification–evaluation cycle. This cycle of purification and evaluation may have to be repeated several times before the desired result is obtained and the synthetic chemist feels confident that the product obtained is the one intended.

The "battery of analytical tests" generally available for the evaluation of synthetic peptides are the tools that the chemist uses to assess homogeneity and covalent structure. The commonly used techniques are analytical high-performance (or high-pressure) liquid chromatography (HPLC), capillary electrophoresis (CE), amino acid compositional analysis, UV spectroscopy, sequence analysis, and mass spectrometry. Table 4-1 indicates the relative contributions of these tools toward the two goals of assessing homogeneity and covalent

Table 4-1 Advantages and limitations of analytical techniques

Technique	Advantages	Limitations
Reverse-phase HPLC	Assesses level of heterogeneity Useful for purification	Does not yield structural information
Capillary electrophoresis	Assesses level of heterogeneity Provides assessment based on a second physical principal	Not useful for preparative scale Does not yield structural information
Amino acid analysis	Accurately quantitates peptide levels Provides mol ratio of amino acids	Some amino acids not quantitated Accuracy decreases for large peptides Does not distinguish heterogeneous species
Ultraviolet spectroscopy	Assists in monitoring tryptophan integrity Useful for monitoring purification	Does not distinguish heterogeneous species Does not yield structural information
Sequence analysis	Determines linear order of amino acids Very good for detecting deletion and addition sequences Detects presence of some adducts	Blocked peptides are invisible Cannot sequence long peptides to the end Not reliable for quantitation May destroy or reverse some adducts
Mass spectrometry	Provides accurate peptide mass Can indicate heterogeneity Can give sequence information	Not quantitative

structure, and summarizes some of their advantages and limitations. Each one makes its own specialized contribution to the evaluation process by providing unique information, and they are generally used in a complementary fashion. Taken as a whole, the information provided by these analytic techniques should be sufficient to allow the investigator to correctly judge the state of the synthetic peptide.

Infrared and nuclear magnetic resonance (NMR) spectroscopy can also be useful tools for the evaluation of synthetic peptides, but are usually necessary only in cases where secondary structures or new chemistries are being investigated. In addition, NMR availability is often limited, particularly for core facilities, and is expensive and time consuming. Since these two techniques are generally not needed for routine evaluation, they will not be discussed here.

Analytical HPLC or CE is usually the first level of assessment. These two techniques give a relatively rapid overall picture or "feel" for the complexity of the synthetic product in that they reveal the relative number of different species present in the preparation, but they lack direct information about covalent structure. Amino acid analysis quantitates the amount of the peptide and informs whether the requisite amino acids are present at reasonable ratios. However, it is less reliable for certain unstable amino acids such as cysteine, methionine, and tryptophan, and the compositional accuracy decreases as the peptide gets larger. UV spectroscopy monitors the integrity of the aromatic amino acids, particularly tryptophan, and is very useful in monitoring peptide purification, but the amount of information that can be obtained about the covalent state of the peptide in general is very limited. Sequence analysis and mass spectrometry come as close as any to addressing both homogeneity and covalent structure in a single step. However, while these techniques give the best assessment of the covalent nature of the peptide, it must be pointed out that they can lead to incorrect conclusions about the sample if they are relied on exclusively. For example, peptides which are blocked to the Edman chemistry at their amino terminus, either as a consequence of the synthesis chemistry employed or because of an unwanted side reaction, will be completely invisible to sequence analysis. Mass spectrometry is not generally found to be a quantitative technique, so conclusions of relative homogeneity based on mass analysis can be misleading. However, it has generally been found that in a mixture of related species, as would result from a single synthesis run, that the apparent relative level of species is fairly indicative of the yield of those species. Mass spectrometry of peptides has advanced considerably since the first edition of this book was published. Today it is routinely used, sometimes exclusively, as the best and most informative method for analysis of synthetic peptides. Nonetheless, the other analytical techniques listed above can still be useful, often as part of a more in-depth evaluation of a peptide preparation because certain problems were identified from the mass spectra, or for the more rigorous evaluation of homogeneity and yield.

In the examples that follow throughout this chapter, there are instances where incorrect conclusions would have been made if they were based only on the information provided by a single analytical technique. Indeed, in many cases, information from several analytical techniques is necessary in order for the correct conclusion to be made.

Homogeneity and Purification

The assessment of homogeneity of a peptide goes hand in hand with the purification process. First of all, homogeneity is not a characteristic that can be proven directly in that there is no single test that, when met, automatically indicates that a peptide is a clean, single species. Rather, homogeneity is demonstrated by an inability to demonstrate heterogeneity, or the presence of more than one species. This implies, of course, that reasonable attempts have been made to demonstrate heterogeneity under conditions where it should be detected if it is present and that good judgment, based on experience with the techniques used, has been exercised. It often follows that if a particular technique or procedure indicates the existence of multiple species, it also provides either the means to separate and purify the individual species or the evidence upon which a separation protocol can be based. For example, if more than one peptide peak is found using analytical reverse phase HPLC, the individual peptides should be able to be isolated one from the other by collecting fractions of the column effluent and thus physically separating the individual peptides into separate fractions. In contrast, if two sequences show up on automated Edman degradation, the sequencing procedure cannot in itself be used to physically separate the two species, but the sequence information obtained may suggest a separation procedure, say one based on charge differences, that will be productive.

Typically synthetic chemists are called on to produce peptides of two general categories, simply referred to as either unpurified or purified. Unpurified peptides consist of either the synthetic product as it comes directly from the synthesis or after only minimal purification is performed, such as desalting (see figure 4-1). Purified peptides, as the name implies, are those which have undergone rigorous purification to homogeneity or near homogeneity. Obviously, the evaluation criteria applied to each of these categories are quite different. The desired state of the peptide and the extent of evaluation needed depend on what the peptide is intended for, or how it will be used. If the peptide is intended to be used as an antigen for producing polyclonal antisera directed against a protein epitope, perhaps a lesser degree of purification may be acceptable than for a peptide which will be used for detailed conformational studies. Although it may ideally be desirable to purify all peptides to a high degree of homogeneity, there are often practical limitations in the amount of time and effort spent, as well as fiscal limitations. Moreover, a core laboratory may get requests for unpurified peptides from an investigator who intends to do his own purification. It must be emphasized, however, that accepting a lesser degree of purification does not mean sacrificing the integrity of the product. The intended covalent structure should surely be present and some minimal criteria, such as mass spectrometry, should be assessed to verify the likelihood of that situation.

Covalent Structure

In most instances, before the covalent structure of a peptide can be unambiguously determined, the peptide must be relatively homogeneous. Although many techniques used for the evaluation of covalent structure can in themselves indicate the presence of multiple species, those techniques, with the possible exception of mass spectrometry, are not generally able to provide a specific or unambiguous analysis of the covalent nature of each species within a mixture.

The two main factors to be evaluated in the assessment of covalent structure of a peptide are the presence of the correct sequence of amino acids and the presence, or preferably absence, of derivatives or adducts of those amino acids. Probably, the most common problem encountered is that of deletion or addition sequences. Deletion sequence peptides are those that contain an internal absence or deletion of an amino acid residue brought about by either incomplete coupling to or incomplete deprotection of the α-amino group at a particular cycle (Tregear et al., 1977). Addition sequence peptides are those that contain two residues of the same amino acid adjacent to each other in the sequence when only one was intended at that particular location. The degree to which deletion and addition sequences occur depends on the conditions of the individual synthesis, but as long as these do not occur too often during a synthesis and do not represent an overwhelming proportion of the total product, the full-length peptide may still be present in appreciable amounts.

Deletion sequences due to incomplete coupling, but not incomplete deblocking, or other causes (see chapter 3) can be largely eliminated if capping is used. Capping is the chemical derivitization of free α-amino groups, usually by acetylation with acetic anhydride, before deprotection of the just added residue takes place. This results in an irreversibly blocked peptide that is no longer able to be elongated. Although capped peptides will represent contaminants that will have to be removed from the final product, they are generally easier to separate from the full-length peptide than are deletion sequence peptides which may differ from the full-length peptide by only one or two amino acids. Addition sequences are less likely but are more problematic and cannot be prevented by a routine procedure such as capping.

Unwanted modifications usually come in two general forms. They are either side-chain-protecting groups that were not completely removed, or random adducts that were produced as a result of un-controlled side reactions during the synthesis or cleavage and depro-tection steps. Some of the more common side reactions encountered are described in chapter 3 which deals specifically with the synthesis chemistry. Although identification of the exact nature of the adduct may help to prevent it in subsequent syntheses, it is not always possible, or necessary, to do so. If the modified peptides do not represent too high a proportion of the sample, which is often the case, they can often be removed during the purification procedure. A common observation in peptide synthesis is that the crude product will contain a variety of species in addition to the intended product. An acceptable synthesis is one in which there is a predominant com-ponent, when analyzed by HPLC or CE, and some number of minor components which can be separated away. The minor components of the synthesis represent side products due to unwanted modifications and inefficiencies in the chemistry. As long as they do not represent an overwhelming proportion of the product, they can be removed without their identity ever needing to be determined. It is only when the unwanted side products become the major species that the chemical protocols or instrument operation needs to be changed.

Residues That Present Potential Problems

There are certain amino acid residues that can be the source of pro-blems in the finished peptide that deserve some consideration. The problems considered here are not meant to include most of those that result from unwanted reactions during the synthesis, cleavage, and deprotection chemistry as discussed in chapter 3. Rather, they are those that may develop subsequently due to the characteristics of certain amino acid residues in the peptide. They may develop during the work-up of the peptide, or during handling or storage, and present a particular problem in the evaluation steps or the application of the peptide. These amino acid residues are cysteine, methionine, Tryptophan, amino-terminal glutamine, and aspartic acid when it is followed by glycine or alanine.

The main problem with tryptophan, methionine, and cysteine is that they are particularly susceptible to oxidation. Of these, methio-nine and cysteine are the least troublesome because their common oxidation products are usually reversible. When possible, it is perhaps advisable to work below neutral pH and with degassed or de-oxygenated solvents when working up peptides that contain these residues.

Methionine

Methionine is fairly easily oxidized to methionine sulfoxide. Methionine sulfoxide can be converted back to methionine by treatment with thiol reagents such as dithiothreitol or N-methyl-mercaptoacetamide (Cullwell, 1987; Houghton and Li, 1979), but β-mercaptoethanol is not as effective for this purpose (Houghton and Li, 1979). Other methods that have been reported to be effective include NH_4-I-dimethylsulfide (Fujii et al., 1987) and $DMF \cdot SO_3$-EDT (Futaki et al., 1990). The later should only be performed while hydroxyl residues are protected. Peptides that contain methionine sulfoxide tend to elute slightly earlier on reverse-phase HPLC than the corresponding peptide containing methionine due to the more polar nature of the sulfoxide. Thus the possibility of the presence of the sulfoxide derivative can often be spotted in this way. Many laboratories routinely treat methionine-containing peptides with a thiol reagent prior to HPLC analysis to check for elution shifts due to the presence of the sulfoxide. This is particularly important if more than one predominant peak is originally present in the chromatogram. Automated Edman-sequencing instruments that incorporate a reducing agent in their reagents may not detect methionine sulfoxide because it will be converted to methionine during the analysis.

Cysteine

Cysteine is prone to the oxidative formation of disulfide bonds which can result in the presence of multiple species. If a single cysteine residue is present in a peptide, a dimer can form as a result. However, if the peptide contains more than one cysteine, disulfide cross-linked oligomeric aggregates can form.

Generally speaking, peptides usually emerge from the deprotection procedure in their fully reduced state. It is only after exposure to air that appreciable amounts of disulfides start to form. The time of air exposure needed can vary from peptide to peptide. If one can avoid the presence of two or more cysteine residues in the same peptide, it is generally advisable to do so. Also, when using ammonium bicarbonate as a solvent, disulfide formation can be diminished initially if it is chilled before adding it to the peptide.

Any disulfides that form can be reduced by treatment with a molar excess (5–100-fold) of dithiothreitol. It should be pointed out, however, that some complex disulfide cross-linked aggregates can prove to be very resistant to reduction. Peptides linked by intermolecular disulfides tend to elute later on reverse-phase HPLC than their monomeric counterparts. The relative elution behavior of peptides with intramolecular cross-links is less predictable.

Tryptophan

Oxidation products of tryptophan are generally not reversible. Common oxidation products of the indoyl ring of trytophan are oxyindoyl, formylkynureninyl, and kynureninyl derivatives. The latter two result in opening of the five-membered ring of tryptophan and shift the absorbance maximum toward shorter wavelengths. Significant oxidation of tryptophan residues can also be detected by mass spectrometry. In Edman sequence analysis, a common oxidation product of tryptophan elutes just before intact PTH-tryptophan on the HPLC. Tryptophan oxidation can result from the conditions under which the peptide was handled. For instance, the same peptide from a common source, can often be observed, by sequence analysis, to be at different levels of tryptophan oxidation by investigators at different locations. One possible cause that has been suggested is the exposure of the peptide to small rust particles from equipment used in processing the sample. However, it is also a matter of some conjecture as to whether the oxidation was pre-existing or occurred during the sequence analysis itself. Thus, the integrity of the Trp residues can best be monitored by mass spectrometry.

Glutamine

When a glutamine residue occurs at the amino-terminus of a peptide, it can cyclize by interaction of the side chain with the α-amino group to form a lactam structure (figure 4-2). This structure, which is often referred to as pyroglutamate, renders the peptide inaccessible to Edman chemistry so that it cannot be sequenced. This is particularly prone to occur at acid pH but can also occur slowly at neutral or basic pH when stored in aqueous solution. This reaction results in a decrease of 17 Da and can easily be seen by mass spectrometry. This same phenomenon is sometimes seen with glutamic acid, but is much less common.

Aspartic Acid-X Sequences

When using Fmoc chemistry to synthesize peptides, sequences of aspartic acid followed by residues with small side chains are particularly prone to cyclic imide formation. Although cyclic imide formation can also occur at Asp and Asn residues in finished peptides

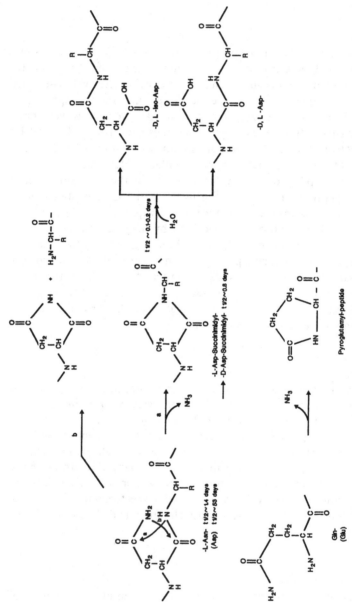

Figure 4-2. Formation of cyclic imides in peptides. The top two reactions depict potential routes for internal imide formation of Asn and Asp. The bottom reaction depicts the potential cyclization of Gln, and occasionally Glu, when they occur at the amino terminus of a peptide.

($t_{1/2} \approx$ 53 days for Asp-Gly and $t_{1/2} \approx$ 1.4 days for Asn-Gly), the problem is particularly evident when Asp residues containing ester based blocking groups, such as O-t-butyl, are used in the synthesis ($t_{1/2} \approx$ minutes). In practice, this is seen most often for Asp-Gly and Asp-Ala sequences when the Asp side chain is ester protected, but can occur with other amino acids following Asp as well. Moreover, the degree to which it occurs increases the earlier in the synthesis that it is present.

The length of the Asp side chain makes it particularly well suited to interact with the peptide backbone by cyclization to a cyclic imide. The cyclic imide can subsequently open up to reform the normal Asp containing peptide with α-peptide linkage or it can also form an iso-Asp peptide with a β-peptide linkage (figure 4-2). In addition, racemization about the α-carbon can also occur. In Fmoc synthesis, piperidine can react with the cyclic imide (succinimidyl ring) to form both α- and β-piperidides. This results in an adduct that is 67 mass units larger than the expected peptide mass and can be easily seen by mass spectrometry.

A second reaction that can occur is chain cleavage due to attack of the main chain carbonyl by the side-chain nitrogen (figure 4-2). This is relatively rare in peptide synthesis because the other reaction is highly favored, however it can become appreciable with bonds such as Asp-Pro where iso-Asp formation is not favored.

Peptide Solubility

The solubility properties of synthetic peptides can have profound effects on their usability and should always be considered before synthesis. In general, it is advisable to incorporate charged amino acids into a synthetic peptide when possible. Peptides composed completely of uncharged amino acids tend to be less soluble in aqueous solutions and their solubility will tend to decrease as their size increases. The presence of His, Lys, and Arg promote solubility in solutions at a pH where their side chains possess a positive charge. Conversely, the presence of Asp and Glu promote solubility in solutions at a pH where their side chains will be negatively charged. A general rule of thumb for solubility in aqueous solution is that the number of charges (net) should be at least 20% of the total number of residues. The contribution of the charge properties of the α-amino and carboxyl groups should also be included. In this regard, it is useful to remember that modification of these end groups, so that they are no longer able to ionize, can significantly effect the solubility properties, especially of small peptides. Long stretches of hydrophobic residues and the presence of polar uncharged side chains are also factors to be considered in assessing solubility. In a "borderline" peptide, their presence can tip the scale one way or the other. However, remember that there are always exceptions and that these general rules may not always be adequate.

Some of the common aqueous solutions used for dissolution of peptides are 0.1% trifluoroacetic acid (TFA), 0.1 N acetic acid, and 0.1 M ammonium bicarbonate. If these are not effective, 25–50% TFA, 30% acetic acid, or the inclusion of acetonitrile or ethanol can help. Aqueous formic acid is another excellent solvent that has been used for peptides, but it has the potential drawback that the peptide may become formylated on primary amino groups if formaldehyde, which is a potential contaminant, is present. In some cases, increasing concentrations of dimethyl sulfoxide (DMSO) may be helpful. Dimethylformamide (DMF) and dichloromethane (DCM) are effective nonaqueous organic solubilizers of synthetic peptides.

Analysis of the Synthetic Product

When the first edition of this book was published, automated peptide synthesizers were relatively new and tBoc-based chemistry was used almost exclusively. It was not uncommon to experience a wide variety of problems throughout the synthesis procedure, and a number of approaches for detecting them were used when these problems occurred. These included resin sampling during the synthesis to determine the extent of coupling efficiency at each step, on-resin sequence analysis, and determination of the peptide-resin mass yield prior to cleavage from the resin. Today, the majority of synthesis is performed using Fmoc-based chemistry. Because of the higher efficiency and fidelity of this chemistry in most situations, the above-mentioned procedures are seldom needed or found useful. However, on-resin sequencing may still be useful on occasion and resin sampling may still be helpful on those occasions when a synthesis proves particularly problematic. These are discussed briefly later.

Today, the two most useful and effective techniques for indicating the initial quality of a synthesis are analytical reverse-phase HPLC and mass spectrometry. These two evaluations are relatively easy to perform and represent the minimum that should always be performed for each synthetic peptide. Analytical HPLC provides a good assessment of the complexity of the unpurified product and mass spectrometry indicates the masses of the species that are present and that can be compared to the expected mass of the desired product.

In fact, these same techniques are used throughout the purification of a peptide to assess the homogeneity of the preparation. In addition to these techniques, amino acid analysis is the best method for quantitation of the peptide product and sequencing by either Edman degradation or tandem mass spectrometry are effective in locating the positions of modifications or deletions. Capillary electrophoresis can be used instead of reverse-phase HPLC for assessing heterogeneity, but there are several disadvantages that lead to its decreased popularity. Although it produces excellent resolution it suffers from a lack of reproducibility and is difficult to use as a preparative method. A recent overview (Angeletti et al., 1997) of a 6-year study of peptide synthesis by the Association of Biomolecular Resource Facilities (ABRF) discusses the advantages and disadvantages of various analytical techniques in more depth and comes to a similar conclusion. This article is recommended reading for anyone making or using synthetic peptides.

Problems can occur either during the synthesis itself or during the procedure used to cleave the peptide from the resin and deprotect the side chains. The ABRF study mentioned above concluded that the majority of problems that they encountered in their studies were a result of the cleavage and deprotection procedures. If a problem does occur, an ability to distinguish at which of these two main steps it manifests itself can be very useful in correcting it. For instance, if a problem occurs during the cleavage or deprotection reaction, it makes no sense to change the synthesis protocol. Resin sampling and on-resin sequencing were traditionally the two main methods used for trying to detect problems during the synthesis itself. However, experience has shown that these techniques can be unreliable and are not generally very revealing. In practice, the general point of origin of most problems are relatively obvious if they are detected during the initial analysis of the cleaved peptide. For instance, deletion or addition products result from problems during synthesis and unremoved or migrated protective groups as well as amino acids modified in other ways generally occur during the cleavage and deprotection process.

Initial Criteria for Evaluation of Synthetic Peptides

Every peptide, whether it will eventually undergo extensive purification or is intended to be used in a realtively crude form, should be subject to some minimal level of characterization. Generally, the minimal criteria that should be present are (1) a predominate peak with reasonable resolution on analytical HPLC or capillary electrophoresis, and (2) a mass consistent with the presence of the desired peptide. If either of these criteria are not met, there is reason to suspect a problem and the peptide should either be evaluated further or resynthesized.

Analytical Electrophoresis and Chromatography

Analytical-scale HPLC and capillary electrophoresis (CE) are the two most common and effective techniques used for the assesment of heterogeneity of synthetic peptides. In practice, they complement each other since their respective modes of separation are based on different physical principles. A detailed discussion of the theory and operation of these two techniques, which is beyond the scope of this chapter, can be found in several excellent reviews and books (CE: Ewing et al., 1989; Sanchez and Smith, 1997. HPLC: Hancock, 1984; Henschen et al., 1985; Mant and Hodges, 1990; Chicz and Regnier, 1990; Mant et al., 1997).

The most common form of depicting the results of either CE or HPLC are measures of absorption of light at a particular wavelength plotted versus elution time or fraction number. The most common wavelengths used for peptides are in the range from 210 nm to 230 nm where the polypeptide bond absorbs strongly. If the peptide contains tryptophan, detection at 280 nm can also be used. However, caution must be excercised when following an elution at 280 nm because peptides that do not contain tryptophan, or where the tryptophan has been extensively oxidized, will not be seen. With both CE and HPLC, a pure peptide is expected to produce a single symmetrical peak. However, it must be kept in mind that that alone is not sufficient criteria for homogeneity. A single symmetrical peak does not guarantee homogeneity (see figure 4-21 below) and a homogeneous sample does not guarantee the intended product (see figure 4-22 below). Furthermore, reversible oxidation of methionine and cysteine (disulfide bonds) can lead to the observation of multiple peaks. These can be misleading but do not, in themselves, necessarily indicate serious chemistry problems. If they are recognized for what they are, they can usually be dealt with successfully.

Finally, as the size of peptides increase, to greater than approximately 50 residues, the expectation that a homogeneous peptide will form a single, sharp symmetrical peak tends to decrease. With peptides of this size, this has been attributed to the onset of the formation of relatively stable, slowly exchanging folded structures which exhibit slightly altered elution characteristics and which will result in a broad peak, often with absorption spikes appearing across its profile (Kent, 1988; Regnier, 1987). Moreover, in the case of HPLC, the onset of these structures can be enhanced or induced by interaction with the chromatography supports themselves (Regnier, 1987), particularly the hydrophobic reverse-phase supports. This can result in the elution profile being broad and irregular. Thus, without independent data, incorrect conclusions regarding heterogeneity can be made.

Capillary Electrophoresis
Capillary electrophoresis, which was first demonstrated in 1981 (Jorgenson and Lukacs, 1981), separates peptides on the basis of

their differential migration in solution in an electric field (Ewing et al., 1989). Like ion exchange HPLC, the separation is a function of the charge properties of the peptide, but the physical basis of separation is different. The separation takes place in a liquid filled capillary tube ($< 100\,\mu m$ in diameter) whose ends are placed in buffers exposed to opposite poles of an electric field. Typical running voltages can be from $5\,kV$ to $30\,kV$ with times from 1 min to greater than 1 hour. Migration takes place due to charge attraction and is recorded when the individual peptide species pass a small window in the capillary which is placed in line with a detector. For peptides, detection is usually by UV absorbance. Actually, migration is due to two components. The first is migration of the charged peptide toward the pole of opposite charge and the second is bulk fluid flow within the capillary that is due to electroendoosmosis. In aqueous solution in fused silica capillaries, electroendoosmotic flow is usually toward the cathode, so, unless the electrophoretic mobility of an anion in the opposite direction is greater than the osmotic flow, theoretically both anions and cations can be analyzed in a single run (Young and Merion, 1990). Since the electroendoosmotic flow is the same rate for all solute molecules (peptides), the separation is a function of the electrophoretic mobility of the individual species. A common set-up for peptide separation uses a pH 2.0 buffer so that all charged peptides are cationic and thus migrate to the cathode. This also minimizes the effect of electroendoosmosis.

CE is often referred to, somewhat misleadingly, as a higher-sensitivity technique when compared to analytical HPLC. While it is true that the actual amount of peptide analyzed is very small when compared to analytical HPLC (nanograms compared to micrograms), it must be remembered that the volume of sample injected into the column is also very small (low nanoliters). Thus, the concentration of peptide in the original sample that is needed for analysis is not much different than that needed for HPLC analysis. Unlike HPLC, which can be scaled up to be used for preparative purposes, CE is mainly useful as an analytical technique. For the analysis of synthetic peptides, the main advantage of CE is that it provides a second method, based on a different physical property, for separation of peptide mixtures and thus the assessment of heterogeneity.

The resolving power of capillary electrophoresis is demonstrated in figure 4-3 which illustrates the separation of several synthetic peptides that differ only slightly in their composition. While peptides that differ in the number of charges they carry are resolved by relatively large distances, it can be seen that even peptides with apparent equal charge can be separated.

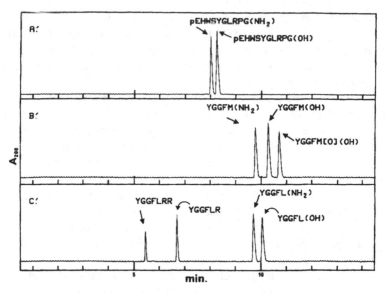

Figure 4-3. Analysis of synthetic peptides by capillary electrophoresis. The capillary was 50 μm × 60 cm and the sample was run at 30 kV, after a 10 s injection, in 100 mM sodium phosphate buffer, pH 2.0. (Reproduced by permission of Dan Crimmins.)

Analytical High Performance Liquid Chromatography
Analytical HPLC separates peptides by their interaction with derivatized insoluble supports, referred to as the stationary phase. In practice, it is completley analogous to conventional "low-pressure" bench-top chromatography. Both conventional chromatography and "high-pressure" or "high-performance" chromatography employ (1) a cylindrical column that contains the derivatized support upon which separation occurs, (2) a means for flowing a modifying solution, referred to as the mobile phase, by which the separation is developed, through the column, and (3) a means to detect the eluting species as they exit from the column. The major difference between high-pressure and low-pressure techniques, and hence the advantage of HPLC, is that HPLC uses columns and pumping systems that can withstand and deliver very high pressures so that very fine particles (3–10 μm mean diameter) can be used as column packing material. The result is superior resolution in relatively short time periods (minutes for HPLC as compared to hours or days for conventional systems). The two general categories of HPLC supports which are used almost exclusively for peptides are ion exchange and reverse phase, with reverse being by far the most widely employed.

Ion-Exchange HPLC

Separation by ion-exchange HPLC, like its conventional low-pressure counterpart, is based on direct charge interaction between the peptide and the stationary phase. The column support is derivatized with an ionic species that maintains a particular charge over a range of pH where a peptide, or mixture of peptides, can exhibit an opposite charge, depending on their amino acid compositions. When the sample is loaded on to the column, the charge interaction binds the peptides to the support. The peptides are eluted from the column by changing either the pH or ionic strength, or both, of the column buffer such that the charge interaction is defeated either through neutralization of the interacting charge groups (pH) or competition with higher concentrations of like charged molecules (ionic strength). Increasing ionic strength with solutions of a salt either in a step-wise fashion or by a gradually increasing gradient of concentration is the most common method. Thus, ion-exchange HPLC, sometimes referred to as "normal-phase" chromatography, binds the peptide at a relatively low ionic strength and elution is accomplished by progressing to conditions of higher ionic strength. Peptides are separated because of their differential interaction with the ionic support and hence elute at different levels of ionic strength. One approach to the ion-exchange separation of synthetic peptides has been the use of strong cation-exchange columns such as sulfoethyl aspartamide columns (Alpert and Andrews, 1988; Crimmins et al., 1988) which separate peptides on the basis of positive charge at acid pH. Sulfoethyl aspartamide SCX columns (obtained from the Nest Group) are particularly useful in this regard because monomeric peptides elute in a monotonic fashion according to their net positive charge at pH 3.0 (Crimmins et al., 1988). An example of peptide separation with this column is shown in figure 4-4. The peptides used in this figure are the same as those in figure 4-3 for CE separation. Note that while the patterns are somewhat different and the order of elution is reversed, both give comparable resolution.

Reverse-Phase HPLC

Reverse phase HPLC is so named because the conditions for elution are essentially the reverse of normal-phase chromatography. Binding is through hydrophobic interaction with the column support and elution is accomplished by decreasing the ionic nature, or increasing the hydrophobicity, of the eluant so that it competes for the hydrophobic groups on the column. Column supports are generally hydrocarbon alkane chains ranging from four (C4) to 18 (C18) carbons in length, or aromatic hydrocarbons such as phenyl groups. Although many different buffer systems and stationary phases have been

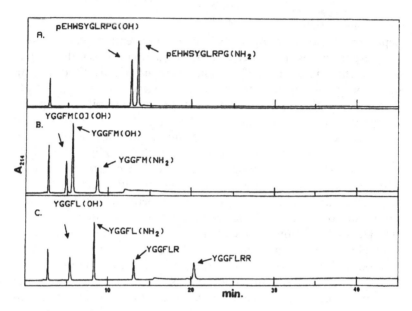

Figure 4-4. Strong cation exchange chromatography of synthetic peptides on a sulfoethyl aspartimide column. The peptide mixtures are the same as those shown in figure 4-3 for CZE separation. The column conditions can be found in Crimmins et al., 1988. (Reproduced by permission of Dan Crimmins.)

reported (Snyder and Kirkland, 1979; Hancock, 1984; Henschen et al., 1985), the most commonly employed system for peptides is a two-buffer system where the column is equilibrated in 0.1% trifluoroacetic acid for sample loading and elution is produced by an increasing gradient of acetonitrile as a water miscible organic modifier. A typical recipe is for reservoir A to contain 0.1% TFA in water and for reservoir B to contain 80% acetonitrile in water that has been made 0.1% in TFA. The columns are usually eluted with a linear gradient run at between 0.5 and 1% B per minute. Most peptides will elute from reverse-phase columns by the time the eluant is 60% acetonitrile. Other buffer recipes and elution protocols can be found in the general references in this section. These include the use of 0.1% heptafluoro-butyric acid, 0.1% phosphoric acid, dilute HCl, and 5–60% formic acid for low pH (pH 2–4); and 10–100 mM ammonium bicarbonate, sodium or ammonium acetate, trifluoroacetic acid/triethylamine, sodium or potassium phosphate, and triethylammonium phosphate for pH 4–8. In addition to acetonitrile, methanol, ethanol, propanol, and isopropanol have been employed as water miscible eluants (Rubinstein, 1979; Mahoney and Hermodson, 1980; Hancock, 1984; Feldhoff, 1991).

It is generally advisable when using a buffer system for the first time to test the compatability of the components by simple mixing experiments. For example, phosphate is insoluble in high concentrations of acetonitrile and the two are potentially disastrous to use together without first determining their solubility limits. Such simple experiments could prevent the loss of an expensive column or a valuable sample.

Again, the basis for separation of peptides is that each peptide interacts differently with the column and will thus elute at a different concentration of organic modifier. The resolution of reverse-phase HPLC can be excellent, but one must be constantly mindful that a single peak does not guarantee homogeneity. This is demonstrated quite convincingly in the chromatograms in figure 4-5 where it can be seen that a single symmetrical peak, generated under one set of HPLC conditions, actually contained two peptides that were separated when the conditions of the chromatography were altered. The two peptides present in this sample differed by only a single glutamic acid residue as a result of a partial deletion during synthesis. In this particular example, the pH of the TFA buffer was raised with triethylamine. Under the basic conditions, the glutamic acid in the peptide becomes negatively charged which in turn alters its interaction with the stationary phase.

A word of caution about pH and reverse-phase columns. Many reverse-phase columns have a silica-based support to which the bonded phase (i.e., the alkane chain) is attached. The bond holding the bonded phase to the silica support is labile to basic pH and these columns should not be used with basic buffers. However, the columns can usually be used at slightly basic pH (< 8), as shown in the example in figure 4-5, for short periods of time as long as they are returned to acidic conditions at the end of the run. Conditions of high pH or exposure to slightly basic pH for extended periods of time should be avoided. Be aware that repeated use at basic pH may gradually destroy the column. Some manufacturers are now offering polymer-based supports that are resistant or unaffected by basic pH.

The most common bonded phases used for the reverse-phase separation of synthetic peptides consist of alkane chains that range in size from 1 to 18 carbon atoms. The most common are C-4, C-8, and C-18. Because elution from the column is a function of the hydrophobicity of the peptide, and since charge decreases hydrophobicity, very highly charged small peptides tend to be retained better by long-chain bonded supports. Conversely, large hydrophobic peptides tend to exhibit better elution properties from short chain bonded supports. However, in practice, they can often be used interchangeably with little significant effect with most peptides.

Carbon load, which refers to the amount of bonded phase present per unit of support, also has an effect on the characteristics of a reverse-phase column. For instance, two different C-18 columns can

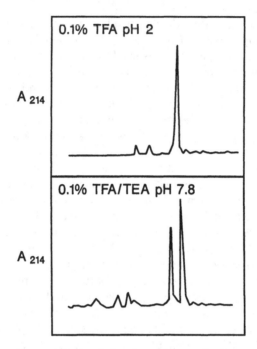

Figure 4-5. Analysis of peptide samples by buffer modulation. Chromatograms are reverse-phase HPLC of the sample depicted in figure 4-21 at two different conditions of pH.

differ significantly in the mass or amount of hydrocarbon chain they contain although they are both C-18 columns. The carbon load not only affects the column's binding capacity, but can also affect retention time. Unfortunately, carbon load is not a parameter often listed by column manufacturers, and usually has to be determined by trial and error or learned by word of mouth.

In addition to the type of bonded phase, particle size and pore size are two other parameters which can have significant effect on resolution of peptides. Particle size or mesh size refers to the actual average size of the particles which make up the support. Modern HPLC columns are generally available in particle size ranges from 3–10 μm. Smaller particle sizes produce better resolution, but they are also more expensive and increase operating pressure, referred to as back-pressure. Commonly, analytical columns (i.e., 4.6 × 250 mm) use 3–10 μm particles, and preparative columns (i.e., 22 × 250 mm utilize 10-μm particles.

The particles used in reverse-phase chromatography are not solid, but rather contain pores (analagous to a honeycomb) which increases the surface area of the particle and hence the interaction with the peptide. Pore size refers to the average size of the pores within the particle itself through which solvent and solute pass during interaction with the stationary phase. In order for a molecule to effectively interact with the particle, it must be able to pass through the pores. For large proteins, pore sizes of 300 Å are more effective. This pore size can also be used effectively for most peptides, but for very small peptides (six residues or less) pore sizes of 80–100 Å generally produce sharper peaks and better resolution.

Column age can also have a profound effect on its performance. A column that exhibits an extremely high degree of resolution when it is new may lose resolving power as it ages. Thus, shoulders will no longer be evident or two similar peptides may co-elute. As a result, heterogeneity may go undetected with a column that is operating at less than optimal efficiency. In this regard, it is helpful to aquire a peptide standard that contains closely eluting peaks that can be used to assess column performance periodically. When a column starts loosing resolving power, it should be replaced.

Purification

The crude product obtained from stepwise solid phase peptide synthesis will often contain a variety of by-products in addition to the desired product. These will usually include a family of peptides containing deletion or addition sequences generated during the synthesis itself and a series of peptides containing chemical modifications usually generated during the cleavage and deprotection steps. The two requirements to obtaining the intended product are that it is present in this mixture in sufficient amount and that the other components can be successfully removed. In general, using modern synthetic procedures, crude peptides of 30–45 residues can be expected to be between 60–80% target peptide, and those of less than 30 residues will be between 80% and 95% target peptide. All that is required then, is an effective purification protocol.

Most peptides under approximately 50 residues in length can be purified using reverse phase HPLC. In cases where reverse phase is not entirely successful, ion exchange HPLC can be an effective and valuable complement to reverse phase techniques.

The power of reverse phase HPLC as a purification tool is illustrated in figure 4-6. It shows the reverse-phase HPLC elution profile of the cleaved product from the synthesis of a 12-residue peptide. It shows the preparative and analytical scale elution profiles of the peptide run on a C-18 reverse phase column with a standard acetonitrile gradient in 0.1% TFA. The analytical scale run (middle) shows the presence of at least four major species and the preparative

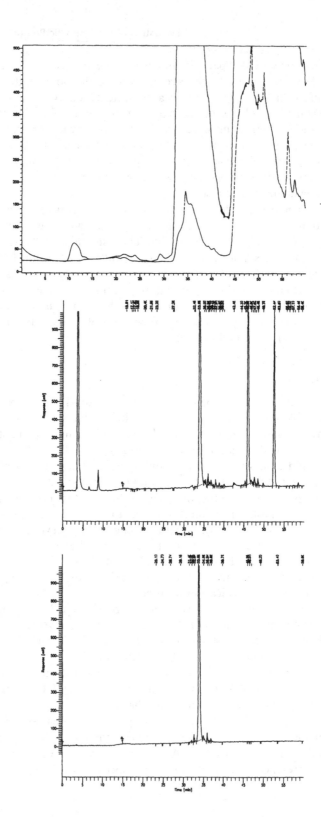

scale run (top), which is monitored at both 214 and 254 nm, suggests that the desired peptide is the one which elutes around fraction 34. The higher ratio of A_{254} signal to A_{214} signal suggests that the later peaks are peptide species that still retain some of the aromatic protecting groups used during synthesis. Aliquots of preparative fractions 32 through 42 were analyzed for homogeneity with the analytical scale column, and fractions 33–37 were pooled to produce the final sample whose analytical profile is shown in the bottom panel. The single symmetrical peak suggests that a homogeneous peptide was obtained from the original sample. This was subsequently verified by mass spectrometry.

The process of fraction-by-fraction analysis of complex patterns found in the crude synthetic products by analytical scale HPLC can be a very effective approach to purification. This systematic approach often allows the retrieval of the desired product from a synthesis that would otherwise appear to be too complex.

Detailed Characterization of Covalent Structure

The previous sections dealt mainly with the assessment of peptide homogeneity and purification. These methods do not provide any information on the covalent nature of the peptide. The determination that the covalent structure of the peptide is correct is no less important than obtaining a peptide free of by-products. After all, some amount of thought went into the design of the peptide for a particular experiment, and there is much to be lost by performing the experiment with the wrong peptide.

Mass spectrometry has become a very valuable tool for the evaluation of synthetic peptides and many synthetic chemists now rely on it as a routine part of their evaluation process because it provides the best analysis of the covalent nature of a peptide in a single technique. Electrospray ionization (ESI) and matrix-assisted laser desorption ionization (MALDI) instruments are now readily

---

Figure 4-6. Purification of a synthetic peptide. Top: The elution profile of 320 mg of a cleaved synthetic peptide run on a 21 × 250 mm C-18 reverse phase column with a linear gradient from 0–60% acetonitrile in 0.01% trifluoroacetic acid/water. The column was monitored at 214 nm (solid line) and at 254 nm (dashed line). Middle: The elution profile of 60 µg of the same peptide preparation run on a 4.6 × 200 mm C-18 reverse column with a linear gradient from 0–60% acetonitrile in 0.01% trifluoroacetic acid/water. Bottom: Analytical HPLC of the pool of preparative fractions 33–37. (Reproduced by permission of John Gorka, Biomolecules Midwest, Inc.)

available and relatively easy to use, and have become the mainstay of the evaluation process.

Determination of the overall mass of a synthetic peptide is the most practical and common utilization of mass analysis and is usually sufficient to evaluate a synthetic peptide effectively and accurately. However, the technology now exists to produce useful information on the actual sequence of a peptide (Hunt et al., 1988; Griffin et al., 1990), and thus pinpoint the location of any unintended modification. Although in practice, evaluation of the mass of the molecular ion is usually sufficient, particularly when used in conjunction with other analytical methods, this section will also discuss sequencing with tandem mass spectrometry.

Peptide-Mass Evaluation

The determination that the mass of a synthetically produced peptide is correct proves in one step that the synthesis was successful and you have obtained the desired product. It may, in fact, be the only way to determine the presence of modifications that are unstable or invisible in other analytical procedures. A list of mass differences that may be encountered in the analysis of synthetic peptides is provided in table 4-2.

While a detailed discussion of the theory and practice of mass spectrometry in the evaluation of synthetic peptides is beyond the scope of this chapter and is covered in detail elsewhere (McCloskey, 1990; Carr and Annan, 1999; Henzel and Stults, 1999), a brief general discussion of the main types of methodologies that have proven to be effective for determination of the mass of synthetic peptides is appropriate.

The development of particle induced desorption techniques opened the way for the determination of the mass of polar, nonvolatile, thermally unstable molecules, such as peptides, without prior chemical derivatization. Prior to the development of these procedures, samples had to be derivatized to affect vaporization without pyrolytic degradation and analysis was limited to molecular sizes of 1000–1500 Da. The first particle-induced desorption technique that was effective for other than the smallest of peptides was fast atom bombardment (FAB) mass analysis (Barber et al., 1981). Although this technique is still available today, it suffers from a severe limitation in its upper mass limit and has been surpassed by more facile techniques. As modern automated synthesizers and chemistry has improved, it has become possible to routinely synthesize peptides with molecular weights above the range of even the best FAB instruments. Since the introduction of FAB, three other types of instruments have been developed that have the advantage of being able to analyze large peptides as well as proteins.

Table 4-2 Mass differences relevant to synthetic peptides

Average mass change	Modification
−79	5′ Dephospho
−18	Dehydration ($-H_2O$)
−18	Pyroglutamic Acid formed from Glutamic Acid
−17	Pyroglutamic Acid formed from Gln
−2	Disulphide bond formation (Cystine)
−1	Amide formation (C-terminus)
1	Deamidation of Asparagine and Glutamine to Aspartate and Glutamate
14	Methylation (N-terminus, N^ε of Lysine, O of Serine, Threonine of C-terminus, N of Asparagine)
16	Hydroxylation (of δ C of Lysine, β C of Tryptophan, C3 or of Proline, β C of Aspartate)
16	Oxidation of Methionine (to Sulfoxide)
16	Sulfenic Acid (from Cysteine)
22	Sodium
28	Ethyl
28	N, N dimethylation (of Arginine or Lysine)
28	Formylation (CHO)
32	Oxidation of Methionine (to Sulphone)
38	Potassium
42	Acetylation (N-terminus, N^ε of Lysine, O of Serine) (Ac)
43	Carbamylation
43	N-Trimethylation (of Lysine)
44	disodium
45	Nitro (NO_2)
56	t-Butyl ester (OtBu) and t-butyl (tBu)
57	Glycyl (-G-, -Gly-)
60	Sodium + potassium
69	Dehydroalanine (Dha)
71	Alanyl (-A-, -Ala-)
71	Acetamidomethyl (Acm)
75	Glycine (G, Gly)
76	β-Mercaptoethanol adduct
76	Phenyl ester (Oph) (on acidic)
80	Phosphorylation (O of Serine, Threonine, Tyrosine and Aspartate, N^ε of Lysine)
80	Sulfonation (SO_3H)
80	Sulfation (of O of Tyrosine)
82	Cyclohexyl ester (OcHex)
83	Homoseryl lactone
83	Dehydroamino butyric acid (Dhb)
85	γ-Aminobutyryl
85	2-Aminobutyric acid (Abu)
85	2-Aminoisobutyric acid (Aib)
86	t-Butyloxymethyl (Bum)
87	Seryl (-S-, -Ser-)
88	t-Butylsulfenyl (StBu)
89	Alanine (A, Ala)

(continued)

Table 4-2 *(Continued)*

Average mass change	Modification
90	Anisyl
90	Benzyl (Bzl) and benzly ester (OBzl)
93	1,2-Ethanedithiol (EDT)
95	Dehydroprolyl
96	Trifluoroacetyl (TFA)
97	*N*-hydroxysuccinimide (ONSu, OSu)
97	Prolyl (-P-, -Pro-)
99	Valyl (-V-, -Val-)
99	Isovalyl (-I-, -Iva-)
100	*t*-Butyloxycarbonyl (*t*Boc)
101	Threoyl (-T-, -Thr-)
101	Homoseryl (-Hse-)
103	Cystyl (-C-, -Cys-)
104	Benzoyl (Bz)
104	4-Methylbenzyl (Meb)
105	Serine (S, Ser) ·
105	Pyridylethylation of cysteine
106	HMP (hydroxymethylphenyl) linker
106	Thioanisyl
106	Thiocresyl
107	Dipthamide (from Histidine)
111	Pyroglutamyl
111	2-Piperidinecarboxylic acid (Pip)
113	Hydroxyprolyl (-Hyp-)
113	Norleucyl (-Nle-)
113	Isoleucyl (-I-, -Ile-)
113	Leucyl (-L-, -Leu-)
114	Ornithyl (-Orn-)
114	Asparagyl (-N-, -Asn-)
114	*t*-Amyloxycarbonyl (Aoc)
115	Proline (P, Pro)
115	Aspartyl (-D-, -Asp-)
117	Succinyl
117	Valine (V, Val)
117	Hydroxybenzotriazole ester (HOBt)
118	Dimethylbenzyl (diMeBzl)
119	Threonine (T, Thr)
119	Cystenylation
120	Benzyloxymethyl (Bom)
120	*p*-Methoxybenzyl (Mob, Mbzl)
121	4-Nitrophenyl, *p*-Nitrophenyl (ONp)
121	Cysteine (C, Cys)
125	Chlorobenzyl (ClBzl)
126	Iodination (of Histidine[C4] or Tyrosine[C3]
128	Glutamyl (-Q-, -Gln-)
128	Lysyl (-K-, -Lys-)
129	*O*-Methyl Aspartamyl
129	Glutamyl (-E-, -Glu-)

(continued)

Table 4-2 (*Continued*)

Average mass change	Modification
130	N^α-(γ-Glutamyl)-Glu
131	Norleucine (Nle)
131	Hydroxyproline (Hyp)
131	$\beta\beta$-Dimethyl Cystenyl
131	Isoleucine (I, Ile)
131	Leucine (L, Leu)
131	Methionyl (-M-, -Met-)
132	Asparagine (N, Asn)
132	Pentoses (Ara, Rib, Xyl)
133	Aspartic Acid (D, Asp)
134	Dmob (Dimethoxybenzyl)
134	Benzyloxycarbonyl (Z)
134	Adamantyl (Ada)
135	*p*-Nitrobenzyl ester (ONb)
137	Histidyl (-H-, -His-)
142	*N*-methyl Glutamyl
142	*N*-methyl Lysyl
143	*O*-methyl Glutamyl
144	Hydroxyl Lysyl (-Hyl-)
145	Methyl Methionyl
146	Glutamine (Q, Gln)
146	Aminoethyl Cystenyl
146	Pentosyl
146	Deoxyhexoses (Fuc, Rha)
146	Lysine (K, Lys)
146	Aminoethyl cystenyl (-AECys-)
147	Methionyl Sulfoxide
147	Glutamic Acid (E, Glu)
147	Phenylalanyl- (-F-, -Phe-)
149	2-Nitrobenzoyl (NBz)
149	Methionine (M, Met)
153	3-Nitrophenylsulphenyl (Nps)
154	4-Toluenesulphonyl (Tosyl, Tos)
154	3-Nitro-2-pyridinesulfenyl (Npys)
155	Histidine (H, His)
156	Arginyl (-R-, -Arg-)
158	3,5-Dibromo
159	Dichlorobenzyl (Dcb)
160	Carboxyamidomethyl Cystenyl
161	Carboxymethyl Cystenyl
161	Methylphenylalanyl
161	Hexosamines (GalN, GlcN)
161	Carboxymethyl cysteine (Cmc)
162	*N*-Glucosyl (*N*-terminus or N^ε of Lysine) (Aminoketose)
162	*O*-Glycosyl (to Serine or Threonine)
162	Hexoses (Fru, Gal, Glc, Man)
163	Methionyl Sulphone
163	Tyrosinyl (-Y-, -Tyr-)

(*continued*)

Table 4-2 (*Continued*)

Average mass change	Modification
165	Phenylalanine (F, Phe)
166	2,4-Dinitrophenyl (DNp)
166	Pentaflourophenyl (Pfp)
166	Diphenylmethyl (Dpm)
167	Phospho Seryl
169	2-Chlorobenzyloxycarbonyl (ClZ)
170	*N*-Methyl Arginyl
172	Ethaneditohiol/TFA cyclic adduct
173	Carboxy Glutamyl (Gla)
174	Acetamidomethyl Cystenyl
174	Acrylamidyl Cystenyl
174	Arginine (R, Arg)
177	Benzyl Seryl
177	*N*-Methyl Tyrosinyl
179	*p*-Nitrobenzyloxycarbonyl (4Nz)
179	2,4,5-Trichlorophenyl
180	2,4,6-trimethyloxybenzyl (Tmob)
180	Xanthyl (Xan)
181	Phospho Threonyl
182	Mesitylene-2-sulfonyl (Mts)
186	Tryptophanyl (-W-, -Trp-)
188	*N*-Lipoyl- (on Lysine)
201	HMP (hydroxymethylphenyl)/TFA adduct
203	*N*-acetylhexosamines (Gal*N*Ac, Glc*N*Ac)
204	Tryptophan (W, Trp)
204	Cystine (Cys)2
212	4-Methoxy-2,3,6-trimethylbenzenesulfonyl (Mtr)
213	2-Bromobenzyloxycarbonyl (BrZ)
222	9-Fluorenylmethyloxycarbonyl (Fmoc)
226	Biotinylation (amide bond to lysine)
226	Dimethoxybenzhydryl (Mbh)
233	Dansyl (Dns)
238	2-(*p*Biphenyl)isopropyl-oxyxarbonyl (Bpoc)
242	"Triphenylmethyl (Trityl, Trt)"
252	Pbf (pentamethyldihydrobenzofuransulfonyl)
252	3,5-Diiodination (of Tyrosine)
266	*O*-GlcNAc-1-phosphorylation (of Serine)
266	2,2,5,7,8-Pentamethylchroman-6-sulphonyl (Pmc)

The first of these was plasma-desorption mass spectrometry (PDMS) (MacFarlane et al., 1974; Cotter, 1988; Fontenot et al., 1991), which is another particle-induced desorption technique. Although PDMS is still a viable technique it has been almost completely surpassed by electrospray-ionization mass spectrometry (ESMS), and matrix-assisted laser desorption ionization (MALDI) mass spectrometry.

Electrospray ionization introduces a peptide in the liquid state and produces a series of multiply charged ions where the charge is carried on protonatable groups. At acid pH, the free α-amino group as well as the side chains of basic residues such as arginine and lysine carry a positive charge. This produces a series of ions differing from each other by a single charge which are then analyzed by a computer to calculate the molecular weight of the polypeptide. It is this multiple charging phenomenon that has allowed direct molecular weight analysis of large peptides and proteins at high sensitivity (picomol level) and accuracy. An example of a peptide mass spectra produced by electrospray ionization is shown in figure 4-7. Note that multiple charging produces multiple signals despite the fact that this is a homogeneous peptide. Each signal, or observed mass (m_0) in the spectrum is equal to the actual mass of the peptide (m) divided by the number of charges (z) carried by the peptide (m/z). That is, $m_0 = [m + m'(z)]/z$, where m is the mass of the peptide, z is the number of charges, and m' is the mass of the charging species. In this case, this is a positive ion spectra where basic groups on the peptide have been protonated $[P + H^+ \rightarrow (PH)^+]$. Thus, the charging species are protons which have a mass of 1. Solving for m produces the relationship, $m = [m_0(z)] - z$. Each observed mass peak is fit into this equation and the values for m are averaged to produce a final peptide mass. Peptides can also be anlyzed in negative ion mode where the acidic groups have been deprotonated (PH $\rightarrow$ P$^-$ + H$^+$). In this case, $m = [m_0(z)] + z$.

In order to perform the calculations described above, referred to as deconvolution of the spectra, the charge state of each of the signals must first be known. On mass spectrometers where the isotopic peaks (resulting from natural abundance of the isotopes) can be resolved, the charge state can be determined from the spacing of the isotopic peaks (figure 4-8). For a singly charged peptide, the isotope peaks are $1\,m/z$ unit apart. For a doubly charged peptide they are $0.5\,m/z$ units apart and so on. The charge state of each signal can also be determined by the relationship of adjacent peaks. If Δm_0 is the difference in observed mass between two adjacent peaks and z_1 and z_2 are the charges on the lesser and greater charged of the two species, respectively, then $z_1 = m_{0(2)}/\Delta m_0$ and $z_2 = m_{0(1)}/\Delta m_0$. In other words, the charged state of an observed signal is equal to the mass of its adjacent signal divided by the difference in mass between the two signals. The charges calculated in this way should be whole numbers. If the charges are not whole numbers, then the two peaks chosen are not truly adjacent peaks for the same peptide. Modern mass spectrometers have software that do these calculations automatically but it is a good idea to become familiar with these simple relationships. The deconvolution of the spectrum in figure 4-7 is shown in figure 4-9 and an example of the manual deconvolution is shown in table 4-3. The result of 4005.6 is in

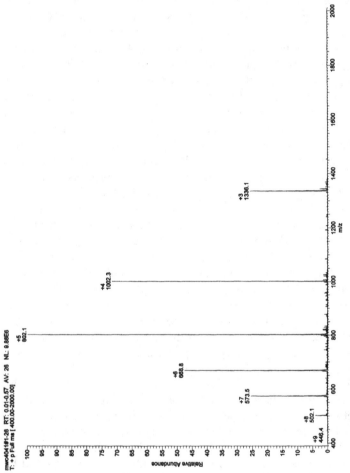

Figure 4-7. An electrospray ionization spectrum of a 33 residue peptide. The peptide was dissolved in 50% acetonitrile/water, 1% acetic acid and injected at 10 μl/min into a Finnigan LCQ ion-trap mass spectrometer.

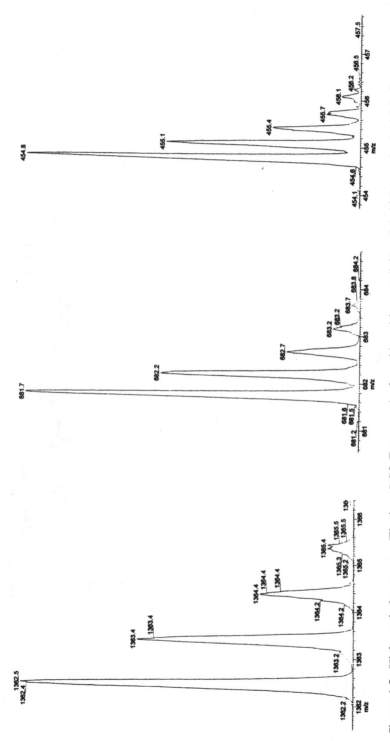

Figure 4-8. High resolution spectra (Finnigan LCQ Zoom scan) of peptides with a +1 (left), +2 (middle), and +3 (right) charge.

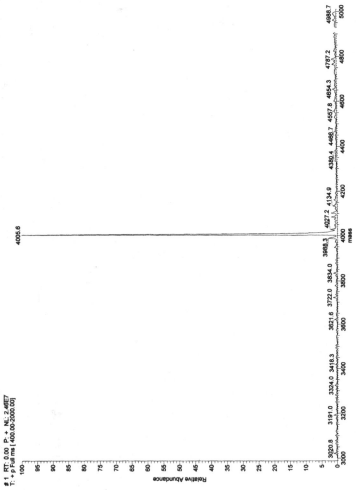

Figure 4-9. Computer-assisted deconvolution of the electrospray data shown in figure 4-7. Deconvolution was performed with Finnigan Bioworks.

Table 4-3 Manual deconvolution of the synthetic peptide spectrum shown in figure 4-7

m_0	z	$[m_0(z)] - z$	Average mass
1336.1	3	4005.3	
1002.3	4	4005.2	
802.1	5	4005.5	4005.3
668.8	6	4006.8	
573.5	7	4007.5	
502.1	8	4008.8	
			4006.5

excellent agreement with the calculated mass of 4005.7 for this 33-residue peptide. Note that in the manual deconvolution shown in table 4-3, greater error can be introduced when multiplying the higher charged peaks by their charge. Therefore, only the lower-charged peaks (+3, +4, and +5 in table 4-3) are routinely used for this procedure.

Matrix-assisted laser desorption ionization mass spectrometry (MALDI-MS) (McCloskey, 1990; Gibson, 1991, and references therein; Henzel and Stults, 1999) employs UV-absorbing matrices which absorb energy from a laser pulse which in turn ionizes peptides and proteins which have been mixed with the solid matrix. The sensitivity of MALDI is in the femtomol range and has the advantage of being less affected by most components commonly found in peptide solutions such as salts. Unlike electrospray, MALDI generally produces a single signal for each peptide that is the singly charged species. However, on occasion a doubly charges species can also be observed, particularly for larger molecules. Figure 4-10 shows a MALDI spectrum of the same peptide shown in the electrospray spectra in figures 4-7 and 4-9. The MALDI spectrum shows the $(M + H)^+$ mass of 4006.6, so the measured mass of the peptide is 4005.6 compared to the calculated mass of 4005.7.

Sequencing Peptides with Mass Spectrometry

Both electrospray and MALDI instruments are capable of producing sequence information for synthetic peptides which can be very helpful in determining the identity or position of an anomoly.

Electrospray Ionization

With electrospray instruments, this is most often accomplished by collision-induced decomposition (CID) of a particular peptide ion.

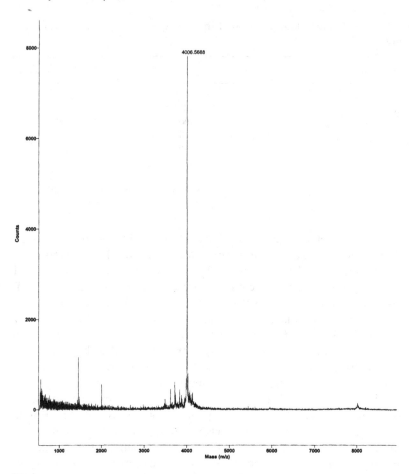

Figure 4-10. MALDI spectrum of the same peptide in figure 4-7. The peptide was mixed with matrix (α-cyano-4-hydroxy cinnamic acid) and the spectrum run on a Perseptive Biosytems DE-Pro spectrometer.

Particularly useful mass spectrometers for this purpose are instruments that employ an ion-trap. The ion-trap allows a particular ion to be easily selected, to the exclusion of all other ions, for collision with an inert gas such as helium which causes it to be broken into smaller fragments. The instrument then determines the mass of each new fragment and produces a spectrum which can be used to determine the sequence of the peptide.

Mass spectrometrists have developed a nomenclature to refer to the peptide fragments that are produced which is illustrated in figure 4-11. With low-energy CID. A peptide can be fragmented in three places along the polypeptide chain. Fragmentation of side chains is usually

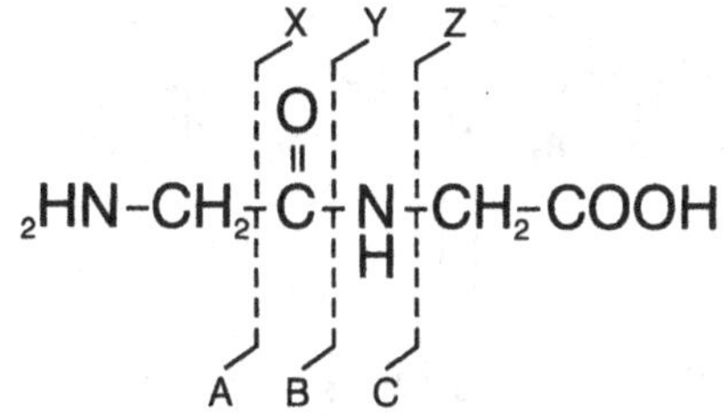

$$_2HN-CH_2-C-N-CH_2-COOH$$

Figure 4-11. Nomenclature for ions produced by collision-induced decomposition.

not observed under these conditions. When this occurs at any one place, two fragments are formed which are named according to the position of fragmentation and whether the charge is retained on the amino terminal portion or the carboxyl-terminal portion. For example, in figure 4-11, if fragmentation occurs between the carbonyl carbon and amide nitrogen, the fragments are referred to as b ions and y ions. B ions retain charge on the amino-terminal portion of the peptide and y ions on the carboxyl-terminal portion. In ion trap instruments, these are the most commonly formed ions. Usually doubly charged ions are selected because they tend to give better fragmention and can produce both b and y ions that retain a single charge. An illustration of a single fragmentation for a doubly charged peptide is shown in figure 4-12. For the parent ion, the charges are found on the amino terminus and the side chain of a lysine residue at the carboxy terminus. In the gas phase, the basicity of the amide bond is similar to that of amino groups. Thus, the protons can migrate along the peptide backbone with relative ease. Protonation weakens the amide bond and makes it more susceptible to being broken upon collision with helium atoms. When this occurs, two fragments are produced, a b ion and a y ion, each with a single charge. This occurs throughout the peptide chain so that a series of fragments are produced that differ by the mass of a particular amino acid residue. This is illustrated in figure 4-13 for a six-residue peptide of sequence Val-Thr-Ala-Asp-Phe-Lys, along with an idealized spectra for this fragmentation.

A residue mass refers to the mass of an amino acid as it occurs in peptide linkage. In other words, it is the free amino acid mass minus water from the loss of the amino-terminal H and carboxyl-terminal OH during formation of the peptide bond. Therefore, each b ion has a mass of the sum of the residue masses plus 1 for the extra proton on the amino terminus. Each y ion has a mass of the sum of the residue masses plus 19 for the extra hydroxyl on the carboxy terminus, the

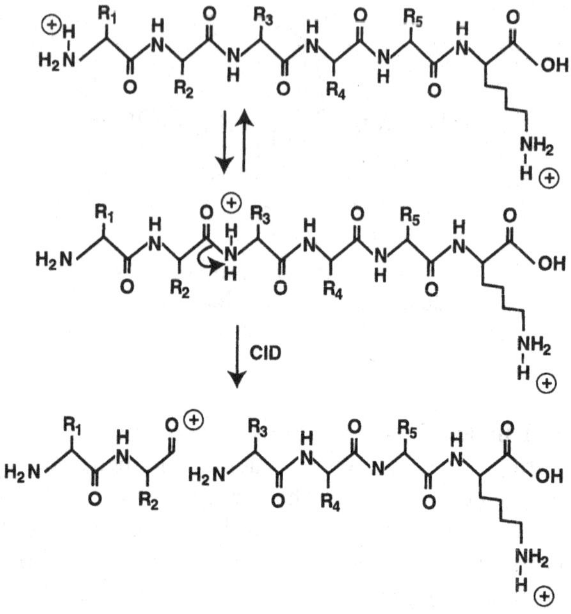

Figure 4-12. Mechanism of CID cleavage in a peptide.
The b ion is on the left and the y ion on the right.

extra proton on the amino terminus and an additional proton that
imparts a positive charge (see figure 4-14). Note that the additional
proton that produces a charge on the y ion can be at the amino
terminus or on a side chain. Within a series, each fragment differs
only by the residue mass of the respective amino acids and each set
of b and y ions are complementary. That is, the sum of their masses
add up to the mass of the parent peptide (i.e., b1 + y5, b3 + y3,
b5 + y1, etc.).

An example of a real spectra generated from a peptide with the
sequence, EGVNDNEEGFFSAR , is shown in figure 4-15. The solution is shown below:

b →	1	2	3	4	5	6	7	8	9	10	11	12	13	14
	130	187	286	400	515	629	758	887	944	1091	1238	1325	1396	1571
	E	G	V	N,	D	N	E	E	G	F	F	S	A	R
	1571	1441	1384	1285	1171	1056	942	813	684	627	480	333	246	175
	14	13	12	11	10	9	8	7	6	5	4	3	2	1 ← y

The fragmentation is produced from the doubly charged ion $m/z =$
786. So, the mass of the peptide is 1570 and the mass of the intact
singly charged peptide would be 1571. Note that the y ion series is

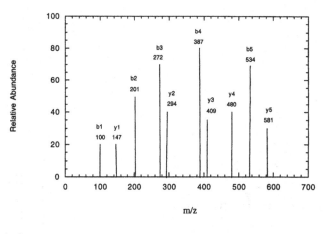

	R_1	R_2	R_3	R_4	R_5	R_6	
H	Val	Thr	Ala	Asp	Phe	Lys	OH
1	99	101	71	115	147	128	$17 = 679 + 2\,H^+ = 681$

b1 $1 + R_1$ **100**   **581** $R_2 + R_3 + R_4 + R_5 + R_6 + 19$ **y5**

b2 $1 + R_1 + R_2$ **201**   **480** $R_3 + R_4 + R_5 + R_6 + 19$ **y4**

b3 $1 + R_1 + R_2 + R_3$ **272**   **409** $R_4 + R_5 + R_6 + 19$ **y3**

b4 $1 + R_1 + R_2 + R_3 + R_4$ **387**   **294** $R_5 + R_6 + 19$ **y2**

b5 $1 + R_1 + R_2 + R_3 + R_4 + R_5$ **534**   **147** $R_6 + 19$ **y1**

Figure 4-13. Idealized spectra for the CID of a 6-residue peptide. The composition of the b and y ions found in the spectra are shown. R_i refers to the residue molecular weight at the ith position.

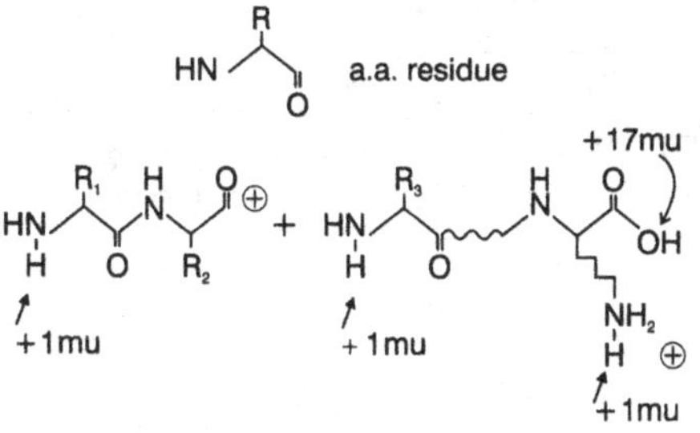

Figure 4-14. Amino acid residue mass and the masses of the b and y ions. In addition to residue mass, the additional mass at the termini must also be taken into account.

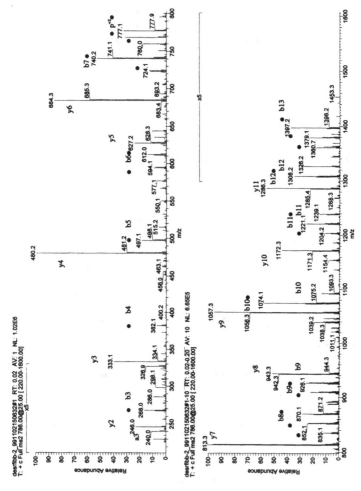

Figure 4-15. Collisional Induced Decomposition spectra from a Finnigan LCQ ion trap instrument of a 14-residue peptide described in the text. The b and y ions are labeled. Dehydrated species (−18) are marked as ●.

clearly present up to y11, but that the b ion series is present mainly as dehydrated (-18) and deammoniated (-17) species. This is probably due to the large number of E, D, Q, and N residues present in the amino terminal part of the peptide. When this is taken into account, almost all of the peaks in the spectra can be accounted for. The peak at 777 is the dehydrated doubly charged parent ion.

MALDI

Peptide sequencing on MALDI instruments is most often done by treating the peptide with an exopeptidase, such as carboxypeptidase, prior to analysis. The peptidase cleaves the peptide, residue by residue, from one end producing a ladder of masses that differ by one amino acid residue. Sequence information can also be obtained from post source decay data.

Peptidase Ladder Sequencing

This approach to peptide sequencing uses an enzyme that will cleave amino acids from the ends of a peptide in a sequential manner (Patterson et al., 1995; Jiménez et al., 1999). When done properly, the result is a ladder of masses that differ by a single residue mass so that the sequence can be read in a straightforward manner. One advantage of this method is that it can be done with a MALDI instrument with only a linear detector. Another advantage is that it is often more successful in identifying the first one or two terminal amino acids than is collision induced fragmentation on electrospray instruments. A disadvantage is that it is not predictable. Since it employs enzymes that differ in their ability to cleave certain amino acids and in the kinetics of amino acid release, a variety of digestion times and enzymes may have to be used to obtain the best results. However, the experiments are easy to perform and the spectra are easy to read. An example is given in figure 4-16 and an easy-to-follow protocol is found in Jiménez et al., 1999.

Post-Source Decay

Sequencing a peptide by post-source decay can only be done in an instrument equipped with a reflector or reflectron which is a device capable of separating fragment ions by flight-time dispersion. A peptide ion that is stable enough to be transported out of the ion source will often not be stable enough to completely survive the flight to the detector and will decompose into fragments in the flight tube. This decomposition process is called post-source decay or metastable decay. When this occurs in the flight tube, all of the fragments will

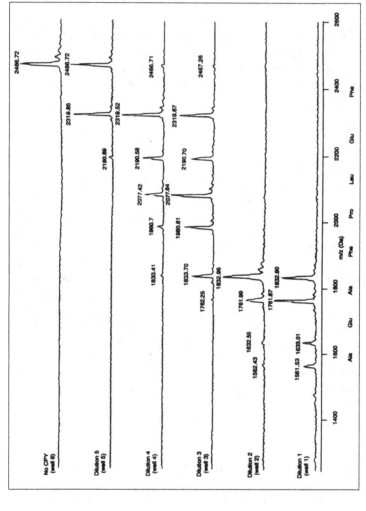

Figure 4-16. Carboxypeptidase Y ladder sequencing of a peptide. The top spectrum is the undiluted peptide. Dilutions 1 to 5 are increasing dilutions of the enzyme. (Reproduced by permission of Keith Waddell, Applied Biosystems, Inc.)

have the same velocity and will arrive at the detector at the same time. Thus, they are all detected as having the same apparent mass with a linear detector. A reflector is capable of separating the in-flight decomposition fragments so that their respective masses can be accurately measured. By changing the reflector voltage during an experiment a complete set of fragments can be seen. Although b and y ions are formed during post-source decay, many other ions can also form from fragmentation in other areas of the amino acid, such as the side chains. This can greatly complicate the spectra and make it difficult, if not impossible, to interpret. Thus, post-source decay analysis is often most productive when it is known what sequence to look for and is not usually suited for analysis of unknown sequences.

Additional Analytical Methods for Synthetic Peptides

Mass spectrometry is a very powerful tool for the analysis of synthetic peptides and as it has become more affordable and widely available it has tended to overshadow other analytical techniques. Nontheless, there are several other techniques that are still useful in particular situations or to answer specific questions that should not be ignored. For instance, resin sampling can still be useful to identify particularly problematic couplings, amino acid analysis is still the best method for peptide quantitation, and Edman sequence analysis can help locate deletion sequences and stable adducts if mass spectrometry fails.

Resin Sampling

Although resin sampling is not commonly used anymore, it can still be useful in determining if particular regions of sequence experience decreased coupling efficiencies. Resin sampling is a procedure by which a small amount of resin is removed from the reaction vessel after the completion of each coupling cycle. Its purpose is to provide a measure of the coupling efficiency at each step. In order for a synthesis to proceed smoothly, the coupling efficiency as each new residue is added must remain high. This is very important because the effect is cumulative. Even if each cycle has a coupling efficiency of 95%, a value considered excellent for most organic reactions, the overall efficiency after 10 couplings is only approximately 60%. At 99% coupling efficiency for each cycle, the overall efficiency after 10 cycles is 90%. If one or more cycles has a significantly reduced coupling efficiency, the outcome of the overall synthesis can be seriously affected, even though all of the other cycles experienced excellent efficiencies. To effectively attempt to correct the problem, it is important to know what cycle is causing the problem. Resin sampling can help in providing that answer. Some instruments perform resin sampling automatically and deliver the resin to a fraction collector. The sampling is done at the end of the coupling step but before the deblocking step. Once collected, the resin then needs to be analyzed manually for the presence of free α-amino groups. This is most commonly done by the Kaiser test (Kaiser et al., 1970, 1980; Sarin et al., 1981) which is a colorimetric ninhydrin analysis specific for primary (ninhydrin) and secondary (isatin) amino groups. The drawbacks of resin sampling analysis are that it consumes resin and it is time consuming to perform on a routine basis. Moreover, it has not been found to be consistently reliable. Within certain sequence contexts, some residues produce a very low color yield with ninhydrin which can lead to incorrect conclusions concerning coupling efficiency. Some laboratories still employ the Kaiser test, but many do not. It is now often reserved for a synthesis where a problem was found during the subsequent evaluation and where one is trying to pinpoint the exact nature of the problem during a resynthesis.

In addition to the Kaiser test, other resin sampling chemistries have also been described but are not as widely used today. These include the utilization of picric acid, 2,4,6-trinitrobenzenesulfonic acid (TNBS), fluorescamine, and chloranil. A description of their use can be found in Stewart and Young (1984). The piperidine derivative of Fmoc, which is formed upon deblocking of the α-amino group of the growing peptide chain, absorbs strongly at 310 nm and can potentially be used to monitor the degree of coupling in Fmoc synthesis (Van Regenmortel et al., 1988). This method has been incorporated as an automated protocol in some synthesizers from PE Biosystems.

Amino Acid Compositional Analysis

Amino acid analysis reveals the content of amino acids in a peptide and quantitates the amount of each. A mole ratio of the amino acids can be calculated from the analysis and compared to the predicted mole ratio based on the intended sequence of the peptide. More importantly, this is the best method for determining the amount of peptide. With the exception of a few amino acids which do not yield quantitatively (see below), the average amount of the amino acids can be directly converted into an accurate quantitation of the peptide. General references for amino acid analysis with excellent descriptions are available (Ozols, 1990; Dunn, 1995; Crabb et al. 1997).

The use of column chromatography for the analysis of amino acids was developed by William Stein and Stanford Moore (Moore and Stein, 1948; Moore et al., 1958; Spackman et al., 1958), for which they won the Nobel Prize in Chemistry in 1972. Their original procedure used ion-exchange chromatography on sulfonated polystyrene resins and post-column detection with ninhydrin. This remained the standard procedure for many years, and is still in routine use today. In fact, it is still considered by many to be the method of choice when quantity is not limiting. A limitation of post-column ninhydrin detection is its limited sensitivity. Even on modern ninhydrin-based instruments, levels below approximately 100 pmols do not quantitate accurately. In addition, post-column ninhydrin analysis requires detection at two wavelengths, 570 and 440 nm, for the detection of both amino and imino (proline and hydroxyproline) acids.

More recently, post-column *o*-phthalaldehyde detection has been used to increase sensitivity to the low picomol range (Roth and Hampai, 1973; Benson and Hare, 1979). However, the imino acid, proline, is not detected with *o*-phthalaldehyde unless it is oxidized with hypochlorite (Bohlen and Mellet, 1979). This procedure also destructively reduces the levels of the other amino acids and thus reduces sensitivity. With the development of reverse-phase HPLC, several, more sensitive precolumn derivitization procedures have been developed. These procedures employ chemical derivatization of the amino acids after hydrolysis, but before chromatography, with a reagent that reacts quantitatively with the free α-amino group of amino acids and which can be detected by fluorescence, UV, or visible absorbance. These include derivitization with *o*-phthalaldehyde (Jones, 1986), dansyl chloride (dimethylaminonapthalene-1-sulfonyl chloride) (Tapuhi et al., 1982; Oray et al., 1983; Marquez et al., 1986), PITC (phenylisothiocyanate) (Heinrikson and Meridith, 1984; Cohen and Strydom, 1988), dabsyl chloride (dimethylaminoazo-benzene-4-sulfonyl chloride) (Chang et al., 1983; Knecht and Chang, 1986), DABITC (dimethylaminoazobenzene-4-isothiocyanate) (Chang, 1983), Fmoc-Cl (9-fluorenylmethyl chloroformate) (Einarsson et al., 1983), a combination of *o*-phthalaldehyde and Fmoc-Cl (Einarsson, 1985; Blankenship et al., 1989), and 6-aminoquinolyl-*N*-hydroxysuccinimidyl carbamate (Cohen and Michaud, 1993).Today several instrument manufacturers offer one or more chemistries as a package. These include Water's Pico-Tag (PITC) and Accu-Tag system, Varian's Amino-Tag (Fmoc-Cl) system, and Agilent Technologies amino acid analysis system. An example of an amino acid analysis of a 24-residue synthetic peptide using the Accu-Tag system is shown in figure 4-17 along with a standard chromatogram. The tabulation of the analysis is shown in table 4-4.

Since all amino acid analysis techniques analyze the free amino acids, any polypeptide, including synthetic peptides, must first be broken down into their individual amino acids by hydrolysis of the peptide bonds. This is commonly done by subjecting the peptide to 6 N HCl in either the liquid or gaseous phase. In the original procedure, the peptide was dissolved in 6 N HCl, the tube was evacuated and sealed, and hydrolysis was accomplished by incubating the tube at

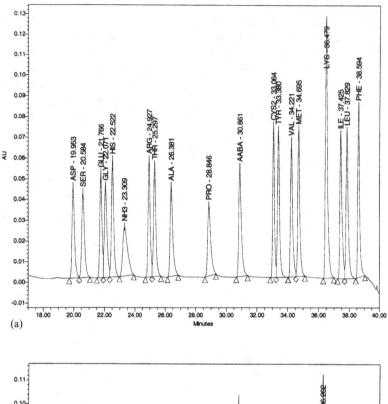

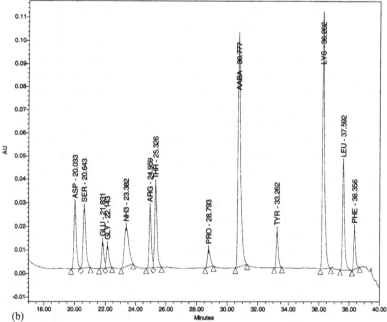

Figure 4-17. Amino acid compositional analysis using the Water's Accu-Tag sytem. (a) Standard mixture of amino acids (250 pmols each). (b) Analysis of a 24-residue synthetic peptide following hydrolysis in 6N HCl.

Table 4-4 Amino acid analysis tabulation of the 24-residue synthetic peptide shown in figure 4-17

Residue	pmol	Mol ratio	Expected
Asp	147.6	2.8	3
Thr	160.5	3.1	3
Ser	148.6	2.9	3
Glu	56.3	1.1	1
Pro	49.5	0.95	1
Gly	51.9	1.0	1
Ala			
Cys			
Val			
Met			
Ile			
Leu	155.7	3.0	3
Tyr	52.3	1.0	1
Phe	52.0	1.0	1
His			
Lys	212.0	4.1	4
Arg	110.9	2.1	2

110 °C for 24 h (Moore and Stein, 1963). More recently the technique of vapor phase hydrolysis has been used (Dupont et al., 1988) where the peptide does not come into direct contact with the liquid HCl. This procedure tends to produce cleaner analyses because nonvolatile contaminants in the liquid HCl no longer contact the peptide. This is more important for high-sensitivity analyses of very low level samples than it is for synthetic peptides where amounts are usually not a problem, but it is routinely used for both with excellent results. Vapor-phase hydrolysis at 150–165 °C for 1 h (Tarr, 1986, Dupont et al., 1988) is also now used routinely with excellent results and with the obvious advantage of saving time.

As mentioned earlier, several amino acids are not stable to hydrolysis in 6 N HCl. Serine, threonine, tyrosine, and methionine are partially destroyed and glutamine, asparagine, cysteine, and tryptophan are usually completely destroyed. Serine, threonine and tyrosine can be fairly accurately determined by carrying out a time course hydrolysis and extrapolating back to zero time. At 110 °C, the times are usually 24, 48, and 72 h, and at 150–165 °C the times are 1, 2, and 3 h. Tyrosine destruction can also be decreased by including 0.1% phenol in the hydrolysis mixture. Phosphoserine, phosphothreonine, and phosphotyrosine readily dephosphorylate under normal hydrolysis procedures but can be quantitated with time course analysis using special procedures (Crabb et al., 1997). Cysteine/cystine and methionine are usually determined after performic acid oxidation (Hirs, 1967) as cysteic acid and methionine sulfone, respectively. Cysteine can also be determined after modification with iodoacetate or 4-vinyl-pyridine (Fontana and Gross, 1986; Hawke and Yuan, 1987, Crabb et al., 1997) as *S*-carboxymethyl cysteine or *S*-pyridylethyl cysteine, respectively.

Hydrolysis with methanesulfonic acid (Simpson et al., 1976; Crabb et al., 1997)), toluenesulfonic acid plus tryptamine (Liu and Chang, 1971), mercaptoethanesulfonic acid (Penke et al., 1974), methanesulfonic acid plus 3-(2-aminoethyl) indole (Simpson et al., 1976), dodecanethiol (Crabb et al., 1997), or in the presence of thioglycolic acid (Matsubara and Sasaki, 1969) have been reported to preserve tryptophan, but in practice the results have been found to be variable. Alkaline hydrolysis (Hugli and Moore, 1972) also preserves tryptophan but destroys other amino acids in the process. It is probably the method of choice for accurate tryptophan determination but due to its relative difficulty, it is seldom done on a routine basis.

The integrity of tryptophan in a polypeptide can be evaluated to some extent by its UV absorbance. Although this does not generally yield accurate quantitation, the absorbance profile of a peptide, scanned from 200 to 320 nm, can provide useful information on the state of the tryptophan in a peptide. An example is shown in figure 4-18 which illustrates a UV scan of a 30-residue peptide containing one tryptophan residue and compares it to a scan where the tryptophan is either missing or destroyed. Intact tryptophan absorbs maximally at 280 nm, while its oxidation products absorb at lower wavelengths and with reduced extinction coefficients. *N*-Formyl tryptophan, which is used in *t*-Boc synthesis absorbs maximally at 300 nm, so it would be readily apparent if deformylation was not successful.

Glutamine and asparagine are completely converted to glutamic acid and aspartic acid, respectively, by acid catalyzed deamidation. Therefore, the values for Glu and Asp derived from amino acid analysis of an acid hydrolyzed sample represent the Glu plus Gln and Asp plus Asn content, respectively.

The rate of peptide bond hydrolysis can also be affected by amino acid sequence. This is most apparent when the two aliphatic side chain β-branched amino acids, Val or Ile, occur adjacent to one another. In this case, hydrolysis of the peptide bond may not be complete until after 72 h of hydrolysis (3 h for high temperature hydrolysis) so values determined at earlier time points will be low.

Amino acid analysis is particularly well suited for the analysis of synthetic peptides because it is more accurate for smaller molecules such as peptides than it is for larger molecules such as proteins. This is due to the inherent error in the procedure. Recent studies (Crabb et al., 1990; Tarr et al., 1991) indicate an average error in practice of

<pre>
 5   10   15   20   25   30
TWKPYDAADLDPTENPFDLLDFNQTQPERC
</pre>

	Expected	AAA Original	AAA Remake
Asp	7	7.0	7.0
Thr	3	2.9	2.5
Ser	0		
Glu	4	5.2	4.5
Pro	4	3.1	3.4
Gly	0		
Ala	2	1.5	1.9
Cys	1	--	--
Val	0		
Met	0		
Ile	0		
Leu	3	2.0	2.9
Tyr	1	0.4	0.8
Phe	2	1.7	1.9
His	0		
Lys	1	0.7	1.0
Arg	1	1.7	1.2
Trp	1	--	--

30

| Yield | | 409mg | 807mg |

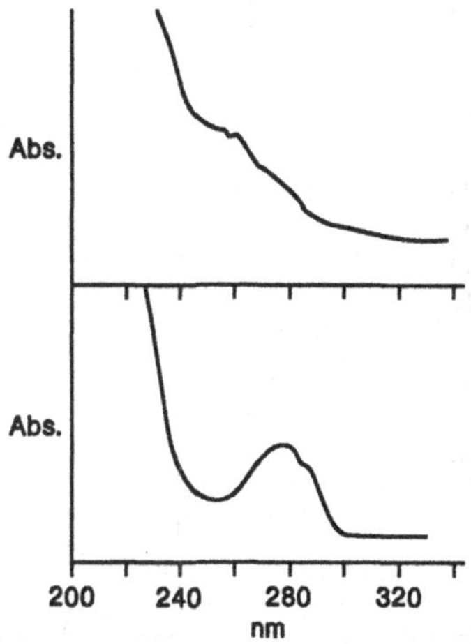

Figure 4-18. Amino acid analysis and UV spectroscopy of a tryptophan-containing peptide. The intended sequence is shown along the top. The amino acid analysis data are on the left and the ultraviolet spectra of the original synthesis (top) and the successful resynthesis (bottom) are on the right.

approximately 10% for nmol level analyses and as much as 16–20% for pmol level analyses. When analyzing a synthetic peptide at the nmol level that contained two aspartic acid residues, the expected error would be approximately 0.2 residues. A calculated value of between 1.8 and 2.2 residues would quite accurately indicate that only two residues of aspartic acid were present. If, however, a protein that contained 20 aspartic acid residues was being analysed, the error would be approximately two residues and give values between 18 and 22 residues per molecule. Synthetic peptides are usually small enough that the inherent error in the analysis is not large enough to cause such an ambiguity. Thus, amino acid analysis of synthetic peptides is expected to be quite accurate and any significant deviation from integral values of residues is usually a sign of heterogeneity in the sample. There are, however, two general types of exceptions which must be taken into consideration when evaluating a peptide by amino acid analysis. The first comes from the fact that not all amino acids survive

Table 4-5 Amino acid analysis[a] of a 16-residue synthetic peptide

Residue	nmol	Mol ratio	Expected
Asp	10.9	2.0	2
Thr	6.6	1.2	1
Ser	4.2	0.8	1
Glu	5.7	1.0	1
Pro	5.5	1.0	1
Gly			
Ala	6.0	1.1	1
Cys	4.2	0.8	1
Val	5.9	1.1	1
Met	5.7	1.0	1
Ile	8.7	1.6	2
Leu			
Tyr			
Phe			
His			
Lys	10.3	1.9	2
Arg	5.1	0.9	1

[a] The peptide was vapor hydrolyzed with 6N HCl containing 0.1% phenol for 1 h at 160°C and analyzed on a Beckman 6300 amino acid analyzer.

the hydrolysis procedure quantitatively and the second from the fact that the hydrolysis can remove modifying groups that may be present, such as side-chain-protecting groups, regenerating the free amino acid.

Although amino acid analysis provides valuable information in the evaluation of a synthetic peptide, it is dangerous to rely on it as the sole characterization. The analysis shown in table 4-5 demonstrates this point. The analysis of this 16-residue peptide looks very reasonable when compared to the expected values. The peptide contains one residue of tryptophan which is not detected, of course, and the two isoleucines are adjacent to one another, so the low Ile value is consistent with the expected slow hydrolysis of the Ile-Ile bond. All other values look good. However, sequence analysis of this peptide revealed that the two aspartic acid residues are not entirely present as free aspartic acid. Rather, the HPLC chromatograms (figure 4-19) of the PTH-amino acids indicate that as much as 60% of the material at the Asp positions elute approximately 6 minutes later than aspartic acid, between the positions for alanine and tyrosine. This example clearly illustrates the presence of unintended modification of the aspartic acid residues that was not detected by amino acid analysis because the adduct was not stable to the hydrolysis procedure and regenerated the free amino acid.

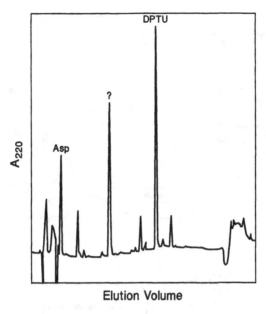

Figure 4-19. Reverse-phase HPLC analysis of PTH-amino acids from a cycle of automated Edman degradation on an Applied Biosystems 477A sequencer. Abbreviations: Asp, aspartic acid; DPTU, diphenylthiourea; ?, unknown amino acid adduct.

Edman Sequence Analysis

Automated Edman degradation (Reim and Speicher, 1997; Smith, 1997), more commonly referred to as amino acid sequence analysis, is obviously well suited for determining the order and identity of amino acids in a synthetic peptide. It can also be very helpful in identifying some, although not all, derivatives of amino acids that may be present for a variety of reasons (see figure 4-19). The first sequence of a polypeptide was determined in 1951 when Sanger determined the sequence of insulin using fluorodinitrobenzene (Sanger and Tuppy, 1951). Repetitive degradation of proteins from the amino-terminal end with phenylisothiocyanate was first reported by Edman in 1950 (Edman, 1950) and remains the chemistry of choice for sequence analysis today (figure 4-20). The automation of Edman chemistry in 1967 (Edman and Begg, 1967), with the introduction of the first "spinning cup" sequencer, marked the beginning of the modern age of sequencing. Later, modifications to the spinning cup sequencer greatly improved the sensitivity of sequence analysis (Hunkapiller and Hood, 1978), and the introduction of polybrene (Hunkapiller and Hood, 1978; Tarr et al., 1978), which helped to

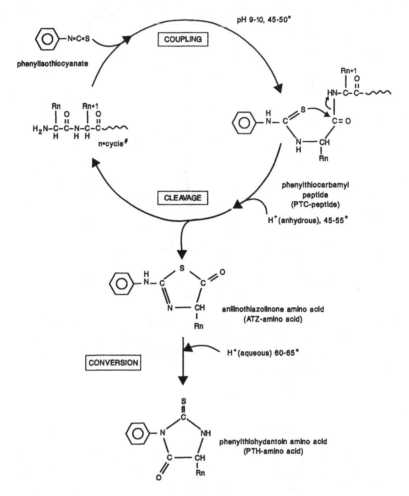

Figure 4-20. Edman Chemistry for the sequential degradation of proteins and peptides. The reaction is carried out in successive cycles with the product of each cycle being a PTH-amino acid. The PTH-amino acid identified at each cycle corresponds to the position from the amino terminus that the amino acid occupied in the peptide.

physically retain the sample in the cup, extended the length of sequence that could be obtained in a single run. Most recently, the development of the "gas-phase" sequencer in 1981 (Hewick et al., 1981) played a major role in advancing the sequential determination of polypeptide structures.

The original spinning cup sequencer of Edman and Begg required protein amounts in the 100 nmol range or greater. Later, improvements in the spinning-cup instrumentation allowed sequence analysis of as little as 1–10 nmols of polypeptide, but the introduction of the gas-phase sequencer, which could routinely analyze 100 pmol of polypeptide, quickly outpaced the utility of the spinning-cup sequencer. Today, the spinning-cup sequencer is no longer available and modern sequencers are now capable of providing very clear analyses on as little as 1 pmol or less of peptide. For a synthetic peptide of 25 residues, this is approximately 2.5 ng of material. Since sensitivity is not a major concern for sequence analysis of synthetic peptides because quantity is seldom a consideration, synthetic peptides are usually analyzed in the 10–100 pmol range.

Sequence analysis, as the name implies, reveals the sequence or order of the amino acids in the polypeptide chain. In this regard, one of the technique's main strengths is in revealing the presence of deletion or addition sequences which are occasional problems encountered in synthetic peptides. It can also be used effectively in revealing the presence of side-chain-blocking groups that failed to be removed during the deblocking step. However, it is more effective in this regard for peptides synthesized by t-Boc chemistry than by Fmoc chemistry due to the acid lability of the blocking groups used for Fmoc synthesis (see section on on-resin sequencing below).

There are also some significant characteristics of sequence analysis which limit its usefulness. First, unless covalent coupling of the peptide to a solid support is employed, it is not always possible to sequence to the C-terminal residue of a peptide. This is especially so with longer peptides but can occur with relatively short peptides as well. This is due mainly to a phenomenon known as washout and is brought about by the repeated washings of the sample in the reaction chamber with solvents in the normal course of the sequencing chemistry. After both the coupling and the cleavage steps of the Edman chemistry, the sample is washed with organic slovents to remove excess reagents, unwanted reaction by-products, and the cleaved ATZ-amino acid itself. Although these washings are formulated to remove only the materials of interest and not the peptide itself, some peptide is invariably lost at each cycle. Thus, at some point, the level of peptide remaining is below the detection sensitivity of the technique and sequence identification stops. Washout is usually more prone to occur with hydrophobic peptides and charged residues near the C-terminus tend to reduce washout of the sequentially shortened peptide.

Failure to sequence to the C-terminus of a peptide can also be caused by rising background due to low-level but repeated internal cleavage of the polypeptide chain. This occurs because the reaction cannot be kept completely anhydrous. It usually takes many successive cycles for the background to build up so that this is usually a problem only with long peptides. In practice, both washout and rising background contribute, but for short peptides, the major cause of cessation of sequencing is washout. For long peptides, sequence identification stops at the point where the effects of the two converge. At a particular point in the sequence analysis, the signal being produced by the sequential degradation of the peptide may still be such that it is still within the detection range of the HPLC. But, if the nonspecific background has risen so high that it gives comparable signal intensity such that it is impossible to distinguish that due to sequence and that due to cumulative background, no additional sequence can be determined.

Loss of peptide due to washout can be alleviated to a large extent if the peptide is covalently coupled through its C-terminal residue to an insoluble support so that washout does not occur. Automated solid-phase sequencing was introduced in 1971 (Laursen, 1971) but it has not gained widespread use and there are no automated instruments presently available which use this approach exclusively. However, chemically derivatized PVDF (polyvinylidene difluoride) membranes are commercially available for this purpose under the brand name Sequelon. They come with either isothiocyanate or arylamine groups covalently attached to the membranes through which attachment of a peptide can be made either through the amino or carboxyl terminus. Attachment through the carboxyl terminus is of course preferable for Edman sequence analysis. Any internal carboxyl side chains will also couple to the membrane and result in reduced yields at these positions. However, the coupling conditions can usually be set up so that some of the carboxyl side chains remain free and a positive identification of the residue can be made. These membranes are also very useful for sequencing radiolabeled phosphorylated peptides (Wang et al., 1988; Aebersold et al., 1990; Sullivan and Wong, 1991). Since the phosphoryl group of phosphoserine and phosphothreonine is eliminated from the amino acid during the cleavage step in Edman sequencing, especially rigorous conditions are required to extract the phosphate ion from the sequencing cartridge and covalent attachment of the peptide is critical for success.

It is always desirable, and sometimes necessary to verify the sequence of a peptide all the way to the C-terminal residue. Occasions where this may be the case is if mass spectrometry is not available or if one suspects for some reason that two or more residues may have been transposed due to an operator error in setting up the synthesizer or a software error in directing the order of addition. If sequence analysis of the intact peptide terminates before reaching the C-terminal residue, another approach is needed. One approach which has been successful is subdigestion of the synthetic peptide and sequencing the individual peptides after separation on HPLC. The smaller peptides generated by this procedure are usually able to be sequenced to their C-terminus, so that all of the sequence of the larger peptide can be verified by analysis of its parts. The major constraint in this approach is the availability of cleavage sites within the peptide such that reasonable sized subpeptides can be generated and recovered. Table 4-6 lists the commonly used cleavage agents and the residues or sequences at which they will cleave. General procedures for using these agents can be found in a number of books and references contained therein (Fontana and Gross, 1986; Wilkinson, 1986; Aitken et al., 1989), and in technical bulletins from the suppliers.

A second disadvantage of sequence analysis is that it relies on the availability of standards for identification. That is, identification of the residue at any particular cycle is dependent on comparison, usually of an HPLC peak, to a known compound or standard. This is obviously not a limitation for the normal amino acids or even for the side-chain-protected amino acids since standard compounds are readily identified and available. However, it can be severely limiting for unexpected or unusual adducts that have formed at some point in the synthesis, such as that presented in figure 4-19. While the presence of an unusual peak may indicate a divergence from normal, it does not provide much useful evidence as to the identity of the adduct. Furthermore, the situation can be further complicated if the unknown adduct co-elutes at the position of a standard amino acid.

Another major limitation of sequence analysis is that peptides that are chemically modified at their amino terminus will not be detected. This is of particular importance when capping chemistry is employed during the synthesis. While capping can be very helpful in the purification process, any capped peptide that is not removed from the sample will be invisible to sequence analysis.

The sequencing chemistry can also reverse some derivatives of amino acids, regenerating the parent amino acid in the process, so that its presence goes undetected. Partial reversal is often seen in on-resin sequencing of peptides generated by t-Boc chemistry. With the protecting groups used for Fmoc synthesis, the reversal can be complete. Another example of this is methionine sulfoxide. The sulfur atom of methionine residues can be oxidzed during synthesis and handling, usually to the sulfoxide. In fact, methionine sulfoxide is sometimes employed directly as a protecting group in the synthesis of a peptide. After synthesis, the sulfoxide can be converted back to methionine by treatment with thiol reagents such as dithiothreitol and N-methylmercaptoacetamide (Houghton and Li, 1979; Cullwell, 1987). However, if the sulfoxide is present in the peptide, its presence will go undetected in sequencing because it is converted back to methionine due to the presence of dithiothreitol in the extraction solvent on many sequencers.

Finally, sequence analysis is not very reliable for quantitation, with some residues being detected only at very low levels or not at all. These include serine, threonine, and cysteine which are degraded to varying

Table 4-6 Common reagents and proteases for cleavage of peptides[a]

Bond cleaved	Agent	Reference	Source[b]
Arg, Lys-X (X ≠ Pro)	Trypsin	Wilkinson, 1986	Worthington, Pierce Boehringer Mannheim
Lys-X (X ≠ Pro)	Endoproteinase Lys-C		Boehringer mannheim
	Lysyl Endoproteinase	Masaki et al., 1981a, b	Wako Chemicals
Arg-X (X ≠ Pro)	Endoproteinase Arg-C (Submaxillary Protease)	Levy et al., 1970 Schenkein et al., 1977	Boehringer Mannheim Pierce, Takara Biochemicals
Glu-X (X ≠ Pro), (Asp-X)[c]	Endoproteinase Glu-C S. aureus V8 Protease	Wilkinson, 1986	Boehringer Mannheim Pierce
Asn-X	Asparaginyl Endopeptidase	Ishii et al., 1990	Takara Biochemicals
X-Asp	Endoproteinase Asp-N	Maier et al., 1986	Boehringer Mannheim
Trp, Phe, Tyr, Leu-X (X ≠ Pro)	Chymotrypsin	Wilkinson, 1986	Worthington Boehringer Mannheim
pGlu-X[d]	Pyroglutamate Aminopeptidase	Shively, 1986	Boehringer Mannheim Pierce, Takara Biochemicals
Acetyl-X[e]	Acylamino Acid Releasing Enzyme	Tsunasawa and Narita, 1982	Pierce, Takara Biochemicals
Met-X	Cyanogen Bromide	Fontana and Gross, 1986	
Asn-Gly	Hydroxylamine	Fontana and Gross, 1986	
Asp-Pro	Dilute Acid	Fontana and Gross, 1986	
Trp-X	Iodosobenzoate BNPS-Skatole N-Chlorosuccinimide	Fontana and Gross, 1986	

[a] This table is not intended to be exhaustive. Only commonly used agents of high specificity ate listed.
[b] For the readers convenience, a limited number of sources are listed. For some, these are the only source; for others, additional sources exist but no attempt has been made to list them all.
[c] Occasional cleavage produced at these sites.
[d] pGlu, cyclized Gln at the N-terminal position.
[e] Removes amino-terminal acetylated residue. Active only on peptides up to 20-30 residues.

degrees by β-elimination during sequencing, and tryptophan, which is sensitive to oxidation either during sequencing or during storage and handling prior to sequence analysis. Although dithiothreitol is often added to the ATZ-amino acid extraction solvent to trap the reactive elimination products of serine, threonine, and cysteine, the yields of these amino acid adducts are still low and variable. In fact, cysteine often must be derivatized prior to sequencing in order to detect it at all. There are many reagents available for this purpose (Gambee et al., 1995). Alkylation with 4-vinylpyridine to the S-pyridylethyl derivative can be performed on the sample after it has been applied to the sequencer cartridge (Andrews and Dixon, 1987; Hawke and Yuan, 1987) and is probably the reagent of choice due to its stability and HPLC elution position. Arginine and histidine also tend to produce low yields due to reduced extraction efficiency of the positively charged amino acids and undesirable interactions with incompletely capped HPLC columns.

In addition to variability in individual PTH-amino acid recovery, the overall initial yield of a sequencing run can also vary. This can be due to several factors, such as the operating efficiency of the sequencer and the chemical properties of the sample itself. Thus, even if only stable PTH-amino acids are considered, the overall yield of the sequencing run itself can vary by several fold.

In spite of all the limitations listed above, sequence analysis is still a useful technique for the evaluation of synthetic peptides. What sequence analysis does best, and that no other technique can do on a routine or available basis, is elucidate the actual linear sequence of amino acids as they occur physically in a peptide. In this regard, sequence analysis is outstanding in revealing deletion and addition sequences.

The role of sequence analysis in finding deletion sequences is illustrated in the example presented in figure 4-21. The product of this synthesis runs as a single major peak on the standard reverse-phase HPLC analysis system, which is 0.1% TFA with a linear acetonitrile gradient, and the amino acid analysis is entirely consistent with the desired product. However, in this case, sequence analysis revealed the presence of two peptides that differed only by a single amino acid. As much as 60% of the peptide was present with the deletion of a single glutamic acid residue at position number 5. Quite surprisingly, this was not well reflected in the amino acid analysis and again illustrates the danger of relying on a single type of analysis. The glutamic acid value may have been more revealing in this case if the peptide contained fewer Glu + Gln residues overall. Subsequent reverse-phase chromatography at basic pH, where the glutamate side chain would now be ionized, easily separated the two peptides (figure 4-5). This example also illustrates how relying on a single set of HPLC

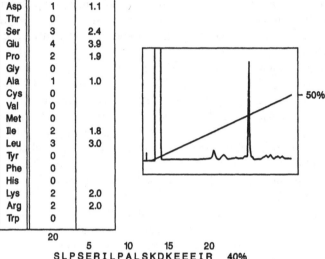

                    5   10   15   20
              S L P S E R I L P A L S K D K E E E I R

	Expected	AAA
Asp	1	1.1
Thr	0	
Ser	3	2.4
Glu	4	3.9
Pro	2	1.9
Gly	0	
Ala	1	1.0
Cys	0	
Val	0	
Met	0	
Ile	2	1.8
Leu	3	3.0
Tyr	0	
Phe	0	
His	0	
Lys	2	2.0
Arg	2	2.0
Trp	0	
	20	

                    5   10   15   20
              S L P S E R I L P A L S K D K E E E I R   40%
              S L P S R I L P A L S K D K E E E I R   60%

Figure 4-21. Evaluation of a 20-residue peptide indicating the presence of a deletion-sequence peptide. The expected sequence is shown at the top and the experimentally determined sequence of the sample is shown at the bottom. The amino acid analysis and reverse-phase HPLC analysis of the sample are also shown.

conditions could be misleading unless other analytic procedures are brought to bear, and it also demonstrates how sequence analysis can complement HPLC as a means of detecting heterogeneity. Mass spectrometry would also have detected the presence of the two peptides if it had been performed. Figure 4-22 illustrates another example of the occurrence of a deletion sequence in a synthetic peptide. In this case, the deletion involves a block of residues rather than just a single residue. This deletion is easily detected by amino acid analysis, but since it occurs at the C-terminus, Edman sequencing may not be entirely effective. Deletions or additions can occur anywhere in the peptide and can be a much more serious problem with longer peptides where sequencing will not reach to the C-terminus.

On-Resin Sequencing

The growing polypeptide chain can also be analyzed by automated Edman degradation while it is still attached to the resin (Kent et al.,

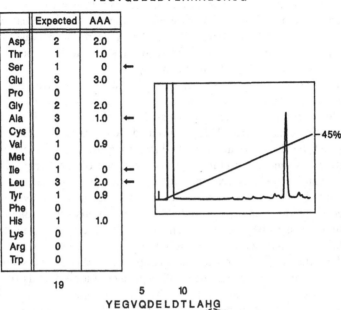

Figure 4-22. Evaluation of a 19-residue peptide indicating the presence of a total deletion sequence of more than one residue. The expected sequence is shown at the top and the experimentally determined sequence of the sample is shown at the bottom. The amino acid analysis and reverse-phase HPLC analysis of the sample are also shown. Arrows indicate obvious deviations from that expected and the indicated residues correspond to the missing sequence shown in the box.

1982). This technique is most applicable to peptides synthesized with t-Boc chemistry, due to the stability of the protecting groups under acidic conditions. This is often done upon completion of the synthesis, but can be done at any time during the synthesis to assess the quality of the growing peptide. This is often referred to as "preview analysis," since deletion sequences will show up as a preview of residue $i + 1$ in cycle i (see also chapter 3). For example, such an analysis might be performed several times during the course of synthesis of a very long peptide. This requires very little resin because of the excellent sensitivity (1 pmol) of modern sequencers. The only requirement is that the resin be removed at a time when the peptide possesses a free α-amino group, such as after removal of the α-amino protecting group but before the beginning of the next coupling cycle. Because the sequencing is being performed in the solid phase while the peptide is covalently coupled to an insoluble bead, sample washout from the sequencer is not a problem. Therefore, the peptide can easily be sequenced to its penultimate residue. The C-terminal residue is not usually seen because it is attached to the resin bead by a bond that is not cleavable by the Edman chemistry.

Since peptides that are sequenced on-resin have not yet been subjected to side chain deprotecting procedures, those amino acids whose side chains contain protecting groups will not elute on an HPLC at the same position as the free PTH-amino acids. Generally, the side-chain-protected PTH-amino acids elute later than the their unprotected counterparts, and most elute well after the last of the normal PTH-amino acids. For this reason, most will not be detected with the standard HPLC program used for routine sequence analysis. For detection of side chain protected PTH-amino acids the gradient and run time needs to be extended (Kent et al., 1982; Kochersperger et al., 1989). Note that with some amino acids, particularly Glu, Asp, and Ser, the side-chain-protecting group is unstable to the sequence chemistry and some regeneration of unprotected amino acids result.

The extended gradients can also be used in routine sequencing of cleaved, deprotected peptides since it gives a better picture of other components which can be present, particularly side chains that were not successfully deprotected. Some of the more common blocking groups that have been detected still attached to cleaved peptides are tosyl-arginine and formyl-tryptophan. A list of the relative elution positions of modified PTH amino acids that may be encountered in the analysis of synthetic peptides is given in table 4-7.

A similar approach can also be used for peptides synthesized with Fmoc chemistry (Kochersperger, et al., 1989), but their are several disadvantages. First, the bond to the resin is not stable under the conditions of the sequencing chemistry, so that the advantage of sequencing in the "solid phase" is lost. Secondly, the side-chain- protecting groups are less stable to the sequencing chemistry and are removed more readily. This is not necessarily a disadvantage for on-resin sequencing but can be a serious drawback in trying to assess the degree of deprotection in cleaved peptides made with Fmoc chemistry.

Storage and Handling of Synthetic Peptides

The manner in which a peptide is handled and stored can be just as important to its integrity and utilization as any of the chemistry used to produce it. In theory, most peptides should be stable at neutral pH in aqueous solution for extended periods of time. However, contamination with microorganisms or metal ions can cause peptide-bond cleavage.

Table 4-7 Relative elution positions of PTH-amino acids on HPLC[a]

Retention time (min)	PTH-amino acid	Notes
5.8	Tyr(P)	stable
6.1	PTC-Gly	stable, partial conversion product to ATZ-Gly
6.2	Asp	stable, slight conversion to Asp derivative
7.1	Asn	~5–10% recovered as PTH-Asp
8.1	PTC-Ala	stable, partial conversion product of ATZ-Ala
8.2	Ser	typical ~0–20% recovered, see Ser derivative
8.8	Gln	~5–20% recovered as PTH-Glu
9.3	Thr	typically 5–25% recovered
9.7	Gly	stable
10.5	Glu	stable, slight conversion to Glu derivative
11.0	Dimethylphenylthiourea (DMPTU)	Edman chemistry by-product
12.5	Cys-acetamidomethyl (Acm)	stable
13.0	Hypro (*cis*)	hydroxyl can be *cis* or *trans*, Edman chemistry produces both
13.8	Ala	stable
14.0	Hypro (*trans*)	hydroxyl can be *cis* or *trans*, Edman chemistry produces both
14.6	His[b]	stable but may not be completely extracted from sequencer
15.1	PTC-Met	stable, partial conversion product of ATZ-Met
15.2	Ser derivative	breakdown product of serine due to Edman Chemistry
16.8	Cys-3-nitro-2-pyridinesulfenyl (Npys)	Sensitive to DTT. Converts to Cys with successive cycles
17.4	Asp-O-allyl (OAl)	partially stable, converts to Asp
17.7	Tyr	stable
18.8	Ser derivative	breakdown product of serine due to Edman Chemistry
19.0	PTC-Val	stable, partial conversion product of ATZ-Val

(*continued*)

Table 4-7 (*Continued*)

Retention time (min)	PTH-amino acid	Notes
19.1	Ser-allyloxyxarbonyl (Aloc)	partially stable, converts to Ser derivatives
19.8	Arg[b]	stable but may not be completely extracted from sequencer
19.9	PTC-Lys	stable, partial conversion product of ATZ-Lys
20.7	PTC-Leu	stable, partial conversion product of ATZ-Ile
20.7	PTC-Ile	stable, partial conversion product of ATZ-Ile
20.8	Tyr-dimethoxyphosphoryl (OP(OMe)₂)	rapidly converted to Tyr(P) in first few cycles
21.1	Pro	stable
21.9	Met	usually stable but susceptible to oxidation
22.3	Val	stable
22.6	Arg-diallyloxycarbonyl (Aloc)₂ derivative	product of conversion of Arg (Aloc)₂, may be Arg (Aloc)
23.3	Thr-allyloxycarbonyl (Aloc)	stable
23.7	Asp derivative	associated with conversion to PTH-Asp
23.9	Diphenylylhiourea (DPTU)	Edman chemistry by-product
23.9	Arg-4-toluenesulfonyl (Tos)	stable
23.3	Glu derivative	associated with conversion to PTH-Glu
24.1	Lys-allyloxycarbonyl (Aloc)	stable
24.7	His-2,4-dinitrophenyl (Dnp)	stable
25.1	Glu-O-allyl (OAl)	stable, slight conversion to Glu derivative
25.3	Trp	typically only partially recovered at inconsistent levels
25.8	Asp-O-*tert*-butyl (O*t*Bu)	very unstable, converts to Asp
26.1	Phe	stable
26.1	Trp-*N*ⁱⁿ-formyl (CHO)	20–40% recovered as PTH-Trp
26.2	Phe-*p*Nitrophenyl	stable
26.2	Cys-allyl (Al)	stable

26.7	Ile	stable
27.1	Lys (ptc)	stable, both amino groups are derivatized with PITC
27.4	Leu	stable
27.5	Norleucine	stable
27.8	His-*tert*butoxymethyl (Bum)	converted to PTH-His with successive cycles
28.2	Ser-benzyl (Bzl)	partially recovered, remainder converted to Ser derivatives
28.5	Cys-allyloxycarbonyl (Aloc)	stable
28.9	Arg-4-methoxy-2,3,6-trimethylbenzenesulfonyl (Mtr)	partially recovered, converted to PTH-Arg with successive cycles
29.2	Arg-mesitylene-2-sulfonyl (Mts)	typically 1–5% recovered as PTH-Arg
29.5	Cys-*tert*-butyl (*t*Bu)	very unstable, converts to Cys
30.0	Asp-O-benzyl (OBzl)	~5–20% recovered as PTH-Asp
30.1	Arg-diallyloxycarbonyl(Aloc)$_2$	partially stable
30.5	Trp derivative	one of multiple Trp degradation products
30.6	His-3-benzyloxymethyl (3-Bom)	stable
30.6	Thr-benzyl (Bzl)	partially stable, products include PTH-Thr and Thr derivatives
31.2	Tyr-allyl (Al)	stable
31.3	Glu-O-benzyl (OBzl)	~10–40% recovered as PTH-Glu
31.7	Tye-*tert*-butyl (*t*Bu)	very stable, converts to Tyr
32.1	Cys-4-methoxybenzyl (Mob)	partially stable, products include Ser derivatives
32.5	Thr-*tert*-butyl (*t*Bu)	very unstable, converts to Thr
33.4	Asp-O-cyclohexyl (OcHex)	~5–20% recovered as PTH-Asp
33.6	Lys-chlorobenzyloxycarbonyl (ClZ)	partially stable, some recovered as PTH-Lys (ptc)
33.6	Lys-2-chlorobenzyloxycarbonyl (2ClZ)	stable, perhaps 1–2% recovered as PTH-Lys (ptc)
33.8	Thr-benzyl (Bzl) derivative	produced from Thr-benzyl (Bzl)
33.9	Lys-2,4-dinitrophenyl (Dnp)	stable
34.0	Trp derivative	one of multiple Trp degradation products
34.3	Hydroxyproline-4-benzyl (4-Bzl)	unstable, converts to Hypro
34.3	Glu-O-cyclohexyl (OcHex)	~10–40% recovered as PTH-Glu

(*continued*)

281

Table 4-7 (*Continued*)

Retention time (min)	PTH-amino acid	Notes
35.1	Cys-4-methylbenzyl (Meb)	Partially stable, products include Ser derivatives
35.4	Hydroxyproline-4-benzyl (4-Bzl)	unstable, converts to Hypro
37.8	Lys-9-fluorenylmethyloxycarbonyl (Fmoc)	converted to PTH-Lys (pts) with successive cycles
38.6	Tyr-2-bromobenzyloxycarbonyl (2-BrZ)	stable, perhaps 1–5% recovered as PTH-Tyr
38.7	Glu-O-9-fluorenylmethyl (OFm)	~10–40% recofered as PTH-Glu
38.9	Trp derivative	one of multiple Trp degradation products

Side-chain-protected amino acids completely or nearly completely unstable to Edman Chemistry

His-allyloxycarbonyl (Aloc)		converts to His
His-tosyl (Tos)		converts to His
His-benzyloxycarbonyl (Z)		converts to His
Asp/Glu-O-tert-butyl (OtBu)		converts to Asp/Glu
Ser/Thr/Tyr/Hypro-tert-butyl (tBu)		converts to Ser/Tht/Tyr/Hypro
Lys/His-tert-butyloxycarbonyl (Boc)		converts to Lys/His
His/Cys/Asn/Gln-triphenylmethyl (Trt)		converts to His/Cys/Asn/Gln
Arg-2,2,5,7,8-pentamethyl-chroman-6-sulfonyl (Pmc)		converts to Arg
Arg-2,2,4,6,7-pentamethyldihydrobenzofuran-5-sulfonyl (Pbf)		converts to Arg
Asn/Gln/Cys-2,4,6-trimethoxybenzyl (Tmob)		converts to Asn/Gln/Cys

[a] Elution positions are relative and may vary somewhat depending on system and column. Some of these derivatives were originally done with different conditions and are now represented on a typical resin-bound sequencing cycle (ABI "REZ" cycle). Table composed from material taken from G. A. Grant, M. W. Crankshaw, and J. Gorka, *Meth. Enzymol.* **289**, 395 (1997), and C. G. Fields, A. Loffet, S. A. Kates, and G. B. Fields, *Anal. Biochem.* **203**, 245 (1992).
[b] Position is variable depending on ionic strength and elution buffer composition.

Moreover, oxygen can adversely affect tryptophan, cysteine, and methionine residues, and the presence of metal ions can cause destruction of tryptophan. Aspartic acid-proline bonds are also extremely sensitive to acid cleavage. Storage of peptides containing this bond for extended times in aqueous solutions that are only slightly acidic can lead to significant hydrolysis. Other peptide bonds with aspartic acid also tend to be more acid labile than most, but not nearly to the degree seen with Asp-Pro. Also in aqueous solution, asparagine-glycine and aspartic acid-glycine bonds, and in some instances aspartic acid N-terminal to other short-side-chain amino acid residues (Ser, Thr, Ala, Asn), can cyclize to form an aspartimide intermediate that can undergo spontaneous ring opening to either regenerate the original peptide or produce a β-aspartyl peptide that possesses two methylene carbons between the backbone amino and carbonyl group of the residue (see figure 4-2). When this occurs, the structure of the peptide is drastically altered. The presence of a β-aspartyl bond can be indicated by sequence analysis. Because of the additional carbon in the backbone, the PTC-peptide is not able to cyclize efficiently, and sequencing will stop when such a structure is encountered.

For the reasons mentioned above, peptides should be stored in solution only for short periods of time. For long-term storage the peptide should be lyophilized from a volatile buffer or solvent and stored dessicated at $-20\,°C$ in either polyethylene, polypropylene or silanized glass containers. When containers of dry peptides are removed from the cold, they should be allowed to come to room temperature before they are opened to minimize condensation of water vapor on the peptide surface. This could contribute to hydrolysis upon the return of the remainder of the peptide to storage.

If a peptide has been stored for an extended period of time, it is not a bad idea to re-evaluate it before use. Even "dry" peptides after lyophilization can contain a significant amount of adsorbed solvent which can contribute to unwanted reactions. An example is the relatively fascile deamidation of Asn and Gln when residual acid is present in the "dry" peptide. Peptides containing amide side chains should either not be dried down from acidic solution or precaution should be taken to assure that all acid is removed from the sample before storage.

If peptides are going to be stored for any length of time after synthesis but before they are worked up, they should be desalted before storage. Many salts and organics can be present as a result of the synthesis, cleavage, and deprotection procedures that can have an adverse effect upon long term storage. Desalting can be accomplished either by standard low pressure gel filtration chromatography or by preparative reverse-phase HPLC.

How Much Is Enough?

How much evaluation is necessary? That is a question that is always asked, both by investigators who are having peptides made for them by someone else and by the synthetic chemists themselves. In an ideal world, all synthetic peptides would be subjected to all possible evaluation techniques, and there would be no need to ask the question. However, most investigators are restrained by matters of cost, time, and instrument availability and are therefore often confronted with this question.

Practically speaking, most peptides which can be purified to a single symmetrical peak on HPLC or CE and which yield a good mass spectrum will be the intended product. The evaluation should be taken at least this far. However, if one procedure indicates a problem while others do not, it is usually not wise to ignore the one in favor of the others. That one procedure is usually trying to tell you something and is sufficient grounds in itself to delve further into the evaluation.

It is also not recommended to rely on a single analytical technique for your evaluation. We have seen examples where apparent single peaks really contained several species and where a reasonable looking amino acid analysis was not truly indicative of the complexity of the sample. Evaluation by sequence analysis alone may reveal only one sequence, and while that sequence may be the expected one, other blocked species can also be present. There is also a common misconception that mass analysis is all that is necessary if it indicates the correct mass ion. It is possible to miss other constituents with this technique for a variety of reasons. The contaminant may produce a weaker signal that is not proportional to its actual presence in the sample if it is not desorbed as efficiently or is outside the mass range of the instrument. Conversely, the major peptide in regard to total sample may give a very weak signal or none at all.

A cumulative 6-year study conducted by the peptide synthesis/mass spectrometry subcommittee of the Association of Biomolecular Resource Facilities (Angeletti et al., 1997) reinforces many of the points made above. While its original intention was to assess the peptide synthesis capabilities in core facilities, it offers many enlightening observations pertaining to the strengths and weaknesses of the methods of peptide evaluation. It is recommended reading for anyone engaged in peptide synthesis and evaluation.

Unfortunately, no one can say in advance how much evaluation needs to be done. Each peptide is different and each synthesis is different. The data need to be evaluated as they are collected in order to make the decision. The answer to the question posed at the beginning of this section can only be "as much as is necessary to convince a knowledgable person that the product is as it should be."

You are the ultimate judge. Hopefully, this chapter, as well as the other chapters in this book, have provided a helpful discussion of the means by which to make that decision. How you apply that knowledge is up to you.

ACKNOWLEDGMENTS   The author gratefully acknowledges the advice, helpful discussions, and materials provided by Mark Crankshaw and the staff of the Washington University Protein Chemistry Laboratory, John Gorka of Biomolecules Midwest, Dan Crimmins of the Department of Pathology at Washington University, and the Association of Biomolecular Resource Facilities.

References

Aebersold, R., Pipes, G. D., Wettenhall, R. E., Nika, H., and Hood, L. E. (1990). Covalent attachment of peptides for high sensitivity solid-phase sequence analysis. Anal Biochem. 187:56–65.

Alpert, A. J., and Andrews, P. C. (1988). Cation exchange chromatography of peptides on (2-sulfoethyl aspartamide)-silica. J. Chromatogr. 443:85–96.

Aitken, A., Geisow, M. J., Findlay, J. B. C., Holmes, C., and Yarwood, A. (1989). Peptide preparation and characterization. In *Protein Sequencing: A Practical Approach*, J. B. C. Findlay and M. J. Geisow, eds., Oxford, IRL Press, pp. 43–68.

Andrews, P. C., and Dixon, J. E. (1987). A procedure for in situ alkylation of cystine residues on glass fiber prior to protein microsequence analysis. Anal. Biochem. 161:524–528.

Angeletti, R. H., Bonewald, L. F., and Fields, G. B. (1997). Six-year study of peptide synthesis. Methods Enzymol. 289:697–717.

Barany, G., Knieb-Cordonier, N., and Mullen, D. G. (1987). Solid phase peptide synthesis: a silver anniversary report. Int. J. Peptide Protein Res. 30:705–739.

Barber, M., Bordoli, R. S., Sedgwick, R. D., and Tyler, A. N. (1981). Fast atom bombardment of solids (F. A. B.): A new ion source for mass spectrometry. J. Chem. Soc. Chem. Commun. No. 7:325–327.

Benson, J. R., and Hare, P. E. (1979). *o*-Phthalaldehyde: fluorogenic detection of primary amines in the picomole range, comparison with fluorescamine and ninhydrin. Proc. Natl. Acad. Sci. USA 72:619–622.

Blankenship, D. T., Krivanek, M. A., Ackermann, B. L., and Cardin, A. D. (1989). High sensitivity amino acid analysis by derivatization with o-phthalaldehyde and 9-fluorenylmethylchloroformate using fluoresence detection: applications in protein structure determination. Anal. Biochem. 178:227–232.

Bohlen, P., and Mellet, M. (1979). Automated fluorometric amino acid analysis: the determination of proline and hydroxyproline. Anal. Biochem. 94:313–321.

Carr, S. A. and Annan, R. S. (1999). Overview of peptide and protein analysis by mass spectrometry. In *Current Protocols in Protein Science*, 16.1.1–16.1.27, New York, John Wiley and Sons.

Chang, J.-Y. (1983). Manual micro-sequence analysis of polypeptides using dimethylaminoazobenzene isothiocyanate. Methods Enzymol. 91:455–467.

Chang, J.-Y., Knecht, R., and Braun, D. G. (1983). Amino acid analysis in the picomole range by precolumn derivatization and high-performance liquid chromatography. Methods Enzymol. 91:41–48.

Chicz, R. M., and Regnier, F. E. (1990). High performance liquid chromatography: Effective protein purification by various chromatographic methods. Methods Enzymol. 182:392–421.

Cohen, S. A. and Michaud, D. P. (1993). Synthesis of a fluorescent derivatizing reagent, 6-aminoquinolyl-N-hydroxysuccinimidyl carbamate. Anal. Biochem. 211:279–287.

Cohen, S. A., and Strydom, D. J. (1988). Amino acid analysis utilizing phenylisothiocyanate derivatization. Anal. Biochem. 174:1–16.

Cotter, R. J. (1988). Plasma desorption spectrometry comes of age. Anal. Chem. 60:781A–793A.

Crabb, J. W., Ericsson, L., Atherton, D., Smith, A. J., and Kutny, R. (1990). A collaborative amino acid analysis study from the association of biomolecular resource facilities. In Current Research in Protein Chemistry: Techniques, Structure, and Function, J. J. Villafranca, ed., San Diego, Calif., Academic Press, pp. 49–61.

Crabb, J. W., West, K. A., Dodson, W. Scott, and Hulmes, J. D. (1997). Amino acid analysis. In Current Protocols in Protein Science, 11.9.1–11.9.42, New York, John Wiley and Sons.

Crimmins, D. L., Gorka, J., Toma, R. S., and Schwartz, B. D. (1988). Peptide characterization with a sulfoethyl aspartimide column. J. Chromatogr. 443:63–71.

Cullwell, A. (1987). Reduction of methionine sulfoxide in peptides using N-methylmercaptoacetamide. Applied Biosystems User Bulletin, Model 430, No. 17.

Dunn, B. (1995). Quantitative amino acid analysis. In Current Protocols in Protein Science, 3.2.1–3.2.3, New York, John Wiley and Sons.

Dupont, D., Keim, P., Chui, A., Bozzini, M., and Wilson, K. J. (1988). Gas-phase hydrolysis for PTC-amino acid analysis. Applied Biosystems User Bulletin, Model 420A, Issue No. 2.

Edman, P. (1950). Method for determination of amino acid sequence in peptides. Acta. Chem. Scand. 4:283–293.

Edman, P., and Begg, G. (1967). A protein sequenator. Eur. J. Biochem. 1:80–91.

Einarsson, S. (1985). Selective determination of secondary amino acids using precolumn derivatization with 9-fluorenylmethylchloroformate and reversed-phase high-performance liquid chromatography. J. Chromatogr. 348:213–220.

Einarsson, S., Josefsson, B., and Lagenkvist, S. (1983). Determination of amino acids with 9-fluorenylmethylchloroformate and revered-phase high-performance liquid chromatography. J. Chromatogr. 282:609–618.

Ewing, A. G., Wallingford, R. A., and Olefirowicz, T. M. (1989). Capillary electrophoresis. Anal. Chem. 61:292A–303A.

Feldhoff, R. (1991). Why not ethanol-based solvents for RP-HPLC of peptides and proteins? In Techniques in Protein Chemistry, Vol. II, J. J. Villafranca, ed., San Diego, Calif., Academic Press, pp. 55–63.

Fontana, A., and Gross, E. (1986). Fragmentation of polypeptides by chemical methods. In *Practical Protein Chemistry: A Handbook*, A. Darbre, ed., New York, John Wiley and Sons, pp. 67–120.

Fontenot, J. D., Ball, J. M., Miller, M. A., David, C. M., and Montelaro, R. C. (1991). A survey of potential problems and quality control in peptide synthesis by the fluorenylmethylcarbonyl procedure. Peptide Res. 4:19–25.

Fujii, N., Otaka, A., Funakoshi, S., Bessho, K., Watanabe, T., Akaji, K., and Yajim, H. (1987). Studies on peptides CLI: Synthesis of cystine-peptides by oxidation of S-protected cysteine peptides with thallium (III) trifluoroacetate. Chem. Pharm. Bull. 26:539–548.

Futaki, S., Yajami, T., Taike, T., Akita, T., and Kitagawa, K. (1990). Sulphur trioxide/thiol: a novel system for the reduction of methionine sulphoxide. J. Chem. Soc. Perkin Trans. 1:653–658.

Gambee, J., Andrews, P. C., DeJongh, K., Grant, G., Merrill, B., Mische, S., and Rush, J. (1995). Assignment of Cysteine and Tryptophan Residues During Protein Sequencing: Results of ABRF 94 SEQ. In *Techniques in Protein Chemistry*, Vol. VI, J. Crabb, ed., San Diego, Calif., Academic Press, pp. 209–217.

Gibson, B. W. (1991). A brief overview of mass spectrometric methods for the analysis of peptides and proteins. In *Techniques in Protein Chemistry*, Vol. II, J. J. Villafranca, ed., Academic Press, San Diego, Calif., pp. 419–425.

Griffin, P. R., Martino, P. A., McCormack, A. L., Shabanowitz, J., and Hunt, D. F. (1990). Protein and oligopeptide sequence analysis on the TSQ-70 triple quadrupole mass spectrometer. In *Current Research in Protein Chemistry: Techniques, Structure, and Function*, J. J. Villafranca, ed., San Diego, Calif., Academic Press, pp. 117–126.

Hancock, W. S., ed. (1984). *CRC Handbook of HPLC for the Separation of Amino Acids, Peptides, and Proteins*, Vol. I and II, Boca Raton, Fla., CRC Press.

Hawke, D., and Yuan, P. (1987). *S*-Pyridylethylation of cystine residues. Applied Biosystems User Bulletin for 470A/477A-120A, Number 28.

Heinrikson, R. L., and Meridith, S. C. (1984). Amino acid analysis by reverse-phase high-performance liquid chromatography: Pre-column derivatization with phenylisothiocyanate. Anal. Biochem. 136:65–74.

Henschen, A., Hupe, K. P., Lottspeich, F., and Voelter, W., eds. (1985). In *High Performance Liquid Chromatography in Biochemistry*, Weinheim, VCH.

Henzel, W. J., and Stults, J. T. (1999). Matrix-Assisted Laser Desorption/Ionization Time-of-Flight Mass Analysis of Peptides. In *Current Protocols in Protein Science*, 16.2.1–16.2.11, New York, John Wiley and Sons.

Hewick, R. M., Hunkapiller, M. W., Hood, L. E., and Dreyer, W. J. (1981). A gas-liquid solid phase peptide and protein sequenator. J. Biol. Chem. 256:7990–7997.

Hirs, C. H. W. (1967). Determining cysteine as cysteic acid. Methods Enzymol. 11:59–62.

Houghton, R. A., and Li, C. H. (1979). Reduction of sulfoxides in peptides and proteins. Anal. Biochem. 98:36–46.

Hugli, T. E., and Moore, S. (1972). Determination of the trypyophan content of proteins by ion exchange chromatography of alkaline hydrolysates. J. Biol. Chem. 247:2828–2834.

Hunkapiller, M. W., and Hood, L. E. (1978). Direct microsequence analysis of polypeptides using an improved sequenator, a non-protein carrier (polybrene), and high pressure liquid chromatography. Biochemistry 17:2124–2133.

Hunt, D. F., Shabanowitz, J., Yates, J. R., Griffin, P. R., and Zhu, N. Z. (1988). In *Analysis of Peptides and Proteins*, C. McNeil, ed., New York, John Wiley and Sons, pp. 151–165.

Ishii, S., Abe, Y., Matsushita, H., and Kato, I. (1990). An asparaginyl endopeptidase purified from jack bean seeds. J. Protein Chem. 9:294–295.

Jiménez, C. R., Huang, L., Qiu, Y. and Burlingame, A. L. (1999). Enzymatic approaches for obtaining amino acid sequence: On-target ladder sequencing. In *Current Protocols in Protein Science*, 16.7.1–16.7.3, New York, John Wiley and Sons.

Jones, B. N. (1986). Amino acid analysis by *o*-phthaldialdehyde precolumn derivatization and reverse phase HPLC. In *Methods of Protein Microcharacterization: A Practical Handbook*, J. E. Shively, ed., Clifton, N.J., Humana Press, pp. 121–151.

Jorgenson, J. W., and Lukacs, K. D. (1981). Zone electrophoresis in open-tubular glass capillaries. Anal. Chem. 53:1298–1302.

Kaiser, E., Colescott, R. L., Bossinger, C. D., and Cook, P. I. (1970). Color test for the detection of free terminal groups in solid-phase synthesis of peptides. Anal. Biochem. 34:595–598.

Kaiser, E., Bossinger, C. D., Colescott, R. L., and Olsen, D. B. (1980). Color test for terminal prolyl residues in the solid phase synthesis of peptides. Anal. Chim. Acta 118:149–151.

Kent, S. B. H. (1988). Chemical synthesis of peptides and proteins. Ann. Rev. Biochem. 57:957–989.

Kent, S. B. H., Riemen, M., LeDoux, M., and Merrifield, R. B. (1982). A study of the Edman degradation in the assessment of the purity of synthetic peptides. In *Methods in Protein Sequence Analysis*, M. Elzinga, ed., Clifton, N.J., Humana Press, pp. 205–213.

Knecht, R., and Chang, J.-Y. (1986). Liquid chromatographic determination of amino acids after gas-phase hydrolysis and derivatization with (dimethylamino) azobenzenesulfonylchloride. Anal. Chem. 58:2375–2379.

Kochersperger, M. L., Blacher, R., Kelly, P., Pierce, L., and Hawke, D. H. (1989). Sequencing of peptides on solid phase supports. Am. Biotech. Lab. 7(3), March: 26–37.

Laursen, R. A. (1971). Solid phase Edman degradation, an automated peptide sequencer. Eur. J. Biochem. 20:89–102.

Levy, M., Fishman, L., and Schenkein, I. (1970). Mouse submaxillary gland proteases. Methods Enzymol. 19:672–681.

Liu, T. Y., and Chang, Y. H. (1971). Hydrolysis of proteins with *p*-tolunesulfonic acid. J. Biol. Chem. 246:2842–2848.

MacFarlane, R. D., Skowronski, R. P., and Torgenson, D. F. (1974). New approach to the mass spectroscopy of non-valatile compounds. Biochem. Biophys. Res. Comm. 60:616–621.

Mahoney, W. C., and Hermodson, M. A. (1980). Separation of large denatured peptides by reverse phase high performance liquid chromatography. J. Biol. Chem. 255:11199–11203.

Maier, G. Drapeau, G. R., Doenges, K. H., and Ponstingl, H. (1986). Generation of starting points for microsequencing with a protease specific for the amino side of aspartyl residues. In *Methods of Protein Sequence Analysis*, K. A. Walsh, ed., Clifton, N.J., Humana Press.

Mant, C. T., and Hodges, R. S. (1990). HPLC of peptides. In *HPLC of Biological Macromolecules*, Chromatographic Science Series, Vol. 51, K. M. Gooding and F. E. Regnier, eds., New York, Marcel Dekker, pp. 301–332.

Mant, C. T., Kondejewski, L. H., Cachia, P. J., Monera, O. D., and Hodges, R. S. (1997). Analysis of synthetic peptides by high performance liquid chromatography. Methods Enzymol. 289:426–469.

Marquez, F. G., Quesada, A. R., Sanchez-Jimines, F., and Nunez de Castro, I. (1986). Determination of 27 dansyl amino acid derivatives in biological fluids by reverse-phase high-performance liquid chromatography. J. Chromatogr. 380:275–283.

Masaki, T., Tanabe, M., Nakamura, K., and Soejima, M. (1981a). Studies on a new proteolytic enzyme from Achromobacter lyticus M497-1 I. Purification and some enzymatic properties. Biochem. Biophys. Acta 660:44–50.

Masaki, T., Fujihashi, T., Nakamura, K., and Soejima, M. (1981b). Studies on a new proteolytic enzyme from Achromobacter lyticus M497-1 II. Specificity and inhibition studies of Achromobacter protease I. Biochem. Biophys. Acta 660:51–55.

Matsubara, H., and Sasaki, R. M. (1969). High recovery of tryptophan from acid hydrolysates of proteins. Biochem. Biophys. Res. Comm. 35:175–181.

McCloskey, J. A., ed. (1990). Mass spectrometry. Methods Enzymol. 193:1–960.

Moore, S., and Stein, W. H. (1948). Photometric ninhydrin method for use in the chromatography of amino acids. J. Biol Chem. 176:367–388.

Moore, S., and Stein, W. H. (1963). Chromatographic determination of amino acids by the use of automatic recording equipment. Methods Enzymol. 6:819–831.

Moore, S., Spackman, D. H., and Stein, W. F. (1958). Chromatography of amino acids on sulfonated polystyrene resins. Anal. Chem. 30:1185–1190.

Oray, B., Lu, H. S., and Gray, R. W. (1983). High-performance liquid chromatographic separation of Dns-amino acid derivatives and application to protein and peptide structural studies. J. Chromatogr. 270:253–266.

Ozols, J. (1990). Amino acid analysis. Methods Enzymol. 182:587–601.

Patterson, D. H., Tarr, G. E., Regnier, F. E., and Martin, S. A. (1995). C-Terminal ladder sequencing via matrix-assisted laser desorption ionization mass spectrometry coupled with carboxypeptidase Y time-dependent and concentration-dependent digestions. Anal. Chem. 67:3971–3978.

Penke, B., Ferenczi, R., and Kovacs, K. (1974). A new acid hydrolysis method for determining tryptophan in peptides and proteins. Anal. Biochem. 60:45–50.

Regnier, F. E. (1987). The role of protein structure in chromatographic behavior. Science 238:319–323.

Reim, D. F., and Speicher, D. W. (1997). *N*-Terminal sequence analysis of proteins and peptides. In *Current Protocols in Protein Science*, 11.10.1–11.10.38, New York, John Wiley and Sons.

Roth, M., and Hampai, A. (1973). Column chromatography of amino acids with fluoresence detection. J. Chromatogr. 83:353–356.

Rubinstein, M., (1979). Preparative high performance liquid partition chromatography of proteins. Anal. Biochem. 98:1–7.

Sanchez, A., and Smith, A. J. (1997). Capillary electrophoresis. Methods Enzymol. 289:469–478.

Sanger, F., and Tuppy, H. (1951). The amino acid sequence in the phenylalanyl chain of insulin. Biochem. J. 49:463–481.

Sarin, V. K., Kent, S. B. H., Tam, J. P., and Merrifield, R. B. (1981). Quantitative monitoring of solid phase peptide synthesis by the ninhydrin reaction. Anal. Biochem. 117:147–157.

Schenkein, J., Levy, M., Franklin, E. C., and Frangione, B. (1977). Proteolytic enzymes from the mouse submaxillary gland. Arch. Biochem. Biophys. 182:64–70.

Shively, J. E. (1986). Reverse Phase HPLC isolation and microsequence analysis. In *Methods of Protein Microcharacterization: A Practical Handbook*, J. E. Shively, ed., Clifton, N.J., Humana Press, pp. 41–87.

Simpson, R. J., Neuberger, M. R., and Liu, T.-Y. (1976). Complete amino acid analysis of proteins from a single hydrolysate. J. Biol. Chem. 251:1936–1940.

Smith, B. J. (ed.) (1997). *Protein Sequencing Protocols*, Clifton, N.J., Humana Press.

Snyder, L. R., and Kirkland, J. J. (1979). *Introduction to Modern Liquid Chromatography*, New York, John Wiley and Sons.

Spackman, D. H., Stein W. H., and Moore, S. (1958). Automatic recording apparatus for use in chromatography of amino acids. Anal. Chem. 30:1190–1206.

Stewart, J. M., and Young, J. D. (1984). *Solid Phase Peptide Synthesis*, Rockford, Ill., Pierce Chemical Co.

Sullivan, S., and Wong, T. W. (1991). A manual sequencing method for identification of phosphorylate amino acids in phosphopeptides. Anal. Biochem. 197:65–68.

Tapuhi, Y., Schmidt, D. E., Linder, W., and Karger, B. L. (1982). Analysis of dansyl amino acids by reverse-phase high performance liquid chromatography. Anal. Biochem. 127:49–54.

Tarr, G. E. (1986). Manual Edman sequencing system. In *Methods of Protein Microcharacterization: A Practical Handbook*, J. E. Shively, ed., Clifton, N.J., Humana Press, pp. 155–194.

Tarr, G. E., Beecher, J. F., Bell, M., and McKean, D. J. (1978). Polyquarternary amines prevent peptide loss from sequenators. Anal. Biochem. 84:622–627.

Tarr, G. E., Paxton, R. J., Pan, Y.-C. E., Ericsson, L. H., and Crabb, J. W. (1991). Amino acid analysis 1990: The third collaborative study from the Association of Biomolecular Resource Facilities (ABRF). In *Techniques in Protein Chemistry*, Vol. II, J. J. Villafranca, ed., San Diego, Calif., Academic Press, pp. 139–150.

Tregear, G. W., van Rietschoten, J., Sauer, R., Niall, H. D., Keutmann, H. T., and Potts, J. T., Jr. (1977). Synthesis, purification, and chemical characterization of the amino-terminal 1–34 fragment of bovine parathyroid hormone synthesized with the solid phase procedure. Biochemistry 16:2817–2823.

Tsunasawa, S., and Narita, K. (1982). Micro-identification of amino-terminal acetyl amino acids in protein. J. Biochem. 92:607–613.

Van Regenmortel, M. H. V., Briand, J. P., Muller, S., and Plaué, S. (1988). In *Synthetic Polypeptides as Antigens*, Amsterdam, Elsevier, pp. 81–82.

Wang, Y. H., Fiol, C. J., DePaoli-Roach, A. A., Bell, A. W., Hermodson, M. A., and Roach, P. J. 1988. Identification of phosphorylation sites in peptides using a gas-phase sequencer. Anal. Biochem. 174:537–47.

Wilkinson, J. M. (1986). Fragmentation of polypeptides by enzymic methods. In *Practical Protein Chemistry: A Handbook*, New York, John Wiley and Sons, pp. 121–148.

Young, P. M., and Merion, M. (1990). Capillary electrophoresis analysis of species variations in the tryptic maps of cytochrome *C*. In *Current Research in Protein Chemistry: Techniques, Structure, and Function*, J. J. Villafranca, ed., San Diego, Calif., Academic Press, pp. 217–232.

5

Applications of Synthetic Peptides

Victor J. Hruby
Terry O. Matsunaga

The tremendous advances in the development of methods for the design and synthesis of peptides, pseudo-peptides and related compounds, as well as the corresponding advances in our understanding of peptide and protein structure, conformation, topography, and dynamics provides unique opportunities to apply designed synthetic peptides for an enormous variety of problems in chemistry, biology, and medicine. In addition, if these advances can be coupled to the advances in molecular biology and the human genome project, on the one hand, and asymmetric synthesis and catalysis, on the other, it should be possible to provide hitherto unavailable, indeed unthinkable, approaches to diverse areas of drug design, behavioral neuroscience, molecular immunology, chemotherapy, and a wide variety of other uses.

Already it is clear that peptide therapy has enormous potential in such diverse areas as growth control, blood pressure management, neurotransmission, hormone action, satiety, addiction, pain, digestion, reproduction, and so forth. Nature has "discovered" that it can control nearly all biological processes by various kinds of molecular recognition, and that peptides and proteins are uniquely suited for this control because of their enormous potential for diversity and their unique physico-chemical properties. This finding may, perhaps, be most readily understood if one recognizes that, considering only the

20 normal eukaryotic amino acids, the number of unique chemical entities for a pentapeptide is 3,200,000 (20^5), for a hexapeptide it is 64,000,000 (20^6), and so on. Considered from this perspective, perhaps it is not unexpected that Nature has "discovered" that peptides and proteins can do it all, from providing structure and motion, to catalysis, to information transduction, to growth and maturation, and so on. The ability of the immune system in higher animals, including humans, to recognize literally millions of foreign materials made by Nature as well as humans, and to get rid of them as part of its survival strategy, is just one example that illustrates the potential of peptide-based drugs, therapeutics, and modulators of biological function.

Despite the enormous potential of peptides and small proteins for these areas, surprisingly little advantage has been taken of the potential of these molecules as drugs and tools for use in basic and clinical research. The reasons given for this are many and include such factors as: (1) the lack of knowledge of most chemists regarding the structures, conformations, and synthesis of peptides and proteins; (2) the putative instability of these compounds in biological systems; (3) the putative lack of bioavailability of these compounds (though they are found and transported everywhere in the body); (4) their high potency and very small concentrations in most biological systems; and (5) the enormous diversity possible. In fact, many of these problems can be overcome or are irrelevant to a particular biological or medical problem. In the case of diversity, this can, in fact, serve as an advantage, since it is possible to construct very large libraries of peptides (10^6–10^{10} and more) by both biological (e.g., Cwirla et al., 1990; Devlin et al., 1990; Scott and Smith, 1990) and chemical (e.g., Furka et al., 1991; Houghten et al., 1991; Lam et al., 1991; Hruby, 1996) methods and to screen them for a very wide variety of functions. These new approaches may have been expected to accelerate the use of peptides and peptide-derived compounds for medical and biological purposes, but full advantage of the possibilities is still a long way from realization.

In this case, we will provide a brief summary of some of the medical and biological applications for synthetic peptides. Much has already been accomplished, but more importantly, current studies are providing a framework for much more comprehensive success in the future. Indeed, it can be anticipated that peptide, protein and peptide-related compounds will be among the most numerous drugs of the future. If this chapter helps you in this process in some small way, we will have accomplished our purpose. (For a few recent reviews or overviews of this area, see for example: Emmett, 1990; Hruby et al., 1990a,b; Hruby and Balse, 2000; Dutta, 1991; Ward, 1991; Rizo and Gierasch, 1992; Goodman and Ro, 1994; Whittle and Blundell, 1994; Sawyer, 1995; Schneider and Kelly, 1995; Böhn and Klebe, 1996; Al-Obeidi et al., 1998; Strand, 1990.)

Structure/Function Studies

One of the principal goals of peptide and protein research is the rational design of peptide ligands whose chemical, physical, and biological properties can be predicted a priori. It is not possible at this time to propose a general set of rules that would apply to structure/function studies for all bioactive peptides acting at all receptors, enzymes, antibodies, membranes, nucleic acids and other acceptors. However, the goal here is to discuss some general approaches in the design of synthetic peptides that can be used to better understand ligand–receptor/acceptor interactions through structure/function studies.

It now seems clear that there are at least three separate steps (structural states) of the ligand–receptor/acceptor complex which occur during the ligand–receptor/acceptor interaction. First, the ligand must be "recognized" by the receptor for binding to take place. In a second step, a change in conformation of the hormone–receptor/acceptor complex must occur that can result in transduction (in the case of an enzyme, a fruitful transition state must be reached) of the biological response through a second messenger. Finally, the hormone (products, etc.) must be released from the receptor so that the process can be repeated. The differentiation between binding and transduction (or inhibitory) states relating to hormone interactions have been well established as being the result of the hormone exhibiting different conformational and topographical features (Rudinger et al., 1972; Meraldi et al., 1975, 1977; Schwyzer, 1977; Walter, 1977; Hruby, 1981a,b).

Ideally, one would like to be able to "know" what a receptor/acceptor is "looking for" for binding, transduction (transition state), and release (turnover) of its complementary binding ligand. In many cases, remarkable progress has been made in these areas. But even with the ability to isolate receptors, there is, currently, no method to design ligands with specific properties a priori. The use of structure/function studies is the classical method for the design of synthetic peptides with specific activities. The ability to perform structure/function studies relies heavily upon the availability of a well-defined ligand–receptor system with sensitive binding assays and bioassays that can answer questions related to selectivity, binding, and transduction for synthetic peptide analogues.

Systematic Modifications

Structure/function studies usually begin with simple modifications of the native ligand for a receptor or receptor class. The most common of these are the alanine scan and the glycine scan in which each residue is systematically replaced by an Ala or Gly. The hormone–receptor system is complex so that, in general, only by individual modifications can one insure that a single substitution is providing the change. This can then be followed by more extensive changes at observed key residues. Thus, a large number of single modifications built upon one another can result in quite sophisticated analogues. In some cases, the modifications in the analogue can be so extensive that it is difficult for an outside observer to see the relation of the analogue structure to that of the endogenous hormone.

In a somewhat related approach, it has been found that many hormones have a minimum sequence that may have full biological potency. Since most peptide analogues are synthesized by either solid-phase or solution-phase synthesis, if a minimum active sequence can be found, then it can greatly simplify the synthesis of all subsequent analogues prepared. A great deal of information can often be gleaned from the kinds of studies mentioned, and, in some cases, the separation of peptide sequences required for binding and transduction can be determined. This, in turn, can lead to the design of agonist and antagonist ligands. Often, assays that relate modes of action through primary and secondary messengers can be examined. In this way, it sometimes is possible to separate an endogenous ligand's affinity for subtypes of multiple receptors. A few examples mostly from our laboratory of some of these approaches will now be discussed.

Deletion Peptides

Deletion peptides of melanin concentrating hormone (MCH), whose native structure is Asp-Thr-Met-Arg-c[Cys-Met-Val-Gly-Arg-Val-Tyr-Arg-Pro-Cys]-Trp-Glu-Val, showed that only residues 5–15 were necessary for full biological activity (table 5-1) (Matsunaga, et al., 1989). Removal of the tryptophan in position 15 lowers the biological potency by more than 10-fold, and this potency continues to decrease with fragments sequentially deleting residues 1–5. These studies suggested the importance of Trp^{15} for interaction with the MCH receptor, with the 5–14 fragment having only 1/300 the potency of the native hormone. This is a common observation, namely that for many peptides, neurotransmitters, hormones, cytokines and other bioactive peptides, N- and/or C-terminal truncation leads to shortened analogues, with full bioactivity but often reduced potency. Once the "minimum active sequence" is found, structural or conformational modifications can restore full potency.

Stereochemical Substitutions

Once a minimum bioactive sequence has been determined, it often is convenient to search for stereochemical requirements of the receptor/acceptor with the systematic substitution of D- for L-amino acids. The

Table 5-1 Relative potencies of MCH fragment
analogues as determined by the in vitro fish (*Synbranchus
marmoratus*) skin bioassay

Peptide	Potency relative to MCH
MCH	1.0
MCH(2–17)	1.0
MCH(3–17)	1.0
MCH(4–17)	1.0
MCH(5–17)	1.0
MCH(1–16)	1.0
MCH(2–16)	1.0
MCH(3–16)	1.0
MCH(4–16)	1.0
MCH(5–16)	1.0
MCH(1–15)	1.0
MCH(2–15)	1.0
MCH(3–15)	1.0
MCH(4–15)	1.0
MCH(5–15)	1.0
MCH(1–14)	0.1
MCH(2–14)	0.07
MCH(3–14)	0.05
MCH(4–14)	0.023
MCH(5–14)	0.014

incorporation of D-amino acids also can provide clues about the importance of certain secondary structures (e.g., β-turns, α-helix), and also may yield further information about specific residues necessary for peptide–receptor interaction as well as contributing stability to analogues against enzymatic breakdown (Hruby, 1982).

In one example from our laboratory, the incorporation of D-Phe at position 7 of [Nle⁴]α-MSH (melanotropin stimulating hormone) resulted in a melanotropin superagonist (Sawyer et al., 1980). The native linear peptide has the active sequence Ac-Ser-Tyr-Ser-Met-Glu-His-Phe-Arg-Trp-Gly-Lys-Pro-Val-NH₂. The substitution was made based on the evidence that heat-alkaline treated α-MSH and [Nle⁴]α-MSH led to prolonged activity with an unusual amount of racemization at the Phe⁷ position. The [Nle⁴, D-Phe⁷]α-MSH analogue was shown to be 60 times more potent than the native hormone with extremely prolonged activity and greatly reduced biodegradation. Other early examples of the use of D-amino acids include LHRH (luteinizing hormone releasing hormone) agonists and antagonists (Coy et al., 1982; Freidinger et al., 1985), substance P antagonists (Piercy et al., 1981), oxytocin antagonists (Melin et al., 1983; Lebl et al., 1985), enkephalin (Morgan et al., 1977) and somatostatin analogues

(Freidinger and Veber, 1984; Torchiana et al., 1978), and many other examples have followed. Indeed a D-amino acid scan is now a standard approach to investigating structural, stereochemical and conformational requirements for bioactivity of native bioactive peptides (e.g., Grieco et al., 2000).

Isosteric and Other Substitutions

Pseudo-isosteric replacements are also quite useful for determining peptide–receptor interactions. Amino acids can be replaced with other amino acids of similar size but different electronic or stereoelectronic properties. Such substitutions would include the use of norleucine (Nle) for the oxidizable methionine, where a methylene unit is used as a replacement for sulfur. This substitution was carried out for α-MSH. The resulting [Nle^4]α-MSH peptide was shown to be 1.5 to 2.2 times more potent than the native hormone having methionine in the fourth position. This slight increase in potency may, in part, be the result of lack of a methionine oxidation during the bioassays (Hruby et al., 1980). Other such substitutions include changes in the aromatic amino acids tyrosine, tryptophan, phenylalanine, and histidine with various modifications of the aromatic moieties. Switching these amino acids might allow study of spatial requirements and/or hydrogen-bonding interactions, π–π, or π–σ interactions and so forth. For example, in oxytocin, the importance of the phenolic OH in the Tyr^2 position was illustrated by the synthesis of [Phe^2]oxytocin (Bodanszky and du Vigneaud, 1959) and [Trp^2]oxytocin (Kaurov et al., 1972), which demonstrate weak agonist activity, while [p-MePhe^2]oxytocin (Rudinger et al., 1972) is an antagonist. Other possibilities include the acidic residues, aspartic acid and glutamic acid, which might be exchanged for their amide counterparts asparagine and glutamine and vice versa. These types of substitutions might have a pronounced impact on the analogue's chemical nature because the net charge on the ligand is changed. Similarly, the basic residues lysine and arginine also are often interconverted to examine how well the amine functional group fits the receptor relative to the guanidine functional group. These types of replacements have been very effective in the design of analogues of enkephalins (Kruszynski et al., 1980; Mosberg et al., 1983a), and somatostatins (Freidinger and Veber, 1984), μ-opioid selective peptides (Kazmierski and Hruby, 1988), and many others.

Constraint

In addition to these more classical approaches, the use of conformational and topographical constraint has been a very powerful approach to peptide ligand design (e.g., for reviews see: Hruby et al., 1982; Kessler, 1982; Toniolo, 1990; Hruby, 1990a; Farlie et al., 1995), including increasing potency, receptor selectivity, providing antagonists, and greatly increasing stability against proteolyte biodegradation. We now provide a few examples of these approaches.

The incorporation of N-methyl amino acids has a tendency to constrain the χ^1 angles of amino acids and provide *cis-trans* isomerism around the amide bond of which they are part. One excellent example is in the case of certain [N-MeNle^3]CCK-8 (CCK = cholecystokinin) analogues, which showed a significant increase in the specificity for the CCK-B (brain) receptor over the CCK-A (peripheral) receptor. It has been suggested from NMR studies that the enhanced selectivity is the result of the brain receptors preference for the *cis* amide bond which is not seen in related [Nle^3]CCK-8 analogues (Hruby et al., 1990b). Another example includes the substitution of N-methyl-D-phenylalanine in the one position of Phe-Cys-Tyr-D-Trp-Orn-Thr-Pen-Thr-NH$_2$ (CTOP). This analogue was shown to be 200 times less potent than [D-Phe^1]CTOP at the μ-opioid receptor, simply because of the change in the χ^1 angle from $-60°$ to $180°$ (Kazmierski et al., 1988). N-Methylalanine was used in place of proline in some bradykinin analogues, decreasing their potency (Filatova et al., 1986). A review of the literature reveals that many similar examples could be cited.

Amino acids that are α-alkylated have also been widely used to examine a receptor's conformational requirements. Besides adding potential additional steric bulk to the peptide, these amino acids, tend to induce β-turns and other secondary structures (De Grado, 1988; Toniolo, 1990). One amino acid that has proved to be extremely useful for this purpose is Aib (aminoisobutyric acid) in the place of glycine.

To study the steric and stereoelectronic nature of a receptor, bulky nonproteinogenic amino acids are often incorporated into peptide analogues. One of the most widely used replacements is the substitution of penicillamine (Pen, 3-mercapto-D-valine or β, β-dimethylcysteine) for cysteine. This substitution has led to the differentiation between agonist and antagonist analogues in oxytocin (table 5-2). A series of analogues that have even larger side chains at the β-carbon also were examined. In these analogues antagonists were obtained only when both hydrogens on the β-carbon were replaced with alkyl groups. Replacement of only one methyl group was not sufficient for antagonism at the uterine receptor. There was little difference in the inhibitory properties observed for β, β-dimethyl or ethyl analogues and bicyclic [Cpp^1]oxytocin. It has been suggested that this antagonism is caused, in part, by the restriction of disulfide interchange between R and S chiralities and to disposition of the χ^1 and χ^2 angles of the Tyr^2 side chain group, and further evidence for this has been

Table 5-2 Activities of some 1-substituted oxytocin analogues

Analogue[a]	Activity U/mg	Reference
Oxytocin	546	Chan and Kelly, 1969
[Mpa[1]]oxytocin	803	Ferrier et al., 1965
[Dmp[1]]oxytocin	$pA_2 = 6.94$	Vavrek et al., 1972
[D,L-β-MeMpa[1]]oxytocin	91	Schulz and du Vigneaud, 1966
[Dep[1]]oxytocin	$pA_2 = 7.24$	Vavrek et al., 1972
[Cpp[1]]oxytocin	$pA_2 = 7.61$	Lowbridge et al., 1979
[Pen[1]]oxytocin	$pA_2 = 6.86$	Vavrek et al., 1972

[a]Mpa: β-mercaptopropanoic acid; Dmp: β,β-dimethyl-β-mercaptopropanoic acid; Dep: β,β-diethyl-β-mercaptopropanoic acid; Cpp: β-cyclopentamethylene-β-mercaptopropanoic acid; Pen: penicillamine.
Definition: pA_2 is defined as the negative logarithm of the molar concentration of a competitive antagonist that reduces the response of the agonists by 50 percent of what it would be in the absence of the antagonist.

Mathematically:

$$E = \frac{E_{max}}{\dfrac{D_2}{a}\left(1 + \dfrac{i}{A_2}\right) + 1} \quad \text{and } pA_2 = -\log A_2$$

E= observed response
E_{max} = maximum response in absence of antagonist
D_2 = half maximum concentration
i = antagonist concentration
a = concentration of agonist

obtained using χ-constrained tyrosine analogues (Liao et al., 1998). The substitution of cysteine by other larger β, β-dialkyl Cys derivatives also has been examined in the development of potent V1 and V2 vasopressin antagonists (Dyckes et al., 1974; Bankowski et al., 1978), δ-selective cyclic enkephalin analogues (Mosberg et al., 1983a,b) and μ-selective somatostatin analogues (Pelton et al., 1985; Kazmierski and Hruby, 1988). The use of other sterically demanding amino acid side chains such as *t*-butyl, adamantyl, naphthyl, and so on, are almost endless, and many such amino acids have been developed and incorporated into peptides.

Ring Size

For cyclic peptides, the optimization of ring size is of great use. The effects of ring size and the necessity of the sulfur atoms in disulfide-containing peptides has been the subject of investigation since the earliest days of bioactive peptide structure–activity studies, and have probably been most completely studied in oxytocin and vasopressin analogues. Replacement of either or both sulfur atoms in oxytocin resulted in analogues with agonist activity (for reviews see Jost, 1977; Hruby and Smith, 1987). Attempts to modify the 20-membered ring size by addition or deletion of single residues or single side-chain carbons or sulfurs led to a 100-fold decrease in potency in all cases. Similar experiments with vasopressin generally resulted in less potent analogues, and the relative effect was largely dependent on the hormone receptor examined. One quite successful exploration of ring size was in the development of somatostatin agonists. Removal of six residues from the ring of somatostatin Ala-Gly-c[Cys-Lys-Asn-Phe-Phe-Trp-Lys-Thr-Phe-Thr-Cys] gave rise to the 20-membered ring analogue D-Phe-c[Cys-D-Trp-Lys-Thr-Cys]-Thr-ol, which is super-active in inhibiting growth hormone release (Bauer et al., 1983). The ring was subsequently reduced to the 18-membered cyclic hexapeptide

$$\boxed{\text{—Pro-Phe-D-Trp-Lys-Thr-Phe—}}$$

which is equipotent to somatostatin in the inhibition of growth hormone, insulin, and glucagon release (Veber et al., 1981). In addition, the use of cyclic hexapeptides and heptapeptides with appropriate constraints has been a critical aspect of conformation-biological activity studies of many bioactive peptides for many years and is an area that will continue to develop rapidly well into the future (for reviews see, for example, Kessler, 1982; Rizo and Gierasch, 1992).

Cyclic Constraints

Conformational features in peptides that are important for receptor interaction often can be stabilized by introducing cyclic constraints into linear peptide ligands (e.g., Hruby, 1982; Kessler, 1982; Hruby et al., 1990a; Marshall, 1993). One early example was the development of α-MSH superagonists. A proposed β-turn around Phe[7] was stabilized by the replacement of Met^4 and Gly^{10} with cysteine residues and cyclization to afford the cyclic peptide c[Cys^4, Cys^{10}]α-MSH, an analogue with much greater potency than the native hormone which also exhibits prolonged activity (Sawyer et al., 1982). In a similar manner, substitution of the Met^5 and Gly^2 positions of enkephalins with D-Pen, followed by cyclization, led to the highly potent and δ-opioid receptor selective cyclic peptide, c[D-Pen^2, D-Pen^5]enkephalin (Mosberg et al., 1983b).

Side-chain cyclizations of polypeptides not involving disulfides generally are not found in nature in higher organisms, but they can be important tools in structure/function studies with synthetic peptides. There are many possibilities for side-chain-to-side-chain cyclizations, including, but not limited to, amides (lactams), esters (lactones), ethers, ketones, and so on. Many possibilities have already been examined (Hruby and Boteju, 1997). Perhaps the most common cyclization, due to its convenience and stability, is the amide bond formed between free carboxylate and free amine groups found on amino acid residue side chains. This type of cyclization has been used quite successfully in a series of oxytocin analogues. The weak

monocyclic agonist [Mpa^1, Glu^5, Cys^6, Lys^8]oxytocin was converted

to the highly potent bicyclic antagonist [Mpa^1, Glu^5, Cys^6, Lys^8]-

oxytocin by amide bond formation between Glu^5 and Lys^8 (Hill, et al., 1990). Not only did this cyclization cause antagonist activity, but it also increased the binding to the receptor over 1000-fold. Cyclizations of a similar manner were also studied in CCK-8. The monocyclic lactam analogue Boc-c[Asp-Tyr(SO$_3$H)-Nle-D-Lys]-Trp-Nle-Asp-Phe-NH$_2$ exhibited a significant decrease in potency at the CCK-A (peripheral) receptor (Charpentier et al., 1987). A monocyclic bis-lactam analogue of CCK-8 was prepared by replacement of Met^{28,31} residues with lysine and cyclizing the N^ε amino groups with succinic acid. This procedure resulted in the highly CCK-B (brain) selective analogue (Rodriguez et al., 1990). Based on computer modeling, α-MSH analogues were prepared incorporating amide cyclizations (Al-Obeidi et al., 1989a,b). This led to the synthesis of the highly potent Ac-c[Nle^4, Asp^5, D-Phe^7, Lys^{10}]α-MSH-(4-10)-NH$_2$ analogue. Amide cyclizations have also been studied in the glucagon (Lin et al., 1992; Trivedi et al., 2000), enkephalin (Schiller et al., 1985), GnRH systems (Struthers et al., 1984), and many others.

Other cyclizations have been reported, including biphenyl-type cyclizations used in an approach to the synthesis of the antitumor agent bovardin (Bates et al., 1984) and diazobridges in enkephalin analogues (Siemion et al., 1980).

Peptidomimetics

More recently, an increasing emphasis of structure/function studies
has been directed toward the synthesis and use of ligands containing
pseudo-peptides and peptidomimetics (e.g. Hruby et al., 1990a).
While it is now possible to stabilize certain conformational features
in peptides with a high degree of predictive confidence, the use of
nonpeptide peptide mimetics as rigid templates for these features of
peptides and proteins in 3-D space and with appropriate dynamic
properties is still problematic. There is also interest in peptides
containing novel amino acids (e.g. β-methyl-2',6'-dimethylphenylala-
nine which can allow peptides to maintain their backbone conformation but predictably bias side-chain conformations of the ligand
(Hruby et al., 1997). These latter considerations are important because
de novo design of nonpeptide peptide mimetics requires careful design
considerations, not only of a proper conformational template but also
the appropriate arrangement of key pharmacophore groups of the
bioactive peptide on the nonpeptide scaffold. This is particularly
important in the design of peptide hormones and neurotransmitters,
cytokines, protease inhibitors and many other ligands that interact
with receptor/acceptor proteins and utilize the side-chain groups as
the primary sites of molecular recognition or for discrimination
between different possible biological receptor/acceptors (selectivity).
While a comprehensive discussion of the design of nonpeptide peptide
mimetics is outside the scope of this chapter, it is critical to know what
the biologically active conformation is, what the structure–activity
relationships are, and what the critical dynamics properties, if any,
of the peptide are if one is to design de novo a true peptide mimetic.
In this regard, the role of peptide chemistry, its many manifestations,
becomes more and more critical to developing a robust approach to
nonpeptide peptide mimetic design (see, for example, Hruby et al.,
1990a; Hirschmann, 1991; Wiley and Rich, 1993; Hruby, 1997;
Sawyer, 1997; Hruby and Balse, 2000). Here we will only discuss
turn design since it often involves use of novel amino acids and/or
pseudo-peptides.

An important aspect of peptide chemistry, particularly with respect
to pharmaceutical development, involves the study of conformational
and topographical aspects of turns. Reverse-turn regions are
important aspects of determining whether a molecule will, or will
not, maintain the desired binding potency and transduction or inhibitory potential. Several examples of peptides associated with turn
regions are (1) the tripeptide RGD sequence found to bind to various
glycoprotein receptors (Ripka et al., 1993a) inhibiting platelet
aggregation, (2) Gramicidin S, a cyclic, naturally occurring peptide
antibiotic whose biological activity has been shown to be related to

its conformation (Ripka et al., 1993b), a variety of peptides used to inhibit HIV proteases (aspartyl proteases) responsible for activity against human immunodeficiency virus (Abbenante et al., 1995) and many more (see above).

The development of peptide turn-mimetics are multi-fold. Rigid turns designed by alterations in the backbone have been developed to aid in the stability of the molecule to enzymatic degradation. Alternatively, stabilizations of conformation and/or topography can aid in imparting selectivity of the compound with respect to a family of receptors. For example, the integrin receptors are found on both platelets (GPIIbIIIa) and on tumor cells ($\alpha_v\beta_3$). A knowledge and design of specific conformational turns may be imperative in selecting for binding to the designated receptor for therapeutic benefit and eliminating binding to other receptors associated with undesirable side effects.

A variety of approaches for synthesizing turn mimetics have been developed. While some investigators focus upon a specific turn conformation comprised of specific amino acid sequences, others have devoted their attention to more global synthetic procedures to develop diverse libraries of turn mimetics. Methodologies directed to the synthesis of turns is diverse as well. For example, an electrochemical methodology has been reported by Moeller, Marshall and coworkers. (Slomczynska et al., 1996). S_NAr chemistry developed in the laboratory of Burgess has provided insight into turn conformations as well (Feng et al., 1998). Alternatively, other investigators have used backbone modifications to help stabilize β-turn conformations. For example, Kessler and coworkers used this approach to help rigidify the RGD conformations (Geyer et al., 1994). This review will elaborate on some of these studies.

Interestingly, the development of diverse libraries of turn mimetics using various amino acids in the i, $i + 1$, $i + 2$, and $i + 3$ positions, can be viewed as being somewhat complementary to the goals and objectives achieved by using the structure/function approach eluded to earlier in the chapter. Although work has progressed to develop diverse libraries, indeed, the turn libraries are only diverse in one aspect, for the objective of these libraries is, in some instances, to stabilize the backbone or conformation of the turn while allowing the topography or side chain to vary. Thus, in essence, the objective is to keep phi, psi, and chi angles of the backbone relatively constant and be able to vary only the topography. Hence, one has reduced the issue of turn optimization by a single variable which, in principle, should aid in understanding the necessary physical properties required for optimal turn mimetic development.

Prior to discussing the synthetic aspects of turn conformations, it should be stated that the evaluation of these mimetics is just as critical to identifying the correct conformation as the synthesis itself. For this reason, strict determination of conformational properties by spectroscopic and molecular modeling techniques can help to confirm the conformational structure of the analogue. At the cornerstone of these techniques is 2-D NMR spectroscopy. In particular, NOE (Nuclear Overhauser Enhancement) spectroscopy in conjunction with correlated spectroscopy and heteronuclear spectroscopy are essential in determining the phi and psi angles of the turn mimetic. In addition, molecular mechanics or molecular dynamics routines in conjunction with NMR constraints can provide insight into structure. Circular dichroism and/or crystal structures (if the compound is crystallizable) can provide more definitive information of the conformation.

Backbone Alteration

Peptidomimetics can be generated by synthetic backbone alterations. The synthesis of pseudo-peptides, whose backbone structure deviates from the conventional amide backbone, has been studied in order to examine conformational effects. Work by Geyer et al., performed a conformational analysis on a cyclic RGD peptide containing a $\psi[CH_2-NH]$ bond. The pseudopeptide cyclo(-Arg^1-Gly^2-Asp^3-D-Phe$^4\psi[CH_2$-NH]Val^5-) was synthesized on solid phase by first incorporating solution-phase reduction of Boc-D-Phe-Val-OH to the corresponding Boc-D-Pheψ [CH$_2$-NH]Val-OH with Lawesson's reagent. The synthesis appeared particularly advantageous because solution-phase syntheses of the dipeptides are virtually free of racemization, and can be crystallized. The pseudo-dipeptide synthesis was then followed by solid-phase coupling to H-Arg(NO$_2$)-Gly-Asp(Obzl)-2-chlorotrityl resin followed by cleavage, cyclization, and deprotection. The compound was compared to its unmodified parent, cyclo (-Arg^1-Gly^2-Asp^3-D-Phe^{4}Val^5-) by solution phase NMR and extensive molecular dynamics simulations. Interestingly, the protonated, amine on the reduced peptide bond exhibited strong hydrogen-bond donor properties, which helped to further rigidify the peptide backbone. This yielded a significantly reduced number of conformers relative to its parent counterpart, exhibiting a γ-conformation with the D-Phe in the $i + 1$ position of the γ-turn and a βII$'$ conformation surrounding the Arg-Gly region. This becomes interesting as the pseudo-dipeptide portion of the analogue induces a shift of the β-II turn, compared to the parent peptide, of two positions in a counterclockwise direction. This work helped to demonstrate how reduction of the backbone can impart both: (1) rigidity to the backbone, and (2) shift the turn region. This study was consistent with previous work by Ma and Spatola (1993), who noted the βII$'$ turn region of another pseudopentapeptide.

On-Resin Formation of Cyclic β-Turn Peptidomimetics
With the development of combinatorial and high throughput parallel methods, the desire to incorporate conformational constraints and, in particular, conformational bias, has emerged. Efficient cyclizations on solid-phase supports with predictable conformations, particularly β-turn conformations, is a priority (Kahn, 1993).

Recently, work by the Burgess group (Feng et al., 1998), has focused on β-turn conformations using S_NAr cyclizations on solid supports for the synthesis of β-turn libraries. Briefly, a Rink amide resin was used as the support followed by synthesis of a prototypical tripeptide followed by coupling to a fluoro-substituted benzoic acid analogue. Cyclization was initiated by side-chain S_NAr reaction of the first residue displacing the fluoro moiety of the benzoyl compound, yielding a macrocyclic analogue. NMR analysis following cleavage was significant for hydrogen bonding of the CO · · · NH being the ring fusion point for 10-membered rings formed by the β-turn analogue.

Work in the Ellman laboratory has produced some interesting libraries of β-turn mimetics as well (Virgilio et al., 1994; 1997a,b). Using a Rink amide resin, the $i + 1$ and $i + 2$ positions were functionalized with an α-bromo acid and α-Fmoc-protected α-amino acid respectively, with an aminoalkylthiol acting as the constraining backbone component. The covalently bound thiol moiety forming the amino alkyl thiol on the analogous $i + 3$ amine, generated the 9- or 10-membered ring. Thus, different combinations of substitutions at the stereocenters on the $i + 1$ and $i + 2$ side chain could be incorporated while still maintaining the putative β-turn conformation. A turn conformation was concluded by virtue of NMR coupling constants and searching the cyclic compounds for lowest energy conformers. Coupling constants between the central amide NH and the Cα-H of the $i + 2$ residue in selected compounds was 9.5 Hz. Incorporating this value to help constrain the backbone φ-dihedral angle, lowest energy conformers were sought using a conformational search analysis. Results indicated that a Type II'-β-turn best fit the conformer search analysis lending support to the conclusion that β-turns were indeed made in these cyclic peptide analogue libraries. A series of 1152 compounds were generated employing the Chiron Mimotope pin apparatus for rapid library synthesis. The β-turns were tested in the fMLF receptor assay which incorporates the fMLF tripeptide in the receptor bound in the plasma membrane of neutrophils. Results indicated binding of some of the compound mixtures with two compounds binding with an IC50 value of 10 μm and 13 μm respectively. Thus, the libraries appeared to generate an ensemble of restricted conformations with possible lead generation.

Shortly thereafter, a similar library based upon the β-turn was developed, however, this time incorporating an additional functionality in the $i + 3$ position as well (Virgilio et al., 1997a, b). This procedure took advantage of a guanidine based resin (acting both as a support and a general base catalyst) which incorporated a disulfide moiety vs. the Rink amide resin used previously. Similarly, a library of 1152 compounds were generated and characterized.

Electrochemical Cyclizations

Another interesting alternative to peptidomimetic turn compounds was employed by Slomczynska et al. (1996) who used electrochemical methodology in situ to induce selective anodic amide oxidations to generate an N-acyliminium cation which is trapped by an intra-molecular hydroxyl group. Experimenting with a variety of solvents, electrodes, and prototype starting materials, Slomczynska and coworkers found that optimal conditions for the cyclization of Boc-Ser-Pro-OMe to a 6,5-bicyclic turn system required dry acetonitrile, tetrabutyl ammonium tetrafluoroborate as a supporting electrolyte, platinum foil electrodes, and a constant current density of $14\,\mathrm{mA\,cm^2}$. Cyclizations were found to be diastereoselective with the S-configuration at the bridgehead being the major isomer. Hydrolysis of the methyl ester with 1.1 equivalent of NaOH yielded a mixture of two diastereomers due to epimerization of the α-carbon of Ser. However, separation could still be achieved using flash chroma-tography with silica gel and a chloroform/methanol/acetic acid (95:5:1, v:v:v) solvent system.

Clearly, this is a novel methodology. However, it is not yet clear as to its utility as a methodology for making turn mimetics since side products can interfere with product isolation, and the dryness of solvent systems are fairly stringent. Nonetheless, the authors provide a very intriguing route to a 6,5-bicyclic turn system.

Use of Peptides as Enzyme Inhibitors and Therapeutic Agents

Enzyme Inhibitors

Inhibition of enzymatic systems remains a prime target for the regula-tion and control of different disease states. Currently, there are numer-ous pharmaceuticals targeted toward the inhibition of enzymes in (1) cancer cells, (2) in bacteria, (3) in the systemic system of humans, and in many other systems. For example, 5-fluorouracil and methotrexate are commonly used as chemotherapeutic agents by the inhibition of dihydrofolate reductase and thymidylate synthetase. Penicillins, native and semi-synthetic, which can be viewed as peptidic in nature, are still considered first-line antibacterials. Similarly, lovastatin, isolated from

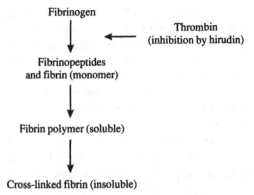

Figure 5-1 Scheme describing the mechanism
of fibrin clot formation and inhibition of
thrombin activity by hirudin.

Aspergillus tereus, is a cholesterol-lowering agent that works by inhibiting the conversion of 3-hydroxy-3 methyl-glutaryl Co-A to mevalonate (Stryer, 1975).

More recently, peptides have become attractive as potential thera-peutic agents for targeting specific enzymes or enzyme systems. For example, hirudin is a 65-amino acid (sulfated on Tyr^{63}) polypeptide inhibitor of the serine protease, thrombin, isolated from the salivary gland of the medicinal leech *Hirudo medicinalis* (Stone and Hofsteenge, 1986). Interestingly, peptides related to hirudin are already being tried experimentally in clinical trials for the prevention of blood clots in patients with rejoined limbs. Mechanistically, hirudin acts as a tight binding inhibitor to thrombin. Thus, once bound, thrombin is no longer able to bind to fibrinogen, thus, preventing the formation of fibrin monomers, which, in turn, would polymerize to form the fibrin (blood) clot (Mann, 1987) (see figure 5-1). Kinetically, it is a slow, tight-binding inhibitor that interacts with thrombin in an apparent two-step fashion. The first step is a slow complexation with thrombin at a remote site while the second step is essentially diffusion controlled $(4.7 \times 10^8 \, mol^{-1} \, s^{-1})$, and involves essentially irreversible complexation to the active site (Stone and Hofsteenge, 1986). Thrombin, being a serine protease enzyme, acts primarily by cleaving fibrinogen to fibrin monomers which later cas-cades into a fibrin clot-forming polymer. For a review of the clotting cascade pathway, the reader is referred to a review by Kayser (1983). Since it is a serine protease, it (as do most enzymes of this class) has a preference for cleavage at the carboxy termini of basic residues, parti-cularly arginine. The synthesis of a peptide affinity label, D-Phe-Pro-Arg-CH$_2$-Cl acted as an irreversible inhibitor of α-thrombin (Sonder and Fenton, 1984; Walker et al., 1985). This covalent peptidyl-thrombin complex was shown to bind covalently at a histidine moiety and consequently blocked the interaction of thrombin with hirudin. Based upon the proposal that the C-terminus may be an important binding region of hirudin, Krstenansky and Mao (1987) synthesized by solid-phase methods, the unsulfated C-terminal fragment N-α-acetyl hirudin$_{45-65}$. They found that this peptide bound to thrombin at a single site with an association rate constant of $10^5\,\mathrm{mol}^{-1}\,\mathrm{s}^{-1}$. In addition, they observed that the synthetic peptide D-Phe-Pip-Arg-p-nitroanilide was still hydrolyzed by the thrombin-hirudin complex. At the same time, the complex was able to inhibit fibrin clot formation. This led to the suggestion that the C-terminus portion of hirudin contained the binding domain to complex to the α-thrombin, yet did not alter the region about the catalytic active site. This work was followed up later by Mao and coworkers (1988) who synthesized smaller C-terminal fragments and determined that the last 10 amino acid residues were sufficient for binding and thrombin mediated clotting. Thus, once the Phe[56] was removed from the C-terminus, no further anticlotting activity was observed. In addition, replacement of Phe[56] by either acidic (Glu), hydrophobic (Leu) or enantiomeric (D-Phe) amino acids, resulted in complete loss of activity. Therefore, the 10 amino acid peptide represented a unique class of noncatalytic site thrombin inhibitors. In addition, binding of the C-terminal fragment induced a conformational change in the thrombin when complexed.

Work by Dennis and coworkers (1990) used site specific mutagenesis to replace Asn[52] with Met hirudin followed by cyanogen bromide cleavage at this single Met site. Their work found, as did Mao and coworkers, that a conformational change in the complex did occur that affects the active site of thrombin. This was shown by a change in the Michaelis constant (K_m) for the substrate D-Phe-Ala-Pip-Arg-p-nitroanilide.

Thus, the work on the hirundin-thrombin complex via synthesis of peptide fragments yielded important information about the mechanism of thrombin-induced coagulation pathways. Similar studies in many areas are likely to lead to many clinically valuable peptide pharmaceuticals in coming years. In the past decade these studies have led to a literal explosion of peptide and nonpeptide inhibitors of thrombin.

Turning toward the pharmaceutical application based upon previous studies, two approaches have evolved in concert leading to the development of an equipotent, stable hirudin analogue for therapeutic application. The first incorporated a solid-phase synthetic approach. This resulted in the development of Hirulog analogues. Hirulogs are a group of synthetic peptides designed to bind bivalently with thrombin at its catalytic site and its anion-binding exosite for fibrinogen recognition (Maraganore, 1993). They are divided into three main components: (1) a tetrapeptide, (D-Phe)-Pro-Arg-Pro, which binds tightly to the active site; (2) a 12-amino acid sequence on the C-terminal (residues 53–64), which binds to the anion-binding exosite; and (3) a linking sequence comprised of a variable number of glycines. This has led to the synthetic design of Hirulog (Biogen) which has been tested in humans.

Parallel work based upon classic structure function studies combined with research in molecular biology has resulted in the development of a recombinant analogue of hirudin, [Leu^1, Thr^3]-63-desulfohirudin (lepirudin [rDNA], Hoechst). The molecule is identical to hirudin with the exception of a recombinant Ile for Leu in position 1 and the absence of a sulfate group on position 63 of Tyr. Thus, research in the area of serine protease inhibitors has resulted in the successful development of a therapeutic anticoagulant.

Peptides as Therapeutic Agents

The development of therapeutic agents based upon peptide design has become an increasingly popular area of research. Previous limitations to the development of peptide therapeutics has been based upon the perception that the stability of peptides and peptide-like compounds in the body precluded their utility as effective therapeutics. However, it has become increasingly clear that peptides and peptide-like compounds can and have been developed as effective therapeutics. Most popular of these areas of research revolve around such agents as (1) cardiovascular agents where angiotensin converting enzyme inhibitors have become a mainstay in managing hypertension; (2) immunological agents where drugs such as cyclosporine have literally made allogeneic heart, kidney, lung, and liver transplants possible to save lives and prolong life; and (3) vascular agents based upon RGD and RGD-like peptides that are now used clinically as antiplatelet drugs to prevent stroke and myocardial infarcts as well as to detect clots in the vasculature (i.e., Reopro). As the review of these topics would be too broad for the scope of this chapter, the author would like to focus on a single example: diagnostic vascular agents.

Diagnostic Vascular Agents

Over the past two decades, the integrins, a family of cell-surface proteins that mediate cell adhesion, have been a popular target for the development of peptide inhibitors. One particular integrin, the Glycoprotein IIb/IIIa receptor found on platelets, has been the focus of extensive research to identify inhibitors of this receptor (Ruoslahti, 1991). The α-IIb component of the GP IIb/IIIa receptors on platelets bind a peptide sequence HHLGGAKQAGDV found on the terminal carboxyl region of the γ-chain of fibrinogen (Hawiger, 1995). These fibrinogen/receptor networks are the basis for the formation of platelet derived clots. Platelets are obviously beneficial for the formation of clots which are an instrumental aspect of the healing process. However, platelet clots may also be somewhat detrimental when vascular injury or tears catalyze the clotting process. These can lead to strokes and/or myocardial infarcts. In addition, invasive procedures such as angioplasties, rotoablation, or stent placement can also provide the nidus for clot formation leading to restenosis of a vessel resulting in an infarct.

Consequently, a therapeutic approach to the prevention of clotting or thrombosis, would entail the development of an antagonist to fibrinogen binding to the GPIIbIIIa receptor. However, in order to do so, a molecule had to be designed to be selective for the GPIIbIIIa receptor versus other integrin receptors found in the body. Work by Scarborough and coworkers (Scarborough et al., 1993) helped to identify analogues of RGDW and KGDW that were more selective for the GPIIbIIIa receptor. Utilizing a cyclic peptide system similar to that of Pierschbacher and Ruoslahti (1987), Scarborough and coworkers identified acetimidated cyclic compounds of KGDW that were potent and highly selective for the GPIIbIIIa receptor. This resulted in cyclo (S,S)-MPR(ε-acetimidyl-Lys)GDWPC-NH$_2$ and cyclo (S,S)-MPR (ε-acetimidyl-Lys)GDWPPen-NH$_2$ which possessed IC$_{50}$ (inhibitory concentrations, 50%) values of $0.55 + 0.02$ m and $0.3 + 0.1$ μm respectively.

Eventually, development of even more potent analogues were developed after identifying in the peptidic model that the spatial proximity of the cationic charge on the Arg versus the anionic charge on the Asp was vital for activity. This eventually led to a variety of compounds based upon tyrosine that resulted in low nanomolar and subnanomolar IC$_{50}$s. Since a detailed review of all these compounds is beyond the scope of this chapter, the author suggests an excellent review of this work by Wang et al. (2000).

Despite the reduction of potent GPIIbIIIa analogues down to a single modified tyrosine analog, peptide analogues based upon RGD have found their way as useful drugs in the diagnostic arena. Diatide Pharmaceuticals based in New Hampshire, has been successful in procuring FDA approval for a Tc99m radiolabeled peptide dimer, Bibapcitide (figure 5-2). The agent has been approved as a radiopharmaceutical for deep vein thrombosis imaging. As can be seen in the structure, the compound is comprised of a dimer of a KGD compound that uses a homolysine analogue that is substituted with sulfur in the γ-position. Once again, this homolysine analogue imparts the correct spatial proximity between the cationic charge of the homolysine

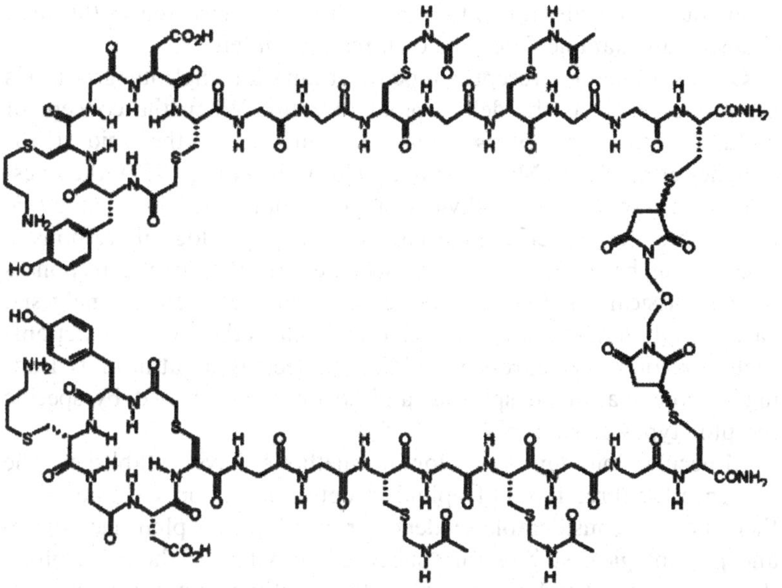

Figure 5-2 Bibapcitide.

analogue and the anionic charge of the aspartic acid to allow the molecule to impart both potency and selectivity for the GPIIb/IIIa receptor.

Receptor Studies: A Brief Overview of the Use of Peptides for Studies of Structure–Activity Relationships

Perhaps the most important applications of synthetic peptides are for the delineation of receptor types and subtypes and for providing analogues (radiolabeled and otherwise) that can be used for examination of receptor system in vivo. Clearly receptor pharmacology, physiology, endocrinology, molecular biology and drug development are complicated by the fact that there are multiple types and subtypes of different receptors. One may assume that each subtype will have a particular functional role in vivo, but the complexities of biological systems, the overlapping redundancies of biological control, and the plasticity of biological systems in response to changes in environment and in disease state all provided further challenges for peptide science. Thus, the synthesis and design of peptides directed towards receptor type and subtype binding and kinetics, as well as toward in vitro and in vivo pharmacology are a critical area of current peptide research, and will continue to be so for many years to come. In this section, we will provide a brief discussion of the many considerations that go with peptide research in this area choosing an example from our own research interests.

One example of a receptor type determination that our group has been involved in is the delta opioid receptor. When the concept of multiple opioid receptors began to surface in the mid 1970s (Hughes et al., 1975; Martin et al., 1976; Lord et al., 1977), the question arose whether these individual types of opioid receptors might be responsible for specific pharmacological physiological responses. Speculation began to arise as to whether the therapeutic responses of the classical alkaloid opiates (e.g., morphine) such as analgesia, vascular dilation, euphoria, and so on as well as the adverse reactions, such as respiratory depression, addiction, decreased gut motility leading to constipation, dysphoria, and so on, were elicited by specific receptor types or subtypes.

Molecular biology and cloning methods have established the existence for three types of opioid receptors μ, δ, and κ. In addition, there now is considerable evidence from extensive pharmacological and physiological studies for "subtype" activities of these receptors, though the structural correlates for these subtype activities have still not been determined. Furthermore there is now considerable evidence, especially in the central nervous system (CNS) for neuroplasticity in response to various stimulae. In such cases the "bioactivities" of the receptors change in response to their endogenous ligands (neurotransmitters, hormones, etc.). Thus ligands that "work" in one circumstance no longer have efficacy in another. Clearly this increases the challenge in designing ligands for their various receptors and receptor "states."

One method of peptide design which has been instrumental in defining these receptor types has been the use of conformational constraints including cyclization (Hruby, 1982; Kessler, 1982; Hruby et al., 1990a, 1991a); (for example, see Mosberg et al., 1983a,b for enkephalins). The rationale for the use of conformational and topographical constraints has been to limit the number of conformations the peptide may assume. Decreasing the number of conformations will increase the likelihood that a peptide ligand will interact primarily at one receptor among several. This technique has been shown to be very fruitful in the discovery of the δ-selective peptide, DPDPE (Mosberg et al., 1983b), the μ-selective peptides, CTOP (Pelton et al., 1985) and TCTAP (Kazmierski and Hruby, 1988; Kazmierski et al., 1988), and the κ-selective dynorphin analogues (Kawasaki et al., 1990).

Prior to describing the design and synthesis of receptor-selective peptides, however, a brief description of the types of bioassays to differentiate between the different receptor moieties is essential. For the example here, we will use studies of the use of opioid receptors and their peptide ligands. As was indicated previously, Lord and coworkers (Lord et al., 1977) demonstrated a difference in affinities of the alkaloid or morphine-like opiates versus the peptidic enkephalins. Utilizing tissues from the guinea pig ileum (GPI) and mouse vas deferens (MVD), Lord was able to show a preference of morphine-like opiates for μ receptors and a preference of enkephalins for δ receptors. This demonstrated that one could study ligands directed towards specific receptor types. It was later discovered that the MVD had approximately 80% δ receptors, the remainder being μ and κ, and the GPI had approximately 30% δ (Leslie et al., 1980). Even though the receptor populations are not homogeneous for δ (MVD) or δ (GPI), one can get a fairly accurate idea of *relative* ligand selectivity by calculating the GPI to MVD ratio of the corresponding IC50 values. Ratio*s* < 1 exhibiting μ selectivity and > 1 exhibiting μ selectivity. Of course, these ratios have to be based on absolute ligand standards which today, we currently use DPDPE (δ) and CTOP (μ) as "standard" δ and μ ligands.

In addition to these bioassays, binding assays generally have utilized membranes from the brain and ^{3}H- or ^{125}I-labeled highly receptor-selective ligands have been developed.

Development of δ-Selective Opioid Receptor Peptides

The prototypical peptide used as the original starting point for δ-selective peptide design begins with the Leu^5-enkephalin and Met^2-enkephalin, since Lord and coworkers (Lord et al., 1977) suggested a preference of both enkephalins for the δ receptor. Initial design principles which address either shortening or lengthening the enkephalin peptapeptides met, in general, with little success with regards to potency (vs. Met^5 or Leu^5 enkephalin) or selectivity. The second design principle which was employed, the substitution of the corresponding L-amino acid for a D-amino acid, initially met with only minor success. Substitution of a D-Ala^2 for Gly^2 in Leu^5 and Met^5 enkephalin did not improve the binding character significantly, but did result in an increase in potency in both the MVD and GPI assay (Kosterlitz et al., 1980). Introduction of a D-Leu in position 5 (DADLE) likewise did not cause any great increase in affinity for the two binding sites; however, it significantly increased the activity in the MVD and reduced it in the GPI such that there was a five-fold increase in preference for the MVD over the GPI. Thus, it became clear that substitution of D-amino acids at positions 2 and 5 were beneficial for δ receptor selectivity.

A major breakthrough in the design of δ-selective peptides evolved from the work of Gacel and coworkers and Fournie-Zaluski et al. (Gacel et al., 1980, Fournie-Zaluski et al., 1981). Starting from structure activity relationships in the enkephalin series, the hexapeptide

Tyr-D-Ser-Gly-Phe-Leu-Thr (DSLET) was prepared. Although no more potent than DADLE in the MVD (0.58 nM $\pm$ 0.20 vs. 0.54 nM $\pm$ 0.09), DSLET was about seven-fold less potent in the GPI versus DADLE (DSLET IC_{50} GPI/MVD = 620). Further modifications based upon DSLET as the prototype eventually led to more δ-selective peptides. Replacement of D-Thr^2 for D-Ser^2 increased the selectivity for the δ receptor even more (Gacel et al., 1988a). This was followed by the development of the O-tert-butylated analogues [D-Ser^2(O-tert-butyl), Leu^5, Thr^6]enkephalin (DSTBULET) and [D-Ser^2(O-tert-butyl), Leu^5-Thr^6 (O-tert-butyl)]enkephalin (BUBU) which further improved upon selectivity (Delay-Goyet et al., 1988; Gacel et al., 1988b).

A design theme that proved to be extremely successful in our laboratory (Hruby, 1982; Hruby et al., 1990a) has been the use of conformational constraints to study receptor selectivity and, even more importantly, receptor-site topography. The concept takes advantage of the fact that conformational constraints not only can help to generate receptor selectivity but also can aid in determining receptor topography, because now the number of possible conformations is substantially reduced compared to their unconstrained counterparts. Hence, a decreased number of degrees of conformational freedom gives one the added advantage of being able to indirectly study receptor topography by examining the conformational properties of the ligand in solution.

Work that led to the development of [D-Pen^2, D-Pen^5]enkephalin (DPDPE) (Mosberg et al., 1983b) was based upon our knowledge of pseudo-isosteric cyclization of side-chain groups in our previous studies with melanotropins. In this case Sawyer et al. (1982), showed that substituting the Met^4 and Gly^{10} residues by Cys^{4,10} moieties followed by subsequent oxidation led to a compound that assumes a similar steric space as its linear counterpart, but now becomes constrained. This led to a superpotent analogue. Similarly, it was realized that the Gly^2 and Met^5 side chains of enkephalin would be pseudo-isosteric if Cys substitutions were made at these positions. From the patent literature, Sarantakis had found that a [D-Cys^2, L-Cys^5]-enkephalin was an effective analgesic in vivo, but no information was known about its selectivity until Schiller and coworkers resynthesized the compound and found it to be potent but not very selective (Schiller et al., 1981). The suggestions from other work that D-amino acids could be utilized in both the 2 and 5 positions encouraged Mosberg et al. to synthesize both [D-Pen^2, D-Cys^5]enkephalin (DPDCE) and [D-Pen^2, L-Cys^5]enkephalin (DPLCE) (Mosberg et al., 1983a). These were found to be the most selective δ ligands to date. A final permutation led to the synthesis of [D-Pen^2, D-Pen^5]enkephalin (DPDPE), which has served as the pharmacological standard for studying δ-opioid receptor properties for many years.

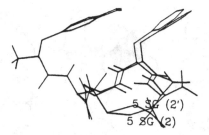

Figure 5-3 Superimposed structures
of the two low energy conformations
found most consistent with ^{1}H NMR
NOESY constraints and vicinal
coupling constant data. A series of
turn structures (type I, Iβ, II, IIβ, III,
IIIβ, IV, IVβ, etc.) were used as
starting points prior to energy
minimization. Note the difference in
the helicity of the disulfide bridge.

The development of DPDPE by Mosberg and coworkers also led to
extensive biophysical studies to help give us a clue to the topography
about the δ-opioid receptor.

Hruby and coworkers (Hruby et al., 1988) conducted extensive 2-D
NMR studies in solution including nuclear Overhauser enhanced
NMR spectroscopy (NOESY) in 90% HOD and pH 3.1 and 6.0.
These studies revealed some rather interesting dipolar interactions
between the aromatic moieties of Tyr[1] and Phe[4] and the β, β-dimethyl
groups of D-Pen[2]. This direct evidence for transannular interactions
led Hruby et al. to suggest that an amphiphilic conformation (Hruby
et al., 1988) may be among the most stable in solution. Starting primarily with coupling constants derived by NMR determinations, combinations of all reasonable phi angles consistent with a 14-membered
ring, were calculated. These starting parameters were in turn energy
minimized and finally compared again with the NMR data. Extensive
energy minimization studies led to finding two pairs of low energy
conformations which differed primarily by the chirality of the disulfide
(figure 5-3). Of the two, one of the low energy conformers satisfied all
of the NMR criteria. The conformation was distinguished by the
existence of a type IV β-turn and the transannular interactions
described previously. In addition, the molecule contained overall
amphipathic topographical features. Thus, because of the constrained
nature of this peptide ligand, one has a general idea of the δ-opioid
receptor topography by virtue of the ligand specific for it. An additional approach using quenched dynamics was conducted by Pettitt

Figure 5-4 CPK models of the nine lowest energy conformers derived
from quenched dynamics studies on [D-Pen^2, D-Pen^5]enkephalin
(Pettitt et al., 1991). Note that all nine conformers exhibit similar
positioning of the aromatic residues above the plane of the backbone
ring, while the carbonyl moieties (oxygens are shaded) are pointed
toward the opposite side of the backbone ring, giving the molecules
their amphiphilic character.

and coworkers (Pettitt, et al., 1991) which yielded a similar set of low
energy conformers that all exhibited the same amphipathic character
which was evident in the earlier work. CPK models of the nine lowest
energy conformers can be seen in figure 5-4.

Further important evidence for the low energy conformations
accessible to DPDPE were provided when an X-ray crystal structure

of DPDPE was finally obtained (Flippen-Anderson et al., 1994) followed shortly thereafter with X-ray crystal structure of analogues of DPDPE (Collins et al., 1996). These important studies provided an opportunity to compare the conformations obtained by NMR, by molecular dynamics, by computational chemistry, and in the crystalline state for both agonist and antagonist biological activity. A major conclusion from these studies is that the 14-membered disulfide ring conformation is the same except the chirality of the disulfide bridge which could be either right-handed or left-handed, a result consistent with the molecular dynamics studies. Thus it appeared that the conformation of the backbone scaffold of the 14-membered ring for DPDPE was the same in both solution and in the crystal and presumably also at the δ-opioid receptor. However, to obtain the biologically active conformation and corresponding pharmacophore in 3-D space it also is critical to know the preferred side-chain conformation χ^1 and χ^2 angles for the key side-chain groups known to be critical for interaction with the δ-opioid receptor, namely Tyr[1] and Phe[4] in DPDPE. Previously, using all four isomers of β-methylphenylalanine (Hruby et al., 1991b) in the 4 position in conjunction with binding and bioassay studies, as well as NMR studies and computational chemistry (Nikiforovich et al., 1993) we were able to conclude that the gauche (-) χ^1 conformation was best for high potency and selectivity at the δ-opioid receptor. However, when we utilized the same approach with all four isomers of β-methyltyrosine in position 1 of DPDPE the biological results were equivocal (Töth et al., 1992). This led us to search for more χ^1 and χ^2 constrained amino acids. Thus we designed and developed an asymmetric synthesis (Qian et al., 1995) of all four isomers of β-methyl-2',6'-dimethyltyrosine and incorporating them in DPDPE (Qian et al., 1994, 1996). Only one of these isomers [(2S,3R)β-methyl-2',6'-dimethyltyrosine]DPDPE had both high potency and high selectivity for the δ-opioid receptor, and both NMR studies in solution and computational studies showed that the trans χ^1 conformation of the Tyr was compatible with the bioactive conformation at the δ receptor. Interestingly, this same analogue was a weak antagonist at the μ-opioid receptor, while the [(2S,3S)TMT[1]]DPDPE analogue was a modestly potent at both the μ and δ receptors (Qian et al., 1996) had considerably better analgesic activity (Bilsky et al., 2000).

These results led to the development of a three-dimensional topographical model in vector space of the biologically active pharmacophore for the interaction of DPDPE with the δ-opioid receptor (Shenderovich et al., 2000). This then provided a de novo approach to the design of a highly potent and receptor selective nonpeptide ligand for the δ-opioid receptor that had similar structure–activity relationships as DPDPE but different structure–activity relationships than other putative nonpeptide peptide mimetic analogues with high binding affinity for the δ-opioid receptor (Liao et al., 1998). Clearly appropriate conformational and topographical constraints of bio-active peptides in conjunction with careful biological assay studies, and rigorous conformational studies using spectroscopic and compu-tational studies can provide new insights into structure–activity relationships for ligands that interact with complex biological systems such as the G-protein coupled receptors (GPCRs) of which the δ receptor is a typical example.

In conclusion, using the δ-opioid receptor as an example, one can see how the design of δ-receptor selective analogues has developed from: (1) lengthening and shortening the sequence; (2) systematic sub-stitution of D-amino acids for native L-amino acids; (3) imposing of conformational constraints by pseudo-isosteric replacement; (4) solution phase NMR; (5) X-ray crystallography; (6) molecular mechanics and molecular dynamics calculations; (7) design and asym-metric synthesis of novel topographically constrained amino acids; (8) their incorporation into the cyclic constrained bioactive peptides; and (9) careful binding and bioassay studies can provide detailed insight into the bioactive conformation of peptides which can be used for de novo design of a nonpeptide mimetic. All have helped to develop a "picture" of the 3-D structural requirements of the δ-opioid receptor that can be tested. Clearly, more work needs to be done to refine the picture. Future endeavors incorporating all the above design principles as well as molecular biology and molecular pharmacology techniques may one day soon allow us to isolate and to view the δ-opioid receptor itself as it interacts with its agonist and antagonist ligands. Efforts in that direction have begun to appear (e.g. Salamon et al., 2000).

The Use of Peptides for Affinity Labeling of Receptors

Establishment of some type of irreversible but specific interaction between a ligand and its biological acceptor, termed affinity labeling, is a major approach to identify, characterize, and even isolate receptor molecules for hormones and neurotransmitters. In the case of enzyme–substrate interactions, this approach helps in identifying the chemical constitution (amino acid residues) of the binding and/or catalytic site(s). The most general approach for establishing irreversible binding between a ligand and its acceptor relies on the establishment of a covalent bond between the two entities. Two main methods for achiev-ing this are chemical-affinity labeling and photoaffinity labeling. While both techniques have found their applicability in numerous peptide–hormone receptor systems, photoaffinity labeling generally is preferred because of its nondependence on proper juxtaposition as well as the

Table 5-3 Reagents for the introduction of maleimido groups in peptides

Reagent	Reference
	Yoshitake et al., 1979
	Kitagawa et al., 1976
	Martin and Papahadjopoulos, 1982

presence or absence of certain specific chemical functionalities at the receptor binding site.

A common technique for affinity labeling of acceptor molecules involve the use of homo- as well as heterobifunctional cross-linking reagents. These reagents are added to the incubation medium after the binding of a ligand with its acceptor/receptor has been established. This technique leads to a covalent cross-linkage between the ligand and the acceptor. A number of these reagents which utilize a variety of the chemical functionality and photoaffinity approaches, have been developed and are available commercially (see, for example, tables 5-3–5-6). The photoreactive and/or chemically reactive groups used in these bifunctional cross-linking reagents are, in principle, the same that are described in the following discussion on the development and use of chemical-affinity and photoaffinity peptide ligands.

Chemical-Affinity Labels

Chemical-affinity labeling involves the establishment of a covalent bond between a ligand and its biological acceptor during their specific binding to each other. Usually, the ligand used in this operation is first modified to bear a chemically reactive group capable of reacting with

Table 5-4 Heterobifunctional reagents for the synthesis of photoaffinity peptides

Reagent	Reference

Reactive for amino group (may also react with SH groups):

Reagent	Reference
N_3—⟨benzene⟩—COBr	Hixson and Hixson, 1975
N_3—⟨benzene⟩—CH_2-COO-⟨benzene⟩—NO_2	Eberle and de Graan, 1985
N_3—⟨benzene, NO_2⟩—CH_2-COO-⟨benzene⟩—NO_2	Eberle and de Graan, 1985
N_3—⟨benzene, NO_2⟩—F N_3—⟨benzene, F⟩—NO_2	Levy, 1973 Smith and Knowles, 1974 Fleet et al., 1969
N_3—⟨benzene⟩—N=C=S	Sigrist and Zahler, 1980
N_3—⟨benzene⟩—$\overset{NH_2^+ \ Cl^-}{\underset{}{C}}$—$OCH_3$	Ji, 1977
N_3—⟨benzene⟩—CH_2—$\overset{NH_2^+ \ Cl^-}{\underset{}{C}}$—$OCH_3$	Ji, 1977
N_3—⟨benzene⟩—COO—N⟨succinimide⟩	Yip et al., 1980 Galardy et al., 1974
N_3—⟨benzene⟩—CH_2—COO—N⟨succinimide⟩	Eberle and de Graan, 1985
N_3—⟨benzene, NO_2⟩—CONH—H_2C—COO—N⟨succinimide⟩	Yip et al., 1980

Table 5-4 (*Continued*)

Reagent	Reference

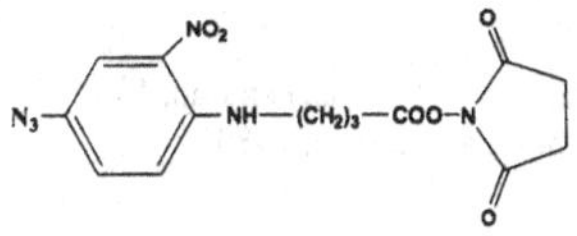

Guire et al., 1977

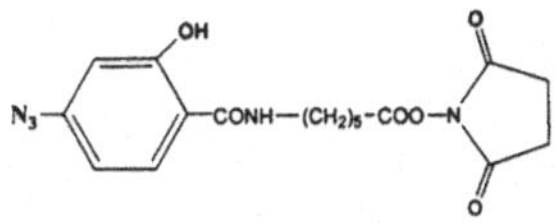

Ji and Ji, 1982

Reactive for carboxy group:

Escher et al., 1979

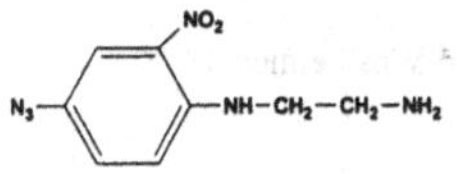

Das and Fox, 1978

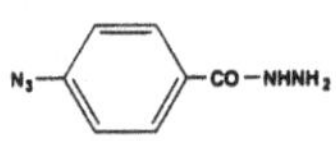

Eberle and de Graan, 1985

Reactive for arginine side-chain functionality:

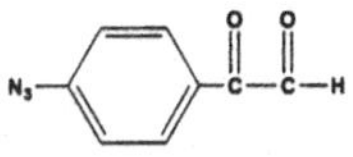

Vanin et al., 1981
Ngo et al., 1981

Reactive for cysteine side-chain functionality:

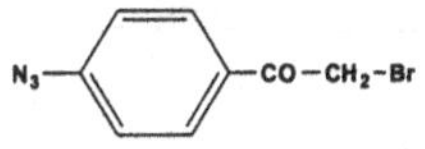

Hixson and Hixson, 1975
Schwartz and Ofengand, 1974

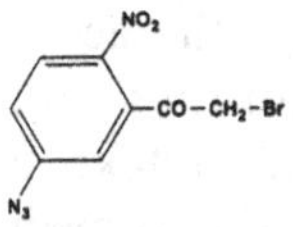

Seela and Rosemeyer, 1977

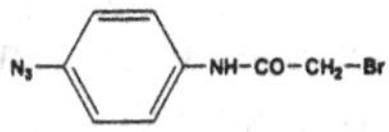

Budker et al., 1974

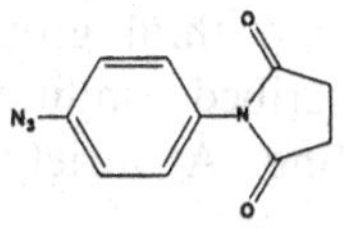

Trommer and Hendrick, 1973

(*continued*)

Table 5-4 (*Continued*)

Reagent	Reference

Reactive for tryptophan side-chain functionality:

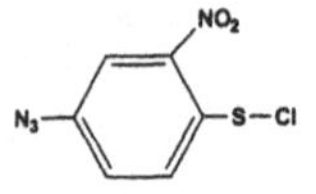

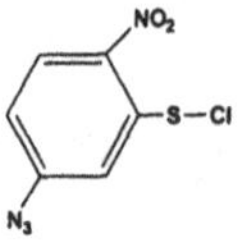

Muramoto and Ramachandran, 1980
Demoliou and Epand, 1980

Reactive for tyrosine/histidine side-chain functionality:

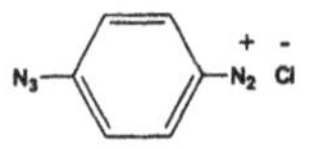

Escher et al., 1979

Carbene generating reagents reactive for amino groups:

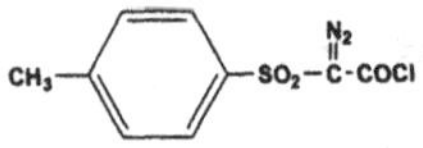

Chowdhry and Westheimer, 1978

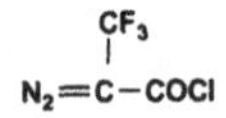

Chowdhry et al., 1976

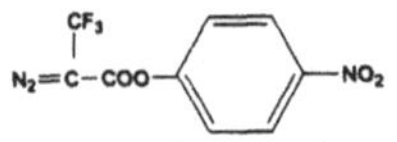

Chowdhry et al., 1976

Burgermeister et al., 1983
Nassal, 1983

any of the chemical functionalities found usually on the acceptor/ receptor molecule. Most importantly, the nature of this reactive group should be the one that allows this reaction to take place under physiologically compatible conditions, that is aqueous medium, ambient temperature, and pH 6–8 values. These requirements limit the type of chemical reactions one can efficiently perform under these conditions with functionalities, such as, amino, carboxyl, thiol, imidazole, phenol, hydroxyl, and so on, that are usually found on the acceptor protein. Among these, the amino and thiol groups, because of their chemical reactivities under the prescribed conditions, are best suited targets for chemical-affinity labeling. A variety of

Table 5-5 Radioactive heterobifunctional reagents for specific derivation of amino groups in the peptides for photoaffinity labeling of their respective receptors

Reagent	Reference
[structure: N₃—benzene ring with OH and *I substituents—COO—N(succinimide)]	Ji and Ji, 1982
[structure: N₃—benzene ring with OH and *I substituents—CONH—(CH₂)₈—COO—N(succinimide)]	Ji and Ji, 1982
[structure: N₃—benzene ring—CONH—CH₂ CONH—C(H)—COO—N(succinimide), with CH₂ linked to benzene ring bearing *I, *I and OH]	Ji and Ji, 1982

*Radioactive iodine atoms may easily be incorporated in these molecules.

derivatives capable of reacting with thiol and amino groups are now known and can be efficiently incorporated in the ligand molecule. The following section provides a general discussion about some of these chemical affinity groups and their incorporation in the peptide-based ligands. Perhaps the most germane point while designing these chemical-affinity ligands, however, is important consideration in designing the site in the ligand molecule at which one can incorporate the chemically reactive group. Because specific biomolecular recognition and binding is of critical importance for any labeling experiment, the site of derivatization must be one that causes little or no change in the recognition and the binding characteristics of the modified ligand.

Thiol Reacting Derivatives

The use of affinity labels reactive toward a thiol group generally is limited to acceptor molecules which possess free thiol groups. Most of the cysteine residues present in a biomolecule exist in their respective oxidative state, forming a disulfide linkage among a pair of cysteine residues. In spite of this, the chemistry for thiol sensitive

Table 5-6 Cleavable heterobifunctional reagents for derivation of the
peptides for reversible labeling of their respective receptors

Reagent	Reference
Amine reactive groups:	
	Ji, 1979 Vanin and Ji, 1981
	Ji, 1979 Vanin and Ji, 1981 Das et al., 1977
	Rinke et al., 1980
	Jaffe et al., 1979 Jaffe et al., 1980
Thiol reactive groups:	
	Vanin and Ji, 1981 Moreland et al., 1982
	Huang and Richards, 1977
	Henkin, 1977

derivatives is well developed due to the nucleophilicity of the thiol
group, and can be performed under mild biocompatible conditions.
As the result, it is the most favored one for conjugating a biomolecule
either with an insoluble matrix for use in affinity chromatography, or
with another biomolecule for use in diagnostic assays, immunization,
antibody, and vaccine production. Two main strategies followed
for reactions with a thiol group are S-alkylation and disulfide bond
formation.

S-Alkylating Reactions

Thiols undergo reaction with α-halo ketones and maleimide groups under physiological conditions. These ketones, particularly the iodo- and bromo-acetyl groups have been used in many instances for alkylating a SH group in a biomolecule. The order of reactivity among α-halo ketones is in the order: iodo $\gg$ bromo $>$ chloro $>$ fluoro. The reactivity of α-iodo as well as α-bromo halo-ketones towards -SH functionality is not entirely specific. While a variety of other functionalities such as water, OH, NH_2, the imidazole ring of histidine, and so on can react with α-iodo as well as α-bromo ketones, the reaction of α-chloroketone with thiols is selective but slow under physiologically compatible conditions. The α-fluoroketones are practically inert. Consequently, α-bromo ketones generally are the most favored chemical affinity alkylating reagents. Bromoacetic acid can be coupled using usual methods of peptide synthesis to either N-terminal or side-chain amino function in a peptide ligand to afford this type of chemical label. For example, such derivatives of α-melanotro-pin have been prepared and used for conjugation with other large molecular weight protein (Eberle et al., 1977).

Unlike α-haloketones the reaction between a thiol and maleimido group is highly specific. Maleimido groups are fully compatible with the methods of peptide synthesis and purification. A variety of reagents containing maleimido groups are available commercially as free carboxylic acid or as active esters like N-hydroxysuccinimide or p-nitrophenyl ester (table 5-3), for facile introduction into peptide ligands.

Disulfide Bridge Forming Reactions

Commercially available 3-(2-pyridyldithio)propionate (PDP) is a commonly used group that is incorporated into peptide or protein ligands through its N-hydroxysuccinimide ester. The resulting highly stable affinity ligand undergoes a very specific reaction at pH 6.0–6.5 with thiols to form a disulfide bond with the ligand. This is the method of choice for conjugating a ligand with another biomolecule or affinity matrix where reversibility is desired. The conjugated ligand can be easily cleaved by reducing the -S-S- bridge with mild reducing agents as dithiothreitol (DTT).

Schwyzer and coworkers (Wunderlin et al., 1985a, b) have established yet another facile technique for making a disulfide bridge between a ligand and another biomolecule. In this approach, as demonstrated by conjugating an α-melanotropin analogue with a bio-molecule, a bromoalkyl group is first incorporated into the ligand by chemical methods of peptide synthesis. The treatment of the resulting stable derivative with sodium thiosulfate yields a "Bunte salt" which is treated in situ with a thiol-containing biomolecule resulting in the establishment of the desired disulfide bridge.

Amine-Reacting Derivatives

Carbonyl compounds such as aldehydes and ketones react with amino groups of lysine residues to form Schiff's bases that are reversible yet stable enough to allow further investigations into affinity labeled products. Imidate, isocyanate, or isothiocyanate groups attached to a ligand are the other most widely used modalities to target amino functionalities on the acceptor molecule. The synthesis of peptide ligands carrying these groups, however, is not straightforward. These groups, therefore, have found more use in bifunctional cross-linking reagents and are being widely utilized in such studies.

Photoaffinity Labels

Photoaffinity labeling of biological receptors/acceptors for peptide and protein hormones, neurotransmitters, and so on, is a valuable means of receptor identification and isolation. In this approach, a chemically stable but photolabile group is conjugated to a potent ligand. After binding this ligand to its receptor/acceptor site, the complex is subjected to photolysis to generate highly reactive carbenes and nitrenes which may undergo hydrogen abstraction, insertion, or cycloaddition reactions with the adjacent chemical functionalities on the receptor molecule, thereby establishing a covalent bond between the receptor and the ligand (Bayley, 1983). Because of their high reactivity, the photoaffinity labels have proved to have advantages over the chemical affinity labels. Prior to activation, they are stable in aqueous solutions, and most importantly, the reactions exhibited upon photo-activation are not dependent on the presence of a nucleophilic center on the receptor for establishing chemical bonds. However, they can offer the disadvantage of nonspecific covalent interactions if they are not receptor bound at the time of photoactivation. The use of a radiolabeled photoaffinity ligand in these studies helps to identify by covalent attachment, the site of ligand–receptor/acceptor interaction.

The five most common groups of photolabels used are α-diazo-ketones, diazirines, α, β-unsaturated ketones, arylazides, and p-nitro-phenyl. The singlet or the triplet state of the photogenerated species as shown in figure 5-5 tends to determine the type of reaction preferred in ligand–receptor cross-linking. Singlet states, in general, exhibit electro-philic reactions such as, insertion, and so on. Triplet states exhibit radical reactions that lead to abstraction of a hydrogen atom (Turro, 1980). Nitrenes, due to longer half-lives than carbenes, tend

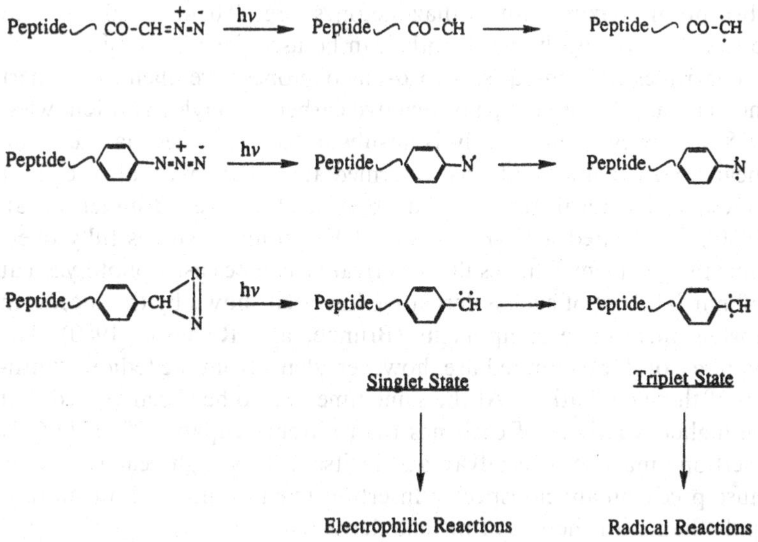

Peptide $\sim$ CO-CH=$\overset{+}{N}$=$\overset{-}{N}$ $\xrightarrow{h\nu}$ Peptide $\sim$ CO-$\overset{\cdot\cdot}{C}$H $\longrightarrow$ Peptide $\sim$ CO-$\overset{\cdot}{C}$H

Peptide $\sim$⟨ ⟩-$\overset{+}{N}$=$\overset{-}{N}$ $\xrightarrow{h\nu}$ Peptide $\sim$⟨ ⟩-$\overset{\cdot\cdot}{N}$ $\longrightarrow$ Peptide $\sim$⟨ ⟩-$\overset{\cdot}{N}$

Peptide $\sim$⟨ ⟩-CH$\overset{N}{\underset{N}{\big\Vert}}$ $\xrightarrow{h\nu}$ Peptide $\sim$⟨ ⟩-$\overset{\cdot\cdot}{C}$H $\longrightarrow$ Peptide $\sim$⟨ ⟩-$\overset{\cdot}{C}$H

Singlet State   Triplet State

↓   ↓

Electrophilic Reactions   Radical Reactions

Figure 5-5 Alpha-diazocarbonyl, aryl azide, and aryl diazirine groups and their respective photolyzed reactive singlet and triplet states. Adapted from Eberle (1983).

to exist in low-energy triplet states which can produce insertion into C-H bonds (McRobbie et al., 1976a,b). In electrophilic reactions the singlet nitrenes prefer an O-H or N-H bond over a C-H bond and thereby are suggested to be less suitable than carbenes for labeling receptor areas rich in lipids or hydrophobic groups (Bayley and Knowles, 1978a,b).

The α-diazoketones that give rise to α-keto carbenes were the first photoaffinity labels used. The active site of chymotrypsin was labeled by a derivative of this class of photolabels by Westheimer and co-worker (Singh, et al., 1962; Chowdhry and Westheimer, 1979). In the case of peptides, however, the use of this type of label has been rather limited. The main reason has been the inherent chemical re-activity of α-diazo carbonyl group. The diazo group is a good leaving group in nucleophilic displacement reactions (e.g., reactions with water, -OH, -SH, or -NH$_2$ functionalities). Further, the α-keto carbene generated upon photolysis of α-diazoketone is likely to undergo Wolff rearrangement reactions thereby lowering the labeling effi-ciency of the photoaffinity labeling of the receptor. Substitution of $^\alpha$C-H with $^\alpha$C-COOEt, or $^\alpha$C-CF$_3$, or $^\alpha$C-SO$_2$C$_6$H$_4$CH$_3$ has been reported to increase the stability of the resulting diazocarbonyl compound as well as suppress the Wolff rearrangement (Vaughan and Westheimer, 1969; Chowdhry et al., 1976; Chowdhry and Westheimer, 1978). This, however, also increases its lability in aqueous medium.

Absorption maxima for α-diazoketones are 250 nm (high α) and 350 nm (low α). Both wavelengths can be used for photolysis.

Diazirines, in comparison to α-diazoketones, are chemically inert and generate the same type of reactive carbenes (Bayley and Knowles, 1978b). However, on photolysis unsubstituted diazirines may generate linear diazo compounds as intermediates that may also exhibit undesired chemical reactivity as described above. Brunner et al. (1980), substituted a C^{α}-H by a C^{α}-CF$_3$ group to successfully overcome this problem. The resulting derivative can be easily photolyzed at 350 nm. The use of such a derivative has been shown by the success in labeling membrane components (Brunner and Richards, 1980). The main reason for its limited use, however, stems from the tedious synthesis of these derivatives. At the same time, it also has been argued that the higher reactivity of carbenes than nitrenes capable of facile C-H insertions may be a disadvantage in itself. This high reactivity may cause predominant nonspecific insertion reaction into cell membrane lipids than with the use of nitrene derivatives.

A few examples on the use of α, β-unsaturated ketones for the photoaffinity labeling of peptides are known. 4-Acetylbenzoyl pentagastrin was covalently coupled to albumin by irradiation at 320nm (Galardy et al., 1974). In this case the photolysis produces a biradical triplet state that has selectivity in the abstraction of C-H hydrogens than O-H hydrogens (Turro, 1965). It has been suggested that this photolabel, due to its high lipophilicity, may considerably alter the binding and biological characteristics of short peptides. That alteration is likely to make this type of photolabel unsuitable for many peptide ligands.

Arylazides are the best-known labels used in combination with peptides. These are stable to most of the acid and base procedures used in peptide synthesis. That makes them compatible with the synthesis of peptide derivatives. These groups, however, are not stable under reducing conditions such as hydrogenation and exposure to thiols. Photolysis of arylazides yield highly reactive nitrenes. The absorption maxima for unsubstituted arylazides is around 260 nm. Introduction of an electron withdrawing group, for example, nitro group, in the position *meta* to the azide group allows one to accomplish photolysis at higher wavelengths.

The *p*-nitrophenyl group has been used as a photolabel in the form of *p*-nitrophenylalanine derivatives incorporated in to peptides. This technique has been successfully used for example, for labeling the active site of α-chymotrypsin (Escher and Schwyzer, 1974), receptor binding sites for angiotensin II (Escher and Guillemette, 1978), and bradykinin (Escher et al., 1981). It, however, failed to label receptors for α-MSH which could be labeled by using a *p*-azidophenyl group containing MSH ligand (Eberle, 1984).

Incorporation of Photolabile Groups in the Peptide Hormones and Their Analogues

Much work on structure–activity relationships for peptide hormones strongly demonstrates that introduction of even a minor structural change for a key residue in the peptide can affect its potency, as well as its selectivity towards a particular receptor subtype. However, for many bioactive peptides, and especially for cyclic conformationally constrained peptides which have two surfaces, specific sites or surfaces exist on these structures where modification of functional groups, even large structural modifications, does not affect bioactivity or potency (e.g., Sharma et al., 1993). These ancillary sites are particularly useful for attaching labeling moieties, with the caveat that their sites of labeling may be somewhat removed from the primary binding site. These factors require careful consideration of the site where a potential group can be incorporated. While not compromising the potency and selectivity towards its biological receptors, the photoaffinity peptide analogues should also exhibit the ability to cross-link with the receptors. In general, the methods of incorporating photolabile groups into peptides can be categorized in two main groups. These categories of; (a) de novo synthesis of a particular photoaffinity peptide analogue, and; (b) derivatization of peptides to photoaffinity peptides are discussed below.

De Novo Synthesis of Photoaffinity Peptides

This approach generally involves introduction of novel amino acids that also can act as photoaffinity labels, such as the p-nitrophenylalanine (Pnp) residue, in the peptide during its total synthesis. The finished peptide containing a Pnp thus acts as a photoaffinity ligand. Alternatively, a Pnp residue, incorporated in a peptide, can be converted to a p-azidophenylalanine (Pap) residue which, as discussed earlier, is a more efficient photolabile group. This conversion can be achieved by hydrogenating the Pnp-containing peptide to give the corresponding p-aminophenylalanine substituted peptide which is then diazotized and treated with sodium azide to form the Pap substituted peptide analogue. As shown in figure 5-6, a more direct method for obtaining a p-amino phenylalanine-substituted peptide derivative is by introducing a benzyloxycarbonyl (Z) protected p-NH_2-phenylalanine residue in the peptide during its synthesis. Photoreactive analogues containing Pap have been reported for many bioactive peptides such as angiotensin II (Escher and Guillemette, 1978), bradykinin (Escher et al., 1981), α-MSH (Eberle and Schwyzer, 1976).

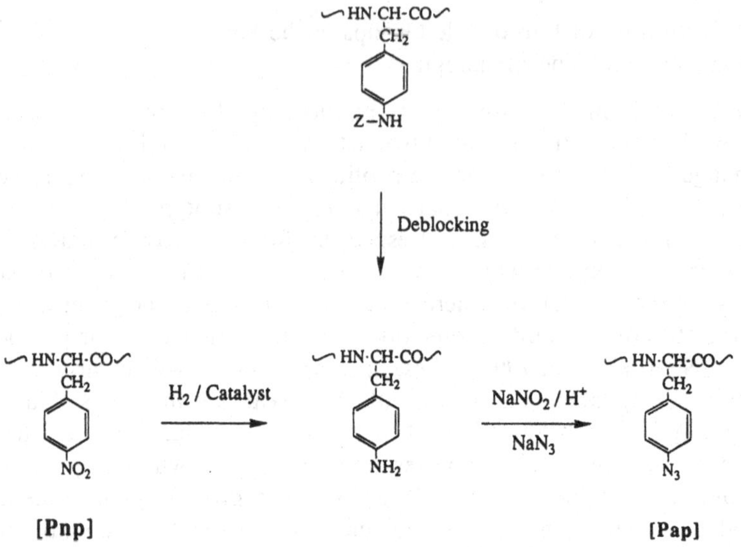

Figure 5-6 Alternate synthetic routes for *p*-azidophenylalanine [Pap] substituted peptides.

Derivatization of Peptides to Yield Photoaffinity Peptides

A number of heterobifunctional reagents containing the photolabile *p*-azidophenyl group have been developed over the last two decades that can be incorporated into finished peptides under rather mild conditions (tables 5-4, 5-5, 5-6). The second chemically reactive species in these reagents is designed to react either with an amino, or a carboxyl functionality that commonly occurs as the *N*- or *C*-terminal, or on certain side-chain functionalities of some amino acids in the peptide. Some reagents that react with the side chain functionalities of Arg, Cys, Trp, His, and Trp residues also have been developed. Most of these reagents (tables 5-4 and 5-5) have been used for constructing photoaffinity labels from peptide ligands and are commercially available from various sources.

Reagents for derivatizing amino groups such as acid chlorides, acid bromides, or *p*-nitro phenyl esters of *p*-azido benzoic acid or *p*-azido phenylacetic acid can lead to quantitative modifications. So do reagents such as fluoro arylazides, *p*-azidophenylisothiocyanates and imidates which, in addition, allow the maintenance of the cationic characteristics of the modified amino group. Some of the reagents, such as those containing an *N*-hydroxysuccinimide ester also provide the choice of varying the length of the spacer between peptide and the photoreactive azidophenyl moiety. The carboxyl modifying reagents contain an amino group that could be coupled to the COOH group on the peptide by any of the well-established methods of peptide bond formation. In this case, however, any amino group that may be present in the peptide itself should be selectively protected prior to this reaction.

An arginine specific reagent, *p*-azidophenyl glyoxal, reacts with the guanidino group at pH 7.0–7.5. The product formed, however, is unstable at alkaline pH regenerating the original guanidino group. Cysteine-specific reagents which contain α-haloketones or maleimide groups, can react quantitatively with the thiol function. However, most cysteine residues are found in peptides in an oxidized disulfide form. The tryptophan-modifying reagent, 2-nitro-4-azido phenyl-sulfenyl chloride reacts at position 2 of the indole ring in Trp. This reagent is also capable of reacting with free Cys (thiol) groups. The azo group in the azido aryl azo derivative reacts exclusively with the tyrosine or histidine ring.

Three carbene generating heterobifunctional reagents that modify an amino functionality also are available. These acid chloride or *p*-nitrophenyl ester reagents are adequately stable for derivatization reactions.

A reversible or cleavable class of heterobifunctional reagents also have been developed. The use of these reagents (table 5-6) offers the possibility of cleaving the peptide ligand from its receptor at will under mild conditions. The receptor thus separated from the ligand can be used in further studies such as generation of receptor specific anti-bodies, examination of structure, or cell-free reconstitution of the receptor. Normally, the cleavable heterobifunctional reagents contain, in addition to the usual two reactive groups, another group such as a disulfide (-S-S-), glycol (-CHOH-CHOH-), or azo (-N=N-) group. These groups can be cleaved respectively by 10–100 mM mercaptoethanol at pH 7–9 (mild reductive conditions), 15 mM sodium periodate for 4–10 h (oxidative conditions), or sodium dithionite (Bayley 1983).

Many other possibilities for affinity labeling a peptide ligand to a protein, nucleic acid, lipid or other receptors/acceptors can be enter-tained. The development of more robust and sensitive methods to study these complexes, especially mass spectrometry, Raman spectro-scopy, and other physical methods, suggests that these applications of peptide chemistry will continue to be used and developed.

Antigenic and Immunogenic Uses of Synthetic Peptides

The use of synthetic peptides in the study of the immune response stems from the ability of peptides to mimic similar antigenic sequences in proteins. Antibodies raised against an antigen in a native protein can bind to the same antigenic sequence in the absence of the rest of the protein. With the advent of the Merrifield solid-phase procedure (Merrifield, 1963), rapid and selective synthesis of pure peptides has become routine, and is now widely used in the study and prediction of protein antigenicity. Primarily synthetic peptides are used to identify key antigenic epitope regions of proteins. This is achieved by initiating the immune response against the native protein, and then testing the binding of the resulting antisera to synthetic fragments of the protein in an appropriate immunoassay, such as the enzyme-linked immuno-sorbent assay (ELISA).

Identification of various protein antigens has far-reaching applications in the fields of biochemistry, immunology, and molecular biology. Short-peptide epitopes can be used in the elucidation of the mechanism of the immune response (Smith, 1989), and allows investigation of the molecular recognition process of the antibody/antigen interaction complex (Hinds et al., 1991). The cross-reactivity of antibodies raised against synthetic epitopes for the native protein has facilitated the development of new generations of synthetic peptide vaccines which are being developed for their enormous potential against viral diseases such as foot and mouth disease, hepatitis, and HIV. In addition, the cross-selectivity of antipeptide antibodies for the native protein has been utilized in the isolation and characterization of gene products. Given the ease of synthesis of short peptides, the major difficulty in any study of this nature is deciding which antigenic fragments to synthesize. The minimum epitope length of a peptide fragment has been indicated to be about five to ten residues, and peptides of those lengths and longer have been used to determine which regions of a macromolecular structure, such as a protein, will be antigenic. Two protocols have been developed to this end. The first of these *epitope mapping* procedures involves the synthesis and testing of all possible peptide fragments in the protein of interest. The second protocol is based on prediction of the antigenic regions from the primary sequence of the protein, and hence involves the synthesis of fewer peptides. The uses of chemical and biological combinatorial synthesis methods have greatly expanded the number and kind of peptide-related ligands that can be used in this approach. In this chapter we will only outline necessary background material. More recent developments which are rapidly expanding are outside the scope of this chapter.

Elucidation of Antigenic Peptide Sequences in Proteins

With the introduction of immunological testing of peptides still bound to the solid support used in their synthesis (Smith et al., 1977), the production of all overlapping peptide fragments in a protein to locate the highly antigenic sequences became attractive (figure 5-7). This

```
           1                                              213
H₂N–T –T –S –A –G –E –S –A –D –P –(201 amino acids)–T–L–COOH
(i)   –T –T –S –A –G –E
   (ii)  –T –S –A –G –E –S
      (iii) –S –A –G –E –S –A
         (iv) –A –G –E –S –A –D
            (v) –G –E –S –A –D –P
```

Figure 5-7 An example of systematic synthesis of all possible
overlapping hexapeptides of the Foot and Mouth Disease Virus
protein VP1, as utilized by Geyson and coworkers (1986).
(Only the first five fragments are indicated.)

systematic protocol was greatly enhanced by the alternative synthetic
procedure of Geysen et al. (1984). They prepared peptides on polyethylene rods utilizing standard Merrifield solid-phase chemistry. The
linker between peptide and support was a polyacrylic acid derivative
compatible with both water and the organic solvent. The polyethylene
rod with attached peptide could then be used as the solid phase in an
ELISA immunoassay. The feasibility of this procedure was demonstrated by Rodda and coworkers (1986) who synthesized all possible
hepta-, octa-, nona-, and decapeptides of the 153-residue protein myoglobin, and assessed the binding of these to polyclonal antibodies
raised against the native protein. In this way, they identified the continuous epitope LKTEAE comprising residues 49–54 of myoglobin as
its strongest antibody-binding fragment.

Although this technique yields a complete systematic epitope mapping it can only be applied to small proteins due to the prohibitive
expense and time required for such a procedure. However, an example
in which a nonselective, systematic, synthetic approach is essential can
be found in the study of the recognition of T helper cells for processed
foreign-protein fragments which are located on the surface of macrophages (Unanue and Allen, 1987). These fragments are produced by
enzymatic degradation or denaturation of the native protein, and they
do not necessarily correspond to the regions of high antigenicity that are
recognized by antibodies. Thus, when studying the immune response
mechanism of T helper cell activation, one cannot easily predict
sequences that are T-cell epitopes (De Lisi and Berzofsky, 1986;
Rothbard and Taylor, 1988), and the most efficient method of finding
the antigenic specificity of T-cell recognition is the synthesis of a complete overlapping series of fragments of the protein. T-cells actually bind
to relatively few antigens due to major histocompatibility complex
restriction, which is thought to suppress autoimmunity. Finnegan
and coworkers (1986) investigated this theory by synthesizing fourteen

20-residue overlapping peptides of staphylococcal nuclease, which is a 149-amino-acid bacterial protein. Binding studies of the overlapping fragments to mouse T-cell clones raised against staphylococcal nuclease indicated only two strongly antigenic sequences, corresponding to segments 81–100 and 91–110. The observation of so few epitopes for mouse T cells on a phylogenetically distant bacterial protein suggested that inhibition of autoimmunity is not a major component in the limited repertoire of T-cell antigens.

Prediction of Protein Antigenic Regions for Antibodies

In predicting an epitope region of a protein, the primary consideration is that an antibody will bind only to the surface without disrupting its overall conformation. Hence, a reasonable assumption is that all highly antigenic regions of a protein are on its surface and are well exposed, for example, secondary structural elements such as amphipathic helices, β-turns, loops and random-coil regions containing high densities of hydrophilic amino acids. When three-dimensional structural information is available from X-ray or NMR, the surface accessibility of the protein can be immediately determined (see, for example Fanning et al. 1986; Thornton et al. 1986). However in the absence of such tertiary structural data, accessible regions can be predicted from knowledge of primary structure according to expected surface properties. (see also chapter 2.)

Use of·Tertiary Structural Information

One of the first and foremost X-ray studies on antigenic interactions was the determination to 2.8 Å-resolution of the 3-D structure of an antibody/antigen complex (Amit et al., 1986). This structure comprised the Fab fragment from a monoclonal antibody bound to its antigenic region on lysozyme. One of the key findings of this work was that two continuous epitopes of lysozyme were in contact with the antibody fragment. These were chain residues 18–27 and 116–129 (see figure 5-3 of Amit et al., 1986). Although this study graphically demonstrated the key epitope-binding sites of an antibody, it also exemplified a major drawback in the use of synthetic peptides as antigens. It has been calculated that the average contact surface area between an antigenic globular protein and antibody is 16–20 Å in diameter (Barlow et al., 1986), and indeed in the aforementioned X-ray structure study the entire antibody/antigen contact area was shown to be approximately 20 × 30 Å. This area constitutes a protein surface that for most cases is too large to contain only one continuous chain of residues. This, indicated that protein antigens are primarily made up of a set of "discontinuous" epitopes. Solid-phase peptide synthesis allows the preparation of linear or "continuous" peptide epitopes only. Thus only partial binding of antibody to linear synthetic peptide epitopes of the native polypeptide can generally be observed relative to the native protein. Despite this drawback, binding between antibodies and continuous synthetic epitopes is substantial enough for most uses and, at least, for those linear sequences corresponding to the regions of highest antigenicity.

Prediction of Antigenicity Based on Primary Structure

Any prediction about regions of high antigenicity from primary structure is concomitant with a prediction about the characteristics of surface regions of a protein. Of the surface characteristics of a protein, the most obvious is hydrophilicity. Globular proteins generally fold, so many of the hydrophobic residues are buried in the protein core, while many of the hydrophilic residues are located at the protein surface in order to maximize good energetic interactions with the aqueous environment. In order to predict protein tertiary structure, relative hydrophilicity (or hydrophobicity) scales have been calculated for the 20 naturally occurring amino acids, such that the average hydrophilicity over six or seven residues in a chain can be estimated. Those segments with the highest average hydrophilicity (lowest hydrophobicity) are then very likely to correspond with exposed surface regions and, hence, are potentially highly antigenic. Several relative hydrophilicity scales have been calculated and com-pared (e.g., Hopp, 1986; van Regenmortel et al., 1988), and generally have a success rate of 65–75% in locating continuous peptide frag-ments of high antigenicity. An example of the Hopp and Woods scale (1981) is plotted for myoglobin in figure 5-8a, together with the known epitopes, and reasonable agreement is observed. A special case of hydrophilicity in proteins is that of amphipathicity. Secondary structural regions are said to be amphipathic if one side of the struc-ture is appreciably more polar than the other. Amphipathic α-helices generally are observed on the surface of proteins, where one half of the helix, as viewed down its axis, contains hydrophilic residues on the protein exterior and the other half comprises hydrophobic residues which interact with the protein core. (This is particularly well represented using a helical wheel diagram—see figure 5-9.) As such, these regions may be exceptionally antigenic. When the hydrophilic residues are presented to the surface in a secondary structural motif, such as an amphipathic α-helix, the hydrophilicity scale described above will only predict a region of intermediate antigenicity, since about half of the residues in the primary sequence are hydrophobic. Hence, based on primary sequence alone, these potentially highly anti-genic sites would be missed. This error has been corrected in some

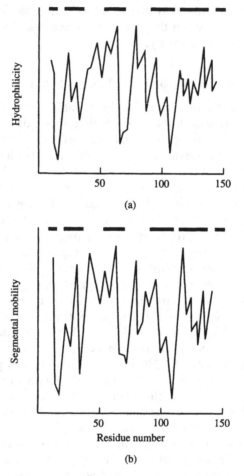

Figure 5-8 (a) Hydrophilicity profiles of
myoglobin calculated with the scale of Hopp
and Woods (1981) plotted against residue
number. The black bars above the profile
indicate regions of greatest antigenicity.
(b) Segmental mobility profile of myoglobin
calculated from the scale of Karplus and
Schulz (1985). Figures adapted from
Van Regenmortel et al. (1988).

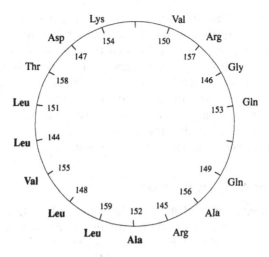

L^{144}RGDLQVLAQKVARTL^{159}

Figure 5-9 Primary sequence and helical wheel representation of the amphipathic α-helix in VP1 protein of Foot and Mouth Disease Virus. The residues constituting the hydrophobic section of the helix are shown in boldface type.

hydrophilicity scales, which incorporate prediction of amphipathic helices from the primary sequence (see, for example, De Lisi et al., 1986; Sette et al., 1986).

The antigenicity of amphipathic α-helices was demonstrated by Pfaff and coworkers (1982). They predicted the presence of seven helices in the foot and mouth disease virus (FMDV) viral protein VP1, and assessed the potential antigenicity of these on their amphipathic character. The fifth helix in the sequence appeared to have the most amphipathic nature (figure 5-9) and indeed a synthetic hexadecapeptide fragment generated from it induced antibodies that neutralized the intact virus. Interestingly, a standard Hopp and Woods (1981) hydrophilicity plot of this hexadecapeptide predicted it to possess limited antigenicity.

Another surface characteristic that can be predicted from primary structure is segmental mobility. The mobility of chains of amino acids can be observed by X-ray or NMR and allows one to calculate an atomic temperature factor (B) which is proportional to the mean square atomic displacement from the equilibrium crystal position. When calculated for the α-carbons of a protein chain and averaged over segments of six to ten residues, a plot of B versus residue number can be made. Since the segments at the surface will be relatively unhindered by the rest of the protein, these will generally have a high B factor and will correspond to relatively accessible antigenic regions. A B-factor scale for the 20 naturally occurring amino acids, parameterized from 31 protein X-ray structures from the Brookhaven Protein Data Bank, has been calculated (Karplus and Schultz, 1985) and an example profile for myoglobin is shown in figure 5-8b. Notably the regions of high segmental mobility correspond well with those predicted to be hydrophilic, and a fair agreement between predicted and experimentally observed antigenicity is seen.

Analogues of Antigenic Peptides

The real potential of synthetic peptides is the ability of the researcher to design novel analogues which can mimic or accentuate certain properties of the original biological sequence, based on conformational information from solution NMR and/or computer aided molecular modeling (Hruby et al., 1990a, 1991a). This approach is widely used in the synthesis of potent and selective analogues of peptide hormones, neurotransmitters, enzyme inhibitors, cytokins, and so on. It has only recently been applied to the conformational analysis of antigenic peptides in studying the molecular recognition processes between antibody and antigen.

In an early example in this area, Dyson and coworkers (1988) studied the immunodominant nonapeptide (YPYDVPDYA) from influenza virus hemagglutinin, noting that the first four residues of this epitope formed a β-turn in aqueous solution as indicated by ^{1}H NMR. Assuming that this conformational motif is important for antigenic recognition, analogues of the pentapeptide YPYDV were prepared, in which positions 3 and 4 were substituted for the other 19 natural amino acids. The β-turn population about Pro[2] and position 3 was assessed by 1- and 2-D NMR in water. In this study the nature of the residues in position 3 was found to be most influential in stabilizing the turn conformation, and in most cases a type II turn was favored. Consistent with this finding, YPGDV was observed to produce the highest turn population, as indicated by the low temperature dependence of the Asp[4] amide proton resonance (table 5-7). The Pro-Gly dipeptide is well known to favor a type II reverse turn (Gierasch et al., 1981). The dependence of antibody binding to this nonapeptide on the type of β-turn conformation about Pro[2]-Tyr[3] was further investigated by Hinds et al. (1991). In these studies, α-alkyl derivatives of Pro[2] were synthesized. In particular the γ-spirolactam (1) and α-Me Pro (2) (figure 5-10) analogues were studied for their ability to stabilize turns when compared to their binding affinity for two antisera for which YPYDV was proven to be the minimum antigenic sequence. The cyclic constraint in 1 forces the backbone to adopt

Table 5-7 Correlation of the relative turn formation of YPYDV and its analogues with the dissociation constant for equilibrium binding to antibody

Sequence	^{1}H NMR $\Delta\delta$ (-p.p.b)	K_d (nM)[a]
YPYDV	4.6	180
YPGDV	3.3	—
1	1.5	no binding observed
2	0.0	3.6

Turns about positions 2 and 3 were determined by observation of the relevant NOEs by ^{1}H NMR. The relative population of the turn is given here by the degree of intramolecular hydrogen bonding between residues 1 and 4 as indicated by the D^{4}NH chemical shift temperature dependence. The lower NH temperature dependencies indicate more stable turns. Hence equilibrium binding increases with turn stability, as long as the amide NH in position 3 remains intact.

[a] Measured against antibody DB 19/1 in Hinds et al. (1991).

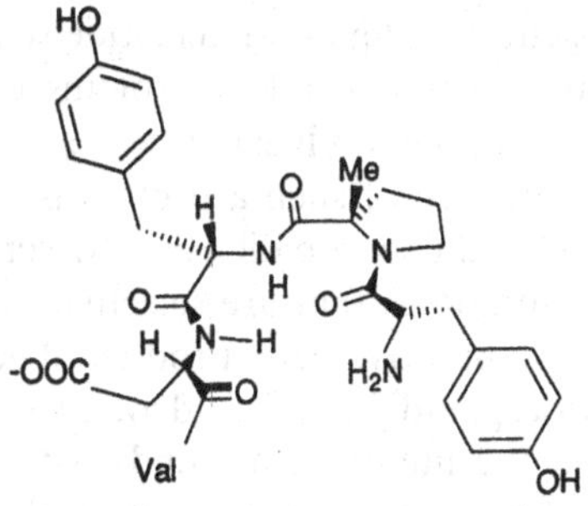

2

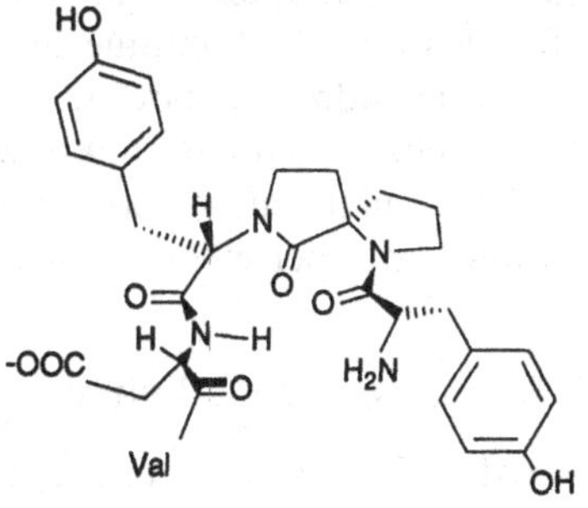

1

Figure 5-10 Analogues of YPYDV tested by Hinds et al. (1991). γ-Spirolactam (1) shown in the constrained type II β-turn conformation; α-MePro derivative (2) shown in a type I turn.

a type II reverse turn as indicated by NOE and the Asp^4 NH temperature dependence (table 5-7). The total lack of binding of 1 to either of the antibodies studied appeared to suggest that a type II turn is not favorable for recognition. However, YP[N-Me]YDV also showed no binding affinity which indicates, that the Tyr^3 NH is critical for formation of the antibody/antigen complex. α-Methylation of Pro^2 in 2 was shown to favor a type I β-turn by gas phase molecular modeling, while ^{1}H NMR indicated the presence of a highly stabilized turn (table 5-7) which was a mixture of both type I and II turns in solution. The enhanced binding of 2 relative to the native pentapeptide sequence was thus attributed to the increased stability of the turn about positions 2 and 3, and this epitope most probably binds in a type I β-turn conformation.

Antigenic Peptides in the Isolation and Evaluation of Gene Products

Rapid advances in gene cloning and nucleotide sequencing have made routine the location of genes encoding proteins which have not been isolated and for which a function has not been assigned. Since the non-processed primary sequence of such a protein is determined from the gene, antigenic regions can be predicted as described above. Hence, raising antipeptide antibodies that are selective for the native protein facilitates isolating the putative gene product. The use of this technique has become widespread in the field of protein purification and characterization and the literature abounds with many variations of its application, especially in the detection of viral proteins and protein intermediates.

This enormous literature undoubtedly will continue to expand and the use of peptides for this purpose continue to grow because of the many questions that can be addressed, both in terms of isolation and functional analysis of proteins which have unknown functions and the analysis of protein folding, maturation, post-translational modification, cellular distribution and other effects which can be examined.

Mimotopes, Peptide Libraries, and the Future

All of the examples cited so far have been based on protein/antigens of known sequence, and the synthesis of the peptide fragments limited to continuous epitopes. An exciting alternative protocol in the search for synthetic-peptide antigens which circumvents these restrictions was proposed by Geysen and coworkers in 1986. They introduced the concept of the "mimotope", which is defined as a molecule able to bind to the combining site of an antibody, but which is not necessarily identical to the native epitope that induced that antibody. A mimotope thus mimics the essential binding properties of the native epitope. Moreover, this definition may be extended to the mimicking of discontinuous epitopes, since a linear mimotope may, in theory, contain the major binding residues from more than one continuous epitope, in the correct topographical configuration for binding.

Applications of Antigenic Peptides: Development of Peptide Vaccines

Over the past two decades, chemists have entertained the use of peptide sequences as potential vaccines for a variety of diseases ranging from viral diseases as hepatitis (Gomara et al., 2000) and AIDS (Cruz et al., 2000; Liao et al., 2000); parasitology (Bueno et al., 2000), and cancer (Appella et al., 1995; Dakappagari et al., 2000; Eisenbach et al., 2000).

However, the development of potentially potent and clinically useful vaccines requires further refinement than the methodology described above. Recent developments in the deconvolution of combinatorial libraries to simpler, more-manageable sets, has aided in the identification of peptide-vaccine analogues with both broader recognition and enhanced potency. Strategies employing iterative and positional scanning techniques, have aided in the identification of more potent vaccines against HIV.

Work by Boehncke et al. (Boehncke et al., 1993; Berzofsky, 1995) addressed the question of increasing both the interaction of a peptide fragment T1 (KQIINMWQEVGKAMYA, residues 428–443 of HIV-1 IIIb gp160 envelope protein), a helper T-cell epitope (Cease et al., 1987), with the major histocompatibility complex (MHC) and thereby increasing immunogenicity. For antigens to be effective, they must be complexed to MHCs on the surface of antigen presenting peptides, and only then can T-cell receptors interact with these antigens to elicit the immune response (Ogasawara, 1999). Boehncke et al. hypothesized that MHC complexing may be effected by regions of the sequence that may seem non-essential but actually provide negative or adverse interactions. Hence, there is the possibility of improving binding to MHC and possibly increasing immunogenicity. Working in collaboration with Richard Houghten and employing positional scanning techniques (Dooley and Houghten, 2000), the investigators were able to identify substitutions for Glu in position 436 of the wild-type HIV sequence that resulted in increased binding potency to that of the wild-type sequence. Elimination of the anionic charge at E436 and substitution with an A or Q increased the potency of the immunogen. The increased potency secondary to elimination of the anionic charge was supported by the fact that E436D substitution led to no increase in potency. Immunogenicity studies with the substituted peptides also confirmed that these agents were more immunogenic at one and two orders of magnitude less peptide versus the wild-type sequence (Berzofsky, 1993).

A second question addressed with regards to more effective vaccines is the issue of MHC polymorphism. Because MHC molecules can vary in sequence in the population, response to epitopes or different antigenic determinants can vary. Thus, it was hypothesized that a slightly larger peptide that incorporated sequences for more than one determinant region (or multideterminant regions) could generate a sequence with overlapping antigenic regions that could form a variety of different MHC types. These "cluster peptides" sequences, known to induce helper T-cell activity, were synthesized and linked to Peptide 18 (P18), a cytotoxic T-cell epitope (Takahashi, et al., 1988) and also part of the V3 loop; a target for neutralizing antibodies (Goudsmit et al., 1988; Palker et al., 1988; Rusche et al., 1988). Results indicated that these multideterminant or cluster peptide constructs were effective at inducing neutralizing antibodies in four different strains of mice, each known to have different MHC types. Control P18 alone did not induce much of an antibody response and failed to induce neutralizing activity.

Thus, these studies demonstrated that the improvement in potency and immunogenicity of peptides could be achieved by new synthetic methodology and cluster design.

Other approaches have been attempted to elicit potent and effective vaccine design. For example, retroinverso (all-D or retroenantio) peptides, that is peptides constructed with all D-amino acids and with reverse sequences relative to the wild-type peptide antigenic sequence, have been designed as vaccines for foot and mouth disease (Muller and Briand, 1998) with varying amounts of effectiveness. Clinical trials are now in progress to test the viability of these vaccines. However, the verdict has not yet been finalized as to their clinical and therapeutic applicability. Perhaps by the publishing of the next edition of *Synthetic Peptides*, the results of these research labors will be known.

If one shifts the emphasis from searching for the native epitope of a protein to finding mimotopes of the native epitope, a totally different approach becomes apparent. Searching for a suitable mimotope requires no previous knowledge of the protein sequence or the native epitopes. Instead, having raised antibodies to the native protein, potent mimotopes may be located by binding to a library of randomized peptides incorporating all the 20 natural amino acids in all positions. However, for a library of all possible hexapeptides (the approximate minimum antigenic length), this would require the synthesis and testing of 20^6 (i.e., 64,000,000) peptides, which until a decade ago was totally impractical. Recently, however, two synthetic procedures for creating such peptide libraries have been presented which offer exciting possibilities for the future. Chronologically the first of these is "light-directed, spatially addressable parallel chemical synthesis" (Fodor et al., 1991). This combines N^α-photolabile protection with photolithography to allow simultaneous multiple-peptide synthesis on a glass microscope slide, using solid phase techniques. This procedure facilitates selective deprotection of the N-terminus by simply masking one region of the slide during light irradiation. Using a computer-controlled binary masking system, 2^n compounds can be synthesized in n steps on the same slide. Since the position of each peptide in the library is known (i.e., the synthesis is spatially addressed), potent mimotopes are easily found in a fluorescence immunoassay simply by the location of the fluorescing regions of the slide. Although this represents, in principle, an efficient procedure of preparing peptide libraries, it relies heavily on an expensive technology that so far not been applied to the preparation of large peptide libraries.

The second method for preparing peptide libraries depends entirely on standard solid-phase chemistry. Thus, it is more attractive to peptide chemists. This method relies on "split synthesis" for randomization in the library (Furka et al., 1991; Houghten et al., 1991; Lam et al., 1991, 1992; figure 5-11). A somewhat modified standard beaded solid support is split into an equal number of aliquots, and a single protected amino acid is coupled to each aliquot. After completing the couplings the aliquots are thoroughly mixed together and split again for deprotection and coupling of the next selection of protected amino acids. This process can be repeated as many times as required, and allows the synthesis of larger libraries than the previously described technique. Having completed the library, the mixed peptide resin will contain beads bearing only one peptide sequence each. Location of potent mimotopes is achieved using assay conditions in which the bead bearing the binding peptide will be stained. This bead then can be removed with the aid of a microscope and micromanipulator and the peptide on the isolated bead is then microsequenced.

Both of these synthetic technologies provide efficient ways to map antigenic peptides. Use of these approaches have greatly enhanced our understanding of the nature of discontinuous epitopes and the molecular mechanisms of antibody/antigen recognition. Possibly of more importance, peptide libraries have facilitated the discovery of mimotope peptide immunogens which can be used as vaccines. With these methodologies vaccine candidates can be directly mapped from monoclonal antibodies known to neutralize biologically important molecules or pathogens, and numerous other possibilities abound, including applications to oncogenic processes in cancer detection and treatment, autoimmune diseases, viral diseases, and related

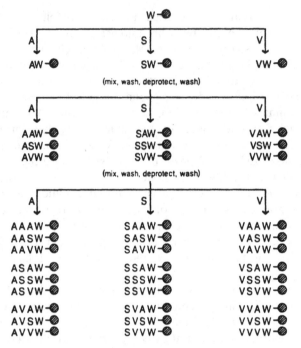

Figure 5-11 Scheme for random peptide synthesis using the split synthesis method for a random tripeptide containing S, A or V with a terminal tryptophan added. Diagram from Lam et al. (1991).

areas where the immune response is of critical importance. In all of these cases, peptide libraries have become invaluable tools in the elucidation of receptors of all types and in the general immunochemistry of the future.

Antisense Peptides

In the quest for understanding the laws of Nature, one of the interesting questions that has been asked is whether there may be a structural or functional relationship between the two peptides that could be coded for individually by each of the two opposing strands of a double-stranded DNA stretch. This new area has been referred to as antisense peptides. The term *antisense peptide* or *complementary peptide*, therefore, refers to a peptide sequence that could *hypothetically* be deduced from the nucleotide sequence that is complementary to the nucleotide sequence coding for a naturally occurring peptide, peptide hormone, neurotransmitter, protein, or the *sense peptide* or protein. Although not studied extensively, some antisense peptides have shown some selectivity in binding with their sense counterparts, and in certain cases have even exhibited surprising biological activity profiles, mediated through the biological receptors for the sense peptide. These initial promising results have not provided any great breakthrough. None the less a brief discussion will be given here.

It appears that, originally, this concept was proposed by Mekler (1969). Jones (1972) independently proposed and tested this concept by synthesizing peptides that were antisense to the C-terminal tetrapeptide of gastrins. Because the gene for this tetrapeptide was not known at that time, these authors had to synthesize various permutations and combinations of the possible antisense peptide sequences. It was observed that one of these peptides exhibited inhibitory activity in a gastric-acid-release assay. A theoretical treatment to this concept was later proposed by Biro (1981a,b,c). A thorough investigation into this concept was initiated by Blalock and coworkers (Bost et al., 1985a,b) in a series of studies aimed at investigating the interaction between the antisense and the sense peptide in solid-phase-binding assays. Later on, Chaiken and coworkers (Shai et al., 1987, 1989) developed high-performance affinity chromatography (HPAC) techniques to investigate the binding between antisense and sense peptides. These two techniques remain the primary methods to study the interactions between a pair of sense and antisense peptides. This section, greatly truncated from the original editon, is aimed at providing a brief overview of this concept, highlighting its potential applications in biomedical research as well as various issues concerning the nature of the binding exhibited by a set of antisense–sense peptides.

Antisense Peptides as Internal Images of the Sense Peptides

It was observed by Blalock and coworkers (Blalock and Smith, 1984; Blalock and Bost, 1986) that, in general, the codons for hydrophobic and hydrophilic amino acids on one strand of DNA may be complemented on the complementary strand, respectively, by the codons for hydrophilic and hydrophobic amino acids. Uncharged (slightly hydrophilic) amino acids on the other hand are complemented by uncharged amino acids (table 5-8). Interestingly, either $5' \rightarrow 3'$ or $3' \rightarrow 5'$ reading of the complementary nucleotide sequence displays the same pattern in hydropathicity of the coded amino acids sequence with respect to the same sense peptide sequence. This similarity in the hydropathic profile of the sequences obtained in either way stems from the fact that the middle base of the triplet codon specifies the hydropathic nature of the coded amino acid and this base remain second in both $5' \rightarrow 3'$ or $3' \rightarrow 5'$ readings. This remarkable pattern in genetic code, termed as

Table 5-8 List of amino acids, their codons, corresponding complementary codons, and antisense amino acids with their respective hydropathic score, as given by Kyte and Doolittle (1982)

Amino acid	Hydropathic value	Codon [5'→3']	Complementary codon [3'→5']	Amino acid encoded by the complementary codon			
				[5'→3'] direction	Hydropathic score	[3'→5'] direction	Hydropathic score
Arg	-4.5	AGA	UCU	Ser	-0.9	Ser	-0.9
		AGG	UCC	Pro	-1.6	Ser	-0.9
		CGA	GCU	Ser	-0.9	Ala	+1.8
		CGC	GCG	Ala	+1.8	Ala	+1.8
		CGG	GCC	Pro	-1.6	Ala	+1.8
		CGU	GCA	Thr	-0.7	Ala	+1.8
Lys	-3.9	AAA	UUU	Phe	+2.7	Phe	+2.7
		AAG	UUC	Leu	+3.7	Phe	+2.7
Asp	-3.5	GAC	CUG	Val	+4.2	Leu	+3.7
		GAU	CUA	Ile	+4.5	Leu	+3.7
Glu	-3.5	GAA	CUU	Phe	+2.7	Leu	+3.7
		GAG	CUC	Leu	+3.7	Leu	+3.7
Asn	-3.5	AAC	UUG	Val	+4.2	Leu	+3.7
		AAU	UUA	Ile	+4.5	Leu	+3.7
Gln	-3.5	CAA	GUU	Leu	+3.7	Val	+4.2
		CAG	GUC	Leu	+3.7	Val	+4.2
His	-3.2	CAC	GUG	Val	+4.2	Val	+4.2
		CAU	GUA	Met	+1.9	Val	+4.2
Pro	-1.6	CCA	GGU	Trp	-0.9	Gly	-0.4
		CCC	GGG	Gly	-0.4	Gly	-0.4
		CCG	GGC	Arg	-4.5	Gly	-0.4
		CCU	GGA	Arg	-4.5	Gly	-0.4
Tyr	-1.3	UAC	AUG	Val	+4.2	Met	+1.9
		UAU	AUA	Ile	+4.5	Ile	+4.5
Ser	-0.9	UCA	AGU	Stop	-	Ser	-0.9
		UCC	AGG	Gly	-0.4	Arg	-4.5
		UCG	AGC	Arg	-4.5	Ser	-0.9
		UCU	AGA	Arg	-4.5	Arg	-4.5
		AGC	UCG	Ala	+1.8	Ser	-0.9
		AGU	UCA	Thr	-0.7	Ser	-0.9
Trp	-0.9	UGG	ACC	Pro	-1.6	Thr	-0.7

	Codon	AA	Value	Codon	AA	Value
Thr −0.7	ACA	Cys	+2.5	UGU	Cys	+2.5
	ACC	Gly	−0.4	UGG	Trp	−0.9
	ACG	Arg	−4.5	UGC	Cys	+2.5
	ACU	Ser	−0.9	UGA	Stop	−
Gly −0.4	GGA	Ser	−0.9	CCU	Pro	−1.6
	GGC	Ala	+1.8	CCG	Pro	−1.6
	GGG	Pro	−1.6	CCC	Pro	−1.6
	GGU	Thr	−0.7	CCA	Pro	−1.6
Ala +1.8	GCA	Cys	+2.5	CGU	Arg	−4.5
	GCC	Gly	−0.4	CGG	Arg	−4.5
	GCG	Arg	−4.5	CGC	Arg	−4.5
	GCU	Ser	−0.9	CGA	Arg	−4.5
Met +1.9	AUG	His	−3.2	UAC	Tyr	−1.3
Cys +2.5	UGC	Ala	+1.8	ACG	Thr	−0.7
	UGU	Thr	−0.7	ACA	Thr	−0.7
Phe +2.7	UUC	Glu	−3.5	AAG	Lys	−3.9
	UUU	Lys	−3.9	AAA	Lys	−3.9
Leu +3.7	CUA	Stop	−	GAU	Asp	−3.5
	CUC	Glu	−3.5	GAG	Glu	−3.5
	CUU	Gln	−3.5	GAC	Asp	−3.5
	UUA	Lys	−3.9	GAA	Glu	−3.5
	UUG	Stop	−	AAU	Asn	−3.5
		Gln	−3.5	AAC	Asn	−3.5
Val +4.2	GUA	Tyr	−1.3	CAU	His	−3.2
	GUC	Asp	−3.5	CAG	Gln	−3.5
	GUG	His	−3.2	CAC	His	−3.2
	GUU	Asn	−3.5	CAA	Gln	−3.5
Ile +4.5	AUA	Tyr	−1.3	UAU	Tyr	−1.3
	AUC	Asp	−3.5	UAG	Stop	−
	AUU	Asn	−3.5	UAA	Stop	−

"anticomplementarity," appears to form the central theme of the concept of antisense peptides. Table 5-9 provides some early examples of peptide systems in which this concept has been applied. As is evident from these results, binding affinities from 10^{-3} M to 10^{-6} M often exists between a sense peptide and an antisense peptide. The ability of an antisense peptide to recognize its sense counterpart and their anticomplementarity has given rise to the suggestion that an antisense peptide may represent an "internal image" of its sense counterpart.

Blalock and coworkers further tested this suggestion by raising antibodies against "HTCA" an antisense peptide for adrenocorticotropin hormone (ACTH) and testing their ability to bind the receptors for ACTH (Bost and Blalock, 1986). It was found that these antibodies, being an image of an image, behaved like ACTH in the receptor-binding assays and were successfully used to purify an ACTH-binding protein or receptor. This concept of receptor purification has since been applied to the purification of binding proteins for ACTH (Bost and Blalock, 1986), Fibronectin (Brentani et al., 1988), vasopressin receptor (Abood et al., 1989), γ-endorphin (Carr et al., 1986), luteinizing hormone releasing hormone, LHRH (Mulchahey et al., 1986), and angiotensin-II (Elton et al., 1988a). It also has been shown that antibodies against a pair of complementary peptides exhibit idiotypic–anti-idiotypic relationships (Smith et al., 1987). The antibodies against synthetic ACTH(1–13) respectively bound the antibodies against "HTCA," the peptide antisense to ACTH.

The selective interactions discussed above gave rise to a suggestion that the DNA strand complementary to the strand coding a peptide hormone might be coding for its receptor. Sequence search analysis performed by Bost et al. (1985a) revealed that regions of complementarity exist between mRNAs for epidermal growth factor (EGF), transferrin (TF), interleukin-2 (IL-2) and their receptors, but that these complementary sequences were small (six amino acids) and were obtained upon $3' \rightarrow 5'$ reading of the nucleotide sequence which is not the usual transcriptional or translational direction during protein synthesis. Though other similar reports have appeared in the literature, the occurrence of such small complementary fragments in nature appears not to be common. Thus the significance of these observations is unclear.

There are many instances reported (and perhaps many others unreported) where no measurable interactions between antisense-sense peptide pairs have been observed (e.g. Rasmussen and Hesch 1987; Eberle and Huber 1991; De Gasparo et al., 1989; Guillemette et al. 1989). For example, Eberle and coworkers failed to raise antibodies against a peptide antisense to human ACTH. Such suggestions of the nonuniversality of this phenomenon, weak unmeasurable interactions, or difficulties in generating antiself antibodies by an animal

Table 5-9 Binding constants for various antisense–sense peptide interactions

Sense peptide	Antisense peptide	Reading direction	Dissociation constant (M)	Assay	Reference
SYSMEHFRWGKPVGKKRRPVKVYP (ACTH)	GVHLHRAPLLAHRLAPAEVFHGVR (HTCA)	5'→3'	0.3×10^{-3}	SPBA[a]	Blalock et al., 1986
			1.9×10^{-6}	SPBA	Bost et al., 1985
	RMRYLVKATPFGHPFFAAGHFHMG	3'→5'	0.3×10^{-9}	SPBA	Blalock et al., 1986
YGGFMTSEKSQTPLVTL (γ-Endorphin)	QRDKGRLALLGGHEPAV	5'→3'	0.2×10^{-3}	SPBA	Carr et al., 1986
QHWSYGLRPG (LHRH)	SRAQSIGPVL	5'→3'	$\sim 1 \times 10^{-4}$	Inhibition of a biological assay	Mulchahey et al., 1986 Also see Bost & Blalock, 1989
RPKPQQFFGLM (Substance P)	AGFGVVKKPNY	3'→5'	6×10^{-5}	SPBA	Bost & Blalock, 1989
KETAAAKFERQHMDSSTSAA (Bovine pancreatic RNase-S-peptide)	RGGSRSVHVLPLKLSRRSFL	5'→3'	4.0×10^{-7}	ZE-HPAC[b]	Shai et al., 1987 Shai et al., 1989
	SRSVHVLPLKLSRRSFL		3.5×10^{-6}		
	RGGSRSVHVLPLK		1.8×10^{-6}		
	VHVLPLKLSRRSFL		8.3×10^{-6}		
	PLKLSRRSFL		2.3×10^{-5}		
	RSVHVLPLK		$>5.9 \times 10^{-5}$		
	VHVLPLKLSRRSFL		1.2×10^{-5}		
	LFSRRSLKLPLVHVSRSGGR	N→C	8.0×10^{-7}		
	SRRSLKLPLVHVSRSGGR	Terminal inversion of the above	1.0×10^{-6}		
	RSLKLPLVHVSRSGGR		6.5×10^{-6}		
	KLPLVHVSRSGGR		2.3×10^{-5}		
	LVHVSRSGGR	Further substitution of underlined amino acids	$>5.9 \times 10^{-5}$		
	KESRRSLKLPLVHVSRSGGR		1.5×10^{-6}		
	EKSRRSLKLPLVHVSRSGGR		1.5×10^{-6}		
	EESRRSLKLPLVHVSRSGGR		3.7×10^{-6}		
	LFPRRPLKLPLVPVPRSGGR		1.2×10^{-5}		
FNLDAEAPAVLSG (Rat liver glycoprotein fragment)	AAQHGGGLGVQVE	5'→3' B.G.S.[c]	920×10^{-6}	ZE-HPAC	Omichinski et al., 1989
	AAQHGRGFGIQIE	5'→3' B.G.S.R.[d]	460×10^{-6}		
	TAKDSGSLSVKVE	Computer optimized sequences	460×10^{-6}		
	AGKDSGSFSVKVE		160×10^{-6}		
	PAKDSGSFSVKVK		59×10^{-6}		
	TAKDSWSFSVKVK		22×10^{-6}		
	AAKDSRGLGIEIE		40×10^{-6}		
	AGKYSWRLRIKIK		0.7×10^{-6}		

(continued)

Table 5-9 (*Continued*)

Sense peptide	Antisense peptide	Reading direction	Dissociation constant (M)	Assay	Reference
GSGSFGTVYKGKWHGDVAVK (c-RAF$_{356\text{-}373}$ sequence)	LYSNISMPLALVHSAKGARA SAKGARA	5'→3'	380×10^{-6} 1500×10^{-6}	ZE-HPAC	Fassina et al, 1989a
	All D-isomers of the above: lysnismplalvhsakgara sakgara FHGHISMPFAFVYSSKAAAA	Computer optimized sequence	380×10^{-6} 1500×10^{-6} 8×10^{-6}		
CYFQNCPRGGKR (Arg^8-Vasopressin-Gly-Lys-Arg)	SQLQVGHGPLAAPWAVLEVA PLAAPWAVLEVA SQLQVGHG	5'→3'	3.5×10^{-3} 5.9×10^{-6e} 7.3×10^{-3} $>30 \times 10^{-3}$	ZE-HPAC	Fassina et al, 1989b

[a] SPBA = solid phase binding assay.

[b] ZE-HPAC = zonal elution—high-performance affinity chromatography in which sense peptide has been immobilized on the column matrix.

[c] B.G.S. = best guess sequence.

[d] B.G.S.R. = best guess sequence for rabbit.

[e] Binding constant for soluble sense and immobilized antisense peptide.

(antisense- peptide-making antireceptor antibodies) have been made to explain these cases. It can, therefore, be concluded that at present the concept of antisense-sense peptides is unclear. The biological relevance, and the mechanistic aspects of the interactions exhibited by these peptide pairs remain the main issues to be understood.

Approaches for the Generation of Antisense Peptides

The generation of an antisense sequence is straightforward in cases where DNA sequence information on a sense peptide is available. A sequence of complementary bases is written opposite the nucleotide sequence of the sense peptide. The entire complementary or antisense base sequence is then read in both $5' \rightarrow 3'$ and $3' \rightarrow 5'$ direction and translated to the peptide sequence according to the genetic code. It has been observed that antisense peptides generated in both the natural $5' \rightarrow 3'$ reading of the nucleotide sequence frame as well as opposite $3' \rightarrow 5'$ direction exhibit the ability to selectively bind the sense peptide. As discussed elsewhere in this section, Table 5-8 describes the amino acids and their corresponding complementary or antisense amino acids when translated in both $5' \rightarrow 3'$ and $3' \rightarrow 5'$ direction.

As is evident from table 5-8, in the absence of knowledge of the nucleotide sequence of a sense peptide, one would be required to test all the various combinations of antisense amino acid sequences resulting due to the degeneracy of the genetic code. None the less, few approaches to generating antisense peptides in such instances have been proposed. One such approach makes a series of educated guesses based on the use of preferred codon usage tables (Hartfield and Rice, 1986; Granthan et al., 1980; Hartfield et al., 1979; Aota et al., 1988) which allow one to assess the probability of a particular codon to be used for each amino acid for a given peptide. In certain instances the frequency of encoding a particular antisense amino acid among a combination of antisense amino acids may also help in making a choice. For example, in $3' \rightarrow 5'$ reading, the arginine codon may give rise to six antisense codons, two of which code for serine and the remaining four for alanine. Alanine may, therefore, be preferred in designating amino acid antisense to arginine. Similarly, in the $5' \rightarrow 3'$ reading frame, arginine may serve as an antisense replacement for proline. The combined use of these approaches may help one to obtain a "best guess" or "consensus" antisense peptide sequence. Examples of this approach have been discussed in the generation of correct antisense peptides to Substance-P read in the $3' \rightarrow 5'$ direction (Bost and Blalock, 1989), and design of a 13-mer peptide antisense to a rat glycoprotein sequence (Omichinski et al., 1989). As is evident from table 5-8, certain anticodons either in $5' \rightarrow 3'$ or $3' \rightarrow 5'$ direction are stop codons. To overcome this problem in generating antisense peptides, Bost and Blalock (1989) have proposed substituting this codon by an amino acid that is suggested by the best-guess or consensus amino acid for that position. Similarly, Shai et al. (1989) replaced the half-Cys residues in a deduced antisense peptide by Ser residues.

Omichinski and coworkers (1989) and Fassina and coworkers (1989a) have successfully demonstrated another approach to the design of antisense peptides. Their approach stems from the observed hydropathy anticomplementarity in the sense-antisense amino acid pairs and consequent suggestion that a hydropathic complementarity may be responsible for the selective interactions between a pair of sense-antisense peptides. As described in table 5-8, hydrophobic and hydrophilic amino acids are, respectively, complemented by hydrophilic and hydrophobic amino acids. Uncharged (slightly hydrophilic) amino acids are complemented by uncharged amino acids. Under this approach the consensus or best-guess sequence obtained earlier or the actual antisense peptide sequence, if known, is subjected to further refinements in order to maximize the existing hydropathic complementarity. The amino acid substitution is made in the antisense peptide so as to maximize a defined "moving average hydropathic score" for it with respect to the sense peptide. In one example, these authors (Omichinski et al., 1989) demonstrated the use of this approach in refining a best-guess antisense sequence for a rat glycoprotein fragment (table 5-9). One of the optimized sequences, AGKYSWRLRIKIK, exhibited much higher binding affinity for the 13-mer sense peptide, FNLDAEAPAVLSG, than the previous best-guess sequence AAQHGGGLGVQVE. In the other example, Fassina and coworkers (1989a) have shown that an optimized sequence antisense to 356–375 peptide sequence of c-raf protein exhibited 47-fold higher binding affinity to immobilized $c\text{-}raf_{356-375}$ peptide than the actual $c\text{-}raf_{356-375}$ antisense peptide (table 5-9). Although not applied yet, these examples strongly suggest this approach, in principle, may allow a de novo design of antisense peptides that are highly selective to a given sense peptide.

Assays for Studying the Interactions Between an Antisense Peptide and Its Sense Counterpart

In general, two types of assays, solid-phase-binding assay and high performance affinity chromatography (HPAC), have been used to study antisense-sense peptide interactions. Both these assays are solid-phase assays designed to overcome the main difficulty one would face in applying the standard techniques of receptor–ligand interactions of high affinity ligands, where the unbound ligand is

removed either by filtration, or centrifugation, or selective adsorption
on an inert insoluble additive like charcoal.

Solid-Phase Binding Assay

Well-established techniques of enzyme-linked immunosorbent assay
(ELISA) were used by Blalock and coworkers (Bost et al., 1985) to
develop the binding assay for studying the interactions between a sense
and antisense peptide. Typically, multi-well microtiter plates are
coated with varying amounts of an antisense peptide (10–50 µg) or
another peptide or protein serving as control for 12–18 hrs. The
wells are then washed and incubated with a fixed concentration of
the radiolabeled sense peptide for 1–2 h. The plates are then washed
with appropriate buffer to remove the unbound radiolabeled sense
peptide and the amount of bound radiolabeled sense peptide
measured. In case antibodies against the sense peptide are available,
radiolabeled sense peptide may be substituted by "cold" sense peptide
and the amounts of the bound sense peptide estimated by the standard
techniques of ELISA using antibodies.

High-Performance Affinity Chromatography
(HPAC) Method

Chaiken and coworkers (Shai et al., 1987) have developed the methods
of high-performance affinity chromatography (HPAC) as a direct
powerful tool to study the interactions between immobilized-sense
peptide and a mobile-antisense peptide. The affinity matrix is prepared
by the standard techniques for reacting the sense peptide with the
activated matrix. Chaiken and coworkers (Shai et al., 1987, 1989)
have employed Accell 78 (Waters Chromatography Division,
Mileford, Mass.) as matrix. This is a silica-based support containing
a six-carbon spacer with terminal *N*-hydroxysuccinimide active ester
for immobilizing molecules through their amino functionalities. The
matrix is reacted with the peptide, then end-capped by aminoethanol
and packed in a glass column. An identical column containing blank
end capped matrix also is prepared. Alternatively, Sharma and coworkers (S. D. Sharma, F. Al-Obeidi, M. E. Hadley and V. J. Hruby,
unpublished data) have used a prepacked activated tresyl column
(SelectiSphere 10, Pierce Chemical Company, Rockford, Illinois) for
immobilizing [Nle$_4$]-α-melanotropin (α-MSH) a tridecapeptide
analogue of α-MSH. This was achieved by pumping a solution
containing a determined amount of α-melanotropin through the
column at pH 7.2. Spectrophotometric evaluations of the eluant
indicated the immobilization of the peptide on the column. The

column was subsequently end capped with aminoethanol to block all the remaining reactive tresyl groups.

The affinity column thus prepared is then used to study the interactions of antisense peptide with the immobilized sense peptide according to experimental and theoretical techniques. Typically, the extent of retardation of different concentrations of a mobile antisense peptide under binding conditions on this column can then be used to calculate equilibrium binding constant for mobile-antisense–immobile sense-peptide complex according to equation 1.

$$\frac{1}{V - V_0} = \frac{K_{M/P}}{[M]_T(V_0 - V_m)} + \frac{[P]}{[M]_T(V_0 - V_m)} \tag{1}$$

where V = experimental elution volume of mobile-antisense peptide, V_0 = unretarded elution volume, V_m = mobile-phase volume, $[M]_T$ = total concentration of matrix-bound sense peptide, $[P]$ = concentration of mobile-antisense peptide in the chromatographed zone, and $K_{M/P}$ = dissociation constant of the complex of mobile-antisense peptide (P) and matrix-bound sense peptide (M).

Alternatively to the above, a fixed concentration of antisense peptide can be made to compete on this column with different concentrations of soluble sense peptide. This affords the binding constants for mobile-antisense–mobile-sense peptide complex (equation 2). In both these methods, called "zonal elution methods," the antisense peptide is applied to the column as a fixed zone of different concentrations.

$$\frac{1}{V - V_0} = \frac{K_{M/P}}{[M]_T(V_0 - V_m)}$$
$$+ \frac{K_{M/P}[L]_T}{K_{L/P}[M]_T(V_0 - V_m)} \tag{2}$$

where $[L]_T$ = total concentration of competing mobile-sense peptide, $K_{L/P}$ = dissociation constant between mobile-antisense peptide and mobile-competing-sense peptide, and the rest of the parameters are as above.

In yet another method, referred to as the "continuous elution method," the solutions containing different concentrations of an antisense peptide are passed through the affinity column and the plateau of maximum absorbance (UV) is observed. The volume of the eluant at which the affinity matrix is half saturated is used to calculate the binding constant of the mobile-antisense–matrix-bound-sense peptide complex according to equation 3.

$$\frac{1}{V - V_0} = \frac{K_{M/P}}{M_T} + \frac{[1/M_T]}{[P]_0} \tag{3}$$

where V = volume at which the affinity matrix is half saturated, V_0 = void volume, $K_{M/P}$ = dissociation constant of complex of the immobilized-sense peptide and antisense peptide, M_T = total amount of immobilized-sense peptide, and $[P_0]$ = initial concentration of mobile antisense peptide.

Biological Activity of Antisense Peptides

In some cases, the antisense peptides have been found to exert biological responses through the biological receptors for the sense peptides. Al-Obeidi and coworkers (1990) assayed a series of antisense peptides corresponding to the β-melanotropin (β-MSH) sequence for melanotropic (MSH-like) activity in frog skin bioassay, and for melanin concentrating hormone (MCH-like) activity in frog skin and fish skin bioassays, respectively. Surprisingly, it was found that some of the antisense peptides exhibited either agonistic or antagonistic activity in these in vitro bioassays suggesting the interaction of these peptides with MSH or MCH receptors. Jones (1972) has reported that after administration of a peptide antisense to C-terminal tetrapeptide of gastrins in conscious dogs some inhibition of gastric juice volume and pepsin output was observed. Bost and Blalock (1989) have also observed a partial inhibition of stress-induced steroid production after in vivo injection of a peptide antisense to ACTH. It could be argued that these inhibitions may result either from the antagonistic action of the antisense peptide on the sense peptide receptors or as a result of interaction between administered antisense peptide and endogenous sense peptide, thereby leaving low levels of sense peptide to elicit its normal biological response.

Antibodies against antisense peptides also may exert biological responses when administered in whole animals. As shown by Görcs and coworkers (1986), intravenous administration of antiserum against a peptide antisense to human placental progonadotropin- releasing hormone, when injected in ovariectomized rats, caused a significant decrease in plasma levels of luteinizing hormone. Since the antibodies against an antisense peptide will behave as the sense peptide in recognition aspects (being an image of an image), this effect has been interpreted in terms of an interference of this antiserum with the actions of endogenous gonadotropin-releasing hormone. Immunocytochemical staining of receptors for gonadotropin-releasing hormone on female rat pituitary cells by this antiserum further supported this inference.

Potential Applications of Antisense Peptides

Though experiments with antisense peptides have shown interesting, even provocative, results from time to time, the early promise has not lead to a systematically useful application of peptide science to problems in biology and medicine, or to develop useful leads for drug candidates in medicinal chemistry. Furthermore, to our knowledge, there is no direct evidence that biological systems synthesize or use antisense peptides, and their applications to peptide hormones, neurotransmitters, enzyme substrates, immunology have not met with any dramatic success. Though this is an interesting area of research, its significance is still not established. It will be interesting to see what further developments may occur in this area.

Summary and Future Perspectives

In this overview of some possible applications of synthetic peptides to modern biology and medicine, we have tried to emphasize the need for both physical/chemical and biological considerations in peptide ligand design, and the application of modern synthetic chemistry including asymmetric syntheses and macrocyclic design and synthesis. Because macromolecular recognition and selectivity are essential features of most biological systems, it is critical that highly purified peptide ligands be utilized to examine these systems, whether the examination is in the form of a binding, in vitro, or in vivo bioassay, or whether it is in the form of a biophysical or biochemical measurement. It is also clear that peptides and their analogues, pseudo-peptides, and peptido-mimetics are likely to be the major drugs of the future. These findings raise many issues in need of further study. As already outlined herein, the discovery process for obtaining peptide ligands that are more or less selective for a particular receptor (or acceptor) type or subtype, a specific antibody or antigen, a specific enzyme, and so on, has been greatly accelerated in recent years by the tremendous advances in synthetic peptide chemistry, combinatorial chemistry which peptide chemists pioneered, and the development of multiple, high-throughput binding and bioassay systems. The following factors all provide powerful tools for future developments: The availability by synthesis of large (billions) peptide libraries that can be used to screen for binding and biologically relevant activities; the ability to design peptide derivatives and analogues with constrained conformations and topographies, and specific two- and three-dimensional properties; the utilization of modern synthetic chemistry to prepare novel amino acids, and designed peptides, pseudo-peptides, and peptide mimetics; the ability to design peptide analogues that are stable to biodegradation and have good bioavailability; the developments of computational methods using modern fast computers to examine the conformations, topographies, and dynamics of peptides. All will be critically important to future developments. Further work is needed in these areas, as well as greatly accelerated efforts to improve methods of peptide delivery in biological systems, to find methods for passing peptides and peptide mimetics through various membrane barriers, to develop further rapid-throughput multiple-assay systems, and to obtain more reliable methods for evaluating peptide conformations and dynamic properties. Most important, there is a need to find ways to facilitate the proper collaboration between chemists, physicists, biologists, and physicians that will be necessary to optimize progress in this area.

The prospects for important scientific advances and for their application to the treatment of disease are exceptionally good. The extent to which these prospects will be realized depends increasingly on comprehensive collaborations to solve these problems. This will require a considerable culture change in the way science is practiced and rewarded, especially in the interactions between chemists and biologists. We hope that this chapter will help facilitate this process.

List of Abbreviations

α-MSH: α-melanotropin, α-melanocyte stimulating hormone

ACTH: adrenocorticotropin hormone

CCK: cholecystokinin

Cp: β-cyclopentamethylene-β-mercaptopropionic acid

CTOP: D-Phe-Cys-Tyr-D-Trp-Orn-Thr-Pen-Thr-NH$_2$

DADLE: [D-Ala^2, D-Leu^5]enkephalin

Dep: β, β-diethyl-β-mercaptopropanoic acid

Dmp: β, β-dimethyl-β-mercaptopropanoic acid

DPDPE: c[D-Pen^2, D-Pen^5]enkephalin

DSLET: Tyr-D-Ser-Gly-Phe-Leu-Thr

DTT: dithiothreitol

EGF: epidermal growth factor

ELISA: enzyme-linked immunosorbant assay

FMDV: foot and mouth disease virus

GPI: guinea pig ileum

HPAC: high-performance affinity chromatography

IL-2: interleukin-2

MAP: multiple antigenic peptide

MCH: melanocyte concentrating hormone

Mpa: β-mercaptoacetic acid

MVD: mouse vas deference

NMR: nuclear magnetic

NOESY: nuclear Overhauser enhancement spectroscopy

OXT: oxytocin

Pap: p-azidophenylalanine

Pen: penicillamine, β, β-dimethyl cysteine

Pip: L-pipecolic acid

TASP: template assembled synthetic protein

TASP: template assembled synthetic protein

TCTAP: D-Tic-Cys-Tyr-D-Trp-Arg-Thr-Pen-Thr-NH^2

TF: transferrin

Thr-ol: Threonal, the reduced form of threonine, 2-amino-1,3-butanedio

Tic: 1,2,3,4-tetrahydroisoquinoline carboxylate

ACKNOWLEDGMENTS This work was supported by grants from the US Public Health Service and the National Institute of Drug Abuse. We thank Cheryl McKinley and especially Margie Colie for extraordinary help in typing and editing this manuscript. We especially thank Drs. Shubh D. Sharma, Nathan Collins and K. C. Russell who helped to write the earlier version of this chapter, much of which still remains in modified form in this chapter.

References

Abbenante, G., March, D. R., Bergman, D. A., Hunt, P. A., Garnham, B., Dancer, R. J., Martin, J. L., and Fairlie, D. P. (1995). Regioselective structural and functional mimicry of peptides. Design of hydrolytically-stable cyclic peptidomimetic inhibitors of HIV-1 protease. J. Am. Chem. Soc. 117:10220–10226.

Abood, L. G., Michael, G. J., Xin, L., and Knigge, K. M. (1989). Interaction of putative vasopressin receptor proteins of rat brain and bovine pituitary gland with an antibody against a nanopeptide encoded by the reverse message of the complementary mRNA to vasopressin. J. Receptor Res. 9:19–25.

Al-Obeidi, F., Hadley, M. E., Pettitt, M. B., and Hruby V. J. (1989a). Design of a new class of superpotent cyclic α-melanotropins based on quenched dynamic simulations. J. Am. Chem. Soc. 111:3413–3416.

Al-Obeidi, F. de L., Castrucci, A. M., Hadley, M. E., and Hruby, V. J. (1989b). Potent and prolonged acting cyclic lactam analogues of α-melanotropin: design based on molecular dynamics. J. Med. Chem. 32:2555–2561.

Al-Obeidi, F. A., Hruby, V. J., Sharma, S. D., Hadley, M. E., and Castrucci, A. M. (1990). Antisense peptides of melanocyte stimulating hormone (MSH): Surprising results. In Peptides: Chemistry, Structure and Biology, J. E. Rivier and G. R. Marshall, eds., Leiden, Escom, pp. 530–532.

Al-Obeidi, F., Hruby, V. J., and Sawyer, T. K. (1998). Peptide and peptidomimetic libraries: Molecular diversity and drug design. Mol. Biotech. 9:205–223.

Amit, A. G., Mariuzza, R. A., Phillips, S. E. V., and Paljak, R. J. (1986). Three-dimensional structure of an antigen–antibody complex at 2.8 Å resolution. Science 233:747–753.

Aota, S., Gojobori, T., Ishibashi, F., Marvyama, T., and Ikamvea, T. (1988). Codon usage tabulated from the gene bank genetic sequence data. Nucleic Acids Res. 16: Suppl. R315–391.

Appella, E., Loftus, D. J., Sakaguchi, K., Sette, A., and Celis, E. (1995). Synthetic antigenic peptides as a new strategy for immunotherapy of cancer. Biomed. Peptides, Proteins, Nucleic Acids 1(3):177–184.

Bankowski, K., Manning, M., Halder, J., and Sawyer, W. H. (1978). Design of potent antagonist of the vasopressor response to arginine-vasopressin. J. Med. Chem. 21:850–853.

Barlow, D. J., Edwards, M. S., and Thornton, J. M. (1986). Continuous and discontinuous protein antigenic determinants. Nature 322:744–751.

Bates, R. B., Gin, S. L., Hassen, H. A., Hruby, V. J., Janda, K. M., Kriek, G. R., Michaud, J-P., and Vine, D. B. (1984). Synthesis of cyclo-N-methyl-L-Tyr-N-Methyl-L-Tyr-D-Ala-L-Ala-O,N-dimethyl-L-Tyr-L-Ala, a cyclic hexapeptide related to the antitumor agent deoxybouvardin. Hetero-cycles 22:785–790.

Bauer, W., Boiner, U., Doepfner, W., Haller, R., Hugeunin, R., Marbach, P., Petcher, T. J., and Pless, J. (1983) Structure–activity relationships of highly potent octapeptide analogues of somatostatin. In *Peptides 1982*, K. Blaha and P. Malon, eds., Berlin, de Gruyter.

Bayley, H. (1983). *Photogenerated Reagents in Biochemistry and Molecular Biology*, Amsterdam, Elsevier.

Bayley, H., and Knowles, J. R. (1978a). Photogenerated reagents for membrane labeling. 1. Phenylnitrene formed within the lipid bilayer. Biochemistry 17: 2414–2419.

Bayley, H., and Knowles, J. R. (1978b). Photogenerated reagents for membrane labeling. 2. Phenylcarbene and adamentylidene formed within the lipid bilayer. Biochemistry 17:2420–2423.

Berzofsky, J. A. (1993). Epitope selection and design of synthetic vaccines: Molecular approaches to enhancing immunogenicity and cross-reactivity of engineered vaccines. Ann. N.Y. Acad. Sci. 690:256–264.

Berzofsky, J. A. (1995). Designing peptide vaccines to broaden recognition and enhance potency. Ann. N.Y. Acad. Sci. 574:161–168.

Bilsky, E. J., Qian, X., Hruby, V. J., and Porreca, F. (2000). Antinociceptive activities of [β-methyl-2',6'-dimethyltyrosine1]-substituted cyclic-[DPen^2, DPen^5]enkephalin and [D-Ala^2, Asp^4]deltorphin analogues. J. Pharmacol. Exp. Ther. 293:151–158.

Biro, J. (1981a). Comparative analysis of specificity in protein–protein interactions. Part I: A theoretical and mathematical approach to specificity in protein–protein interactions. Med. Hypotheses 7:969–979.

Biro, J. (1981b). Comparative analysis of specificity in protein–protein interactions. Part II: The complementary coding of some proteins as the possible source of specificity in protein–protein interactions. Med. Hypotheses 7:981–993.

Biro, J. (1981c). Comparative analysis of specificity in protein–protein interactions. Part III: Models of the gene expression based on the sequential complementary coding of some pituitary proteins. Med. Hypotheses 7:995–1007.

Blalock, J. E., and Smith, E. M. (1984). Hydropathic anti-complementarity of amino acids based on the genetic code. Biochem. Biophys. Res. Commun. 121:203–207.

Blalock, J. E., and Bost, K. L. (1986). Binding of peptides that are specified by complementary RNAs. Biochem. J. 234:679–683.

Bodanszky, M., and du Vingeaud, V. (1959). Synthesis of a biologically active analog of oxytocin, with phenylalanine replacing tyrosine. J. Am. Chem. Soc. 81:6072–6075.

Boehncke, W.-H., Takeshita, T., Pendleton, C. D., Sadegh-Nasseri, S., Racioppi, L., Houghten, R. A., Berzofsky, J. A., and Germain, R. N. (1993). The importance of dominant negative effects of amino acids side chain substitution in peptide-MHC molecule-restricted murine cytotoxic T lymphocytes. J. Immunol. 150:331–341.

Böhm, H.-J., and Klebe, G. (1996). What can we learn from molecular recognition in protein–ligand complexes for the design of new drugs. Angew. Chem. Int. Ed. Engl. 35:2589–2614.

Bost, K. L., and Blalock, J. E. (1986). Molecular characterization of a corticotropin (ACTH) receptor. Mol. Cell. Endocrinol. 44:1–9.

Bost, K. L., and Blalock, J. E. (1989). Preparation and use of complementary peptides. Methods Enzymol. 168:16–28.

Bost, K. L., Smith, E. M., and Blalock, J. E. (1985a). Regions of complementarity between the messenger RNAs for epidermal growth factor, tranferrin, interleukin-2 and their respective receptors. Biochem. Biophys. Res. Commun. 128:1373–1380.

Bost, K. L., Smith, F. M., and Blalock, J. E. (1985b). Similarity between the corticotropin (ACTH) receptor and a peptide encoded by an RNA that is complementary to ACTH mRNA. Proc. Natl. Acad. Sci. USA 82:1372–1375.

Brentani, R. R., Ribeiro, S. F., Potocnjak, P., Pasqualini, R., Lopes, J. D., and Nakaie, C. R. (1988). Characterization of the cellular receptor for fibronectin through a hydropathic complementarity approach. Proc. Natl. Acad. Sci. USA 85:364–367.

Brunner, J., and Richards, F. M. (1980). Analysis of membranes photolabeled with lipid analogues. Reaction of phospholipids containing a disulfide group and a nitrene or carbene precursor with lipids and with gramicidin A. J. Biol. Chem. 255:3319–3329.

Brunner, J., Senn, H., and Richards, F. M. (1980). 3-Trifluoromethyl-3-phenyl diazirine. A new carbene containing group for photolabeling reagents. J. Biol. Chem. 255:3313–3318.

Budker, V. G., Knorre, D. G., Kravchenko, V. V., Lavrik, O. I., Nevinsky, G. A., and Teplova, M. (1974). Photoaffinity reagents for modifications of amino acyl-tRNA synthetases. FEBS Lett. 49:159-162.

Bueno, E. C., Vaz, A. J. Machado, L. D., Livramento, J. A., and Mielle, S. R. (2000). Specific *Taenia crassiceps* and *Taenia solium* antigenic peptides for neurocysticercosis immunodiagnosis using serum samples. J. Clin. Microbiol. 38(1):146–151.

Burgermeister, W., Nassal, M., Wieland, T., and Helmreich, E. J. M. (1983). A carbene-generating photoaffinity probe for beta-adrenergic receptors. Biochem. Biophys. Acta 729:219–228.

Carr, D. J. J., Bost, K. L., and Blalock, J. E. (1986). An antibody to a peptide specified by an RNA that is complementary to γ-endorphin mRNA recognizes an opiate receptor. J. Neuroimmunol. 12:329–337.

Cease, K. B., Margalit, H., Cornette, J. L., Putney, S. D., Robey, W. G., Ouyang, C., Streicher, H. Z., Fischinger, P. J., Gallo, R. C., Delisi, C., and Berzofsky, J. A. (1987). Helper T cell antigenic site identification in the AIDS virus gp120 envelope protein and induction of immunity in mice to the native protein using a 16-residue synthetic peptide. Proc. Natl. Acad. Sci. USA 84:4249–4253.

Chan, W. Y., and Kelly, N. (1969). A pharmacologic analysis on the significance of the chemical functional groups of oxytocin on its oxytocic activity and on the effect of magnesium on the in vitro and in vivo oxytocic activity of neurohypophyseal hormones. J. Pharmacol. Exp. Ther. 156:150–158.

Charpentier, B., Durieux, C., Menant, E., and Roques, B. P. (1987). Investigation of peripheral cholecystokinin receptor homogeneity by cyclic and related linear analogues of CCK_{26-33}: Synthesis and biological properties. J. Med. Chem. 30:962–968.

Chowdhry, V., and Westheimer, F. H. (1978). p-Toluenesulfonyl diazo-acetates: Reagents for photoaffinity labeling. Bioorg. Chem. 7:189–205.

Chowdhry, V., and Westheimer, F. H. (1979). Photoaffinity labeling of biological systems. Annu. Rev. Biochem. 48:293–325.

Chowdhry, V., Vaughan, R., and Westheimer, F. H. (1976). 2-Diazo-3,3,3, trifluoropropionyl chloride: Reagents for photoaffinity labeling. Proc. Natl. Acad. Sci. USA 73:1406–1408.

Collins, N., Flippen-Anderson, J. L., Haaseth, R. C., Deschamps, J. R., George, C., Kövér, K., and Hruby, V. J. (1996). Conformational determinants of agonist versus antagonist properties of [D-Pen^2,D-Pen^5]enkephalin (DPDPE) analogs at opioid receptors. Comparison of X-ray crystallographic structure, solution 1H NMR data, and molecular dynamic simulations of [L-Ala^2]DPDPE and [D-Ala^3]DPDPE. J. Am. Chem. Soc. 118:2143–2152.

Coy, D. H., Horvath, A., Nekola, M. V., Coy, F. J., Erchogyi, J., and Schally, A. V. (1982). Peptide antagonists of LH-RH: Large increases in antiovulatory activities produced by basic D-amino acids in the six position. Endocrinology 110:1445–1447.

Cruz, L. J., Quintana, D., Iglesias, E., Garcia, Y., Huerta, V., Garay, H. E., Duarte, C., and Reges, O. (2000). Immunogenicity comparison of a multiantigenic peptide bearing V3 sequences of the human immunodeficiency virus type 1 with TAB9 protein in mice. J. Peptide Sci. 6(5):217–224.

Cwirla, S. E., Peters, E. A., Barrett, R. W., and Dower, W. J. (1990). Peptides on phage: A vast library of peptides for identifying ligands. Proc. Natl. Acad. Sci. USA 87:6378–6382.

Dakappagari, N. K., Douglas, D. B., Triozzi, P. L., Stevens, V. C., and Kaumaya, T. P. (2000). Prevention of mammary tumors with a chimeric Her-2 B-cell epitope peptide vaccine. Cancer Res. 60(14):3782–3789.

Das, M., and Fox, C. F. (1978). Molecular mechanism of nitrogen action: Processing of receptor induced by epidermal growth factor. Proc. Natl. Acad. Sci. USA 75:2644–2648.

Das, M., Miyakawa, T., Fox, C. F., Pruss, R. M., Aharonov, A., and Herschman, H. R. (1977). Specific radiolabeling of a cell surface receptor for epidermal growth factor. Proc. Natl. Acad. Sci. USA 74: 2790–2794.

De Gasparo, M., Whitebread, S., Einsle, K., and Heussen, C. (1989). Are the antibodies to a peptide complementary to angiotensin II useful to isolate the angiotensin II receptor? Biochem. J. 261:310–311.

De Grado, W. F. (1988). Design of peptides and proteins. Adv. Protein Chem. 39:51–124.

De Lisi, C., and Berzofsky, J. A. (1986). T-cell antigenic sites tend to be amphipathic structures. Proc. Natl. Acad. Sci. USA 82:7048–7059.

Delay-Goyet, P., Sequin, C., Gacel, G., and Roques, B. P. (1988). [^{3}H][D-Ser^2 (O-tert-butyl), Leu^5]enkephalyl-Thr^6 and [D-Ser^2 (O-tert-butyl), Leu^5]enkephalyl-Thr^6 (O-tert-butyl). Two new enkephalin analogs with both a good selectivity and a high affinity toward δ-opioid binding sites. J. Biol. Chem. 263:4124–4130.

Demoliou, C. D., and Epand, R. M. (1980). Synthesis and characterisation of a heterobifunctional photoaffinity reagent for modification of tryptophan residues and its application to the preparation of a photoreactive glucagon derivative. Biochemistry 19:539.

Dennis, S., Wallace, A., Hofsteenge, J., and Stone, S. R. (1990). Use of fragments of hirudin to investigate thrombin-hirudin interactions. Eur. J. Biochem. 188:61–66.

Devlin, J. J., Panganiban, L. C., and Devlin, P. E. (1990). Random peptide libraries: A source of specific protein binding molecules. Science 249:404–406.

Dooley, C. T., and Houghten, R. A. (2000). New opioid peptides, peptidomimetics, and heterocyclic compounds from combinatorial libraries. Biopolymers: Peptide Sci. 51(6):91–110.

Dutta, A. D. (1991). Design and therapeutic potential of peptides. Adv. Drug Res. 21:145–286.

Dyckes, D. F., Nestor, Jr., J. J., Jr., Ferger, M. F., and du Vigneaud, V. (1974). [1-β-Mercapto-β, β-diethylpropanoic acid]-8-lysine-vasopressin, A potent inhibitor of 8-lysine-vasopressin and oxytocin. J. Med. Chem. 17:250–252.

Dyson, H. J., Rance, M., Houghten, R. A., Lerner, R. A., and Wright, P. E. (1988). Folding of immunogenic peptide fragments of proteins in water solution. J. Mol. Biol. 201:161–201.

Eberle, A. N. (1983). Photoaffinity labeling of peptide hormone receptors. J. Receptor Res. 3:313–316.

Eberle, A. N. (1984). Photoaffinity labeling of MSH receptors in Anolis melanophores: Irradiation techniques and MSH photolabels for irreversible stimulation. J. Receptor Res. 4:315–329.

Eberle, A. N., and de Graan, P. N. E. (1985). General principles for photoaffinity labeling of peptide hormone receptors. Methods Enzymol. 109:129–156.

Eberle, A. N., and Huber, M. (1991). Antisense peptides: Tools for receptor isolation? Lack of antisense MSH and ACTH to interact with their sense peptides and to induce receptor-specific antibodies. J. Receptor Res. 11:13–43.

Eberle, A. N., and Schwyzer, R. (1976). Syntheses von [D-alanin^1, 4'-azido-3',5'-ditritio-L-phenylalanine^2, norvalin^4]-α-melanotropins als "Photoaffinits-probe" für Hormon-Rezeptor-Wechselwirkungen. Helv. Chim. Acta 59:2421–2431.

Eberle, A. N., Hhbscher W., and Schwyzer, R. (1977). Synthesis von radioaktiv markierten Bromacetyl- und Diazoacetyl-α-Melanotropin-Derivaten zum Studium von kovalenten Hormone-Makromolekhl-Komplexen. Helv. Chim. Acta 60:2895–2910.

Eisenbach, L., Bar-Haim, E., and El-Shami, K. (2000). Antitumor vaccination using peptide based vaccines. Immunol. Lett. 74(1):27–34.

Elton, T. S., Dion, L. D., Bost, K. L., Oparil, S., and Blalock, J. E. (1988a). Purification of an angiotensin II binding protein by using antibodies to a peptide encoded by angiotensin II complementary RNA. Proc. Natl. Acad. Sci. USA 85:2518–2522.

Elton, T. S., Oparil, S., and Blalock, J. E. (1988b). The use of complementary peptides in the purification of an angiotensin II binding proteins. J. Hypertension 6:S404–S407.

Emmett, J., ed. (1990). *Comprehensive Medicinal Chemistry*, Vol. 3, Oxford, Pergamon Press.

Escher, E., and Guillemette, G. (1978). Photoaffinity labeling of the angiotensin II receptor. 3. Receptor inactivation with photolabile hormone analogue. J. Med. Chem. 22:1047–1050.

Escher, E., and Schwyzer, R. (1974). *p*-Nitrophenylalanine, *p*-azidophenylalanine, *m*-azidophenylalanine, and *o*-nitro-*p*-azido-phenylalanine as photoaffinity labels. FEBS Lett. 46:347–350.

Escher, E. H. F., Robert, H., and Guillemette, G. (1979). 4-Azidoaniline, a versatile protein and peptide modifying agent for photo affinity labeling. Helv. Chim. Acta 62:1217–1222.

Escher, E., Laczko, E., Guillemette, G., and Regoli, D. (1981). Biological activities of photoaffinity labeling analogues of kinins and their irreversible effects on kinin receptors. J. Med. Chem. 24:1409–1413.

Fanning, D. W., Smith, J. A., Rose, G. D. (1986). Molecular cartography of globular proteins with application to antigenic sites. Biopolymers 25:863–882.

Farlie, D. P., Abbenante, G., and March, D. R. (1995). Macrocyclic peptidomimetics-forcing peptides into bioactive conformations. Curr. Med. Chem. 2:654–686.

Fassina, G., Roller, P. P., Olson, A. D., Thorgeirsson, S. S., and Omichinski, J. G. (1989a). Recognition properties of peptides hydropathically complementary to residues 356–375 of the c-raf protein. J. Biol. Chem. 264:11252–11257.

Fassina, G., Zamai, M., Brigham-Burke, M., and Chaiken, I. M. (1989b). Recognition properties of antisense peptides to Arg[8]-vasopressin/bovine neurophysin II biosynthetic precursor sequence. Biochemistry 28:8811–8818.

Feng, Y., Wang, Z., Jin, S., and Burgess, K. (1998). S_NAr cyclizations to form cyclic peptidomimetics of β-turns. J. Am. Chem. Soc. 120:10768–10769.

Ferrier, B. M., Jarvis, D., and du Vigneaud, V. (1965). Deamino-oxytocin. Its isolation by partition chromatography on sephadex and crystallization from water and its biological activities. J. Biol. Chem. 240:4264–4266.

Filatova, M. P., Kri, N. A., Komarova, O. M., Orfkchovich, V. N., Reiss, S., Liepinya, I. T., and Nikiforovich, G. V. (1986). Synthesis and studies of conformationally restricted analogues of peptide inhibitors of the angiotensin-converting enzyme. Bioorg. Khim. 12:59–70.

Finnegan, A., Smith, M. A., Smith, J. A., Berzofsky, J., Sachs, D. H., and Hades, R. J. (1986). The T-cell reportoire for recognition of a phylogenetically distant protein antigen: Peptide specificity and MHC restriction of staphylococcal nuclease-specific T cell clones. J. Exp. Med. 164:897–903.

Fleet, G. W. J., Porter, R. R., and Knowles, J. R. (1969). Affinity labeling of antibodies with aryl nitrene as reactive group. Nature 224:511–512.

Flippen-Anderson, J. L., Hruby, V. J., Collins, N., George, C., and Cudney, B. (1994). X-ray structure of [D-Pen^2,D-Pen^5]enkephalin, A highly potent, delta opioid receptor selective compound: Comparisons with proposed solution conformations. J. Am. Chem. Soc. 116:7523–7531.

Fodor, S. P., Read J. L., Pirrung, U. C., Stryer, l., Lu, A. T., Salas, D. (1991). Light directed, spatially addressable parallel chemical synthesis. Science 251:767–771.

Fournie-Zaluski, M.-C., Gacel, G. W., Maigret, B., Premilat, S., and Roques, B. P. (1981). Structural requirements for specific recognition of α or β opiate receptors. Mol. Pharmacol. 20:484–491.

Freidinger, R. M., and Veber D. F. (1984). Design of novel cyclic hexapeptide somatostatin analogs from a model of the bioactive conformation. In *Conformationally Directed Drug Design: Peptides and Nucleic Acids as Templates and Targets,* J. A. Vida and M. Gordon, eds., ACS Symposium Series 251, Washington, D.C., American Chemical Society.

Freidinger, R. M., Colton, C. D., Randall, W. C., Pitzenberger, S. M., Veber, D. F., Saperstein, R., Brady, E. J., and Arison, B. H. (1985). A cyclic hexapeptide LH-RH antagonist. In *Peptides: Structure and Function*, C. M. Deber, V. J. Hruby, and K. D. Kopple, eds. Rockford, Ill., Pierce Chemical Co.

Furka, A., Sebestyen, F., Asgedom, M., and Debo, G. (1991). General method for rapid synthesis of multicomponent peptide mixtures. Int. J. Peptide Protein Res. 37:487–493.

Gacel, G. W., Fournie-Zaluski, M.-C., and Roques, B. P. (1980). Tyr-D-Ser-Gly-Phe-Leu-Thr, a highly preferential ligand for δ opiate receptors. FEBS Lett. 18(2):245–147.

Gacel, G. W., Daugé, V., Breuze, P., Delay-Goyet, P., and Roques, B. P. (1988b). Development of conformationally constrained linear peptides exhibiting a high affinity and pronounced selectivity for δ opioid receptors. J. Med. Chem. 31:1891–1897.

Gacel, G. W., Zajac, J. M., Delay-Goyet, P., Daugé, V., and Roques, B. P. (1988a). Investigation of the structural parameters involved in the μ and δ opioid receptor discrimination of linear enkephalin-related peptides. J. Med. Chem. 31:374–383.

Galardy, R. E., Craig, L. C., Jamieson, J. D., and Printz, M. P. (1974). Photoaffinity labeling of peptide hormone binding sites. J. Biol. Chem. 249:3510–3518.

Geyer, A., Muller, G., and Kessler, H. (1994). Conformational analysis of a cyclic RGD peptide containing a PSI[CH$_2$-NH] bond: A positional shift in backbone structure caused by a single dipeptide mimetic. J. Am. Chem. Soc. 116:7735–7743.

Geysen, H. M., Meloen, R. H., and Bartelling S. J. (1984). Use of peptide synthesis to probe viral antigens for epitopes to a resolution of a single amino acid. Proc. Natl. Acad. Sci. USA 81:3998–4002.

Geysen, H. M., Rodda, S. J., and Mason, T. J. (1986). A priori delineation of a peptide which mimics a discontinuous antigenic determinant. Mol. Immunol. 23:709–715.

Gierasch, L. M., Deber, C. M., Madison, V., Niu, C. H., and Blout, E. R. (1981). Conformations of (X-L-Pro-Y)$_2$ cyclic hexapeptides. Prefered β-turn conformers and implications for β-turns in proteins. Biochemistry 20:4730–4736.

Gomara, M. J., Riedemann, S., Vega, I., Ibarra, H., Ercilla, G., and Haro, I. (2000). Use of linear and multiple antigenic peptides in the immunodiagnosis of acute Hepatitis A virus infection. J. Immunol. Methods 234(1–2):23–34.

Goodman, M., and Ro, S. (1994). Peptidomimetics for drug design. In *Burger's Medicinal Chemistry and Drug Discovery*, Vol. I, M. E. Wolff, ed., New York, Wiley-Interscience, pp. 803–861.

Goudsmit, J., Debouck, C., Meloen, R. H., Smit, L., Bakker, M., Asher, D. M., Wolff, A. V., Gibbs, C. J., and Gajdusek, D. C. (1988). Human immunodeficiency virus type 1 neutralization epitope with conserved architecture elicits early type-specific antibodies in experimentally infected chimpanzees. Proc. Natl. Acad. Sci. USA 85:4478–4482.

Görcs, T. J., Gottschall, P. E., Coy, D. H., and Arimura, A. (1986). Possible recognition of the GnRH receptor by an antiserum against a peptide encoded by nucleotide sequence complementary to mRNA of a GnRH precursor peptide. Peptides 7:1137–1145.

Granthan, R., Gautier, C., and Gouy, M. (1980). Codon frequencies in 119 individual genes confirm consistent choices of degenerate bases according to genome type. Nucleic Acids Res. 8:1893–1912.

Grieco, P., Balse, P. M., Weinberg, D., MacNeil, T., and Hruby, V. J. (2000). D-Amino acid scan of γ-melanocyte-stimulating hormone: Importance of Trp[8] on human MC3 receptor selectivity. J. Med. Chem. 43:4998–5002.

Guillemette, G., Boulay, G., Gaynon, S., Bosse, R., and Escher, E. (1989). The peptide encoded by angiotensin II complementary RNA does not interfere with antiotensin II action. Biochem. J. 261:309.

Guire, P., Fligen, D., and Hodgson, J. (1977). Photochemical coupling of enzymes to mammalian cells. Pharmacol. Res. Commun. 9:131–141.

Hartfield, D., and Rice, M. (1986). Amino acyl-*t*RNA (anticodon): Codon adaptation in human and rabbit reticulocytes. Biochem. Int. 13:835–842.

Hartfield, D., Mathew, C. R., and Rice, M. (1979). Aminoacyl-transfer RNA populations in mammalian cells: Chromatographic profiles and patterns of codon recognition. Biochim. Biophys. Acta 564:414–423.

Hawiger, J. (1995). Platelets in thrombosis and rethrombosis. In *Molecular Cardiovascular Medicine*, E. Haber, ed., New York, Scientific American, Inc., pp. 157–175.

Henkin, J. (1977). Photolabeling reagents for thiol enzymes. Studies on rabbit muscle creatine kinase. J. Biol. Chem. 252:4293–4297.

Hill, P. S., Smith, D. D., Slaninova, J., and Hruby, V. J. (1990). Bicyclization of a weak oxytocin agonist produces a highly potent oxytocin antagonist. J. Am. Chem. Soc. 112:3110–3113.

Hinds, M. G., Welsh, J. H., Biennoud, D. M., Fisher, J., Glennie, M. G., Richards. N. J. R., Turner, D. L., and Robinson, J. A. (1991). Synthesis, conformational properties and antibody recognition of peptides containing β-turn mimetics based on α-alkyl proline derivatives. J. Med. Chem. 34:1777–1789.

Hirschmann, R. (1991). Medicinal chemistry in the golden age of biology—Lessons from steroid and peptide research. Angew. Chem. Int. Ed. Engl. 30:1278–1301.

Hixson, S. H., and Hixson, S. S. (1975). p-Azidophenacyl bromide, a versatile photolabile bifunctional reagent. Reaction with glyceraldehyde-3-phosphate dehydrogenase. Biochemistry 14:4251–4254.

Hopp, T. P. (1986). Protein surface analysis. Methods for identifying antigenic determinants and other interaction sites. J. Immunol. Methods 88:1–10.

Hopp, T. P., and Woods, K. R. (1981). Prediction of protein antigenic determinants from amino acid sequences. Proc. Natl. Acad. Sci. USA 78:3824–3839.

Houghten, R. A., Pinilla, C., Blondelle, S. E., Appel, J. R., Dooley, C. T., and Cuervo, J. H. (1991). Generation and use of synthetic peptide combinatorial libraries for basic research and drug discovery. Nature 354:84–86.

Hruby, V. J. (1981a). Structure and conformation related to the activity of peptide hormones. In Perspectives in Peptide Chemistry, A. Eberle, R. Geiger and T. Wieland, eds., Basel, S. Karger.

Hruby, V. J. (1981b). Relation of conformation to biological activity in oxytocin, vasopressin, and their analogues. In Topics in Molecular Pharmacology, Vol. 1, A.S.V. Burgen and G.C.K. Roberts, eds., Amsterdam, Elsevier/North-Holland Biomedical Press.

Hruby, V. J. (1982). Conformational restrictions of biologically active peptides via amino acid side chain groups. Life Sciences 31:189–199.

Hruby, V. J. (1996). Synthesis of peptide libraries for lead structure screening. In The Practice of Medicinal Chemistry, C. G. Wermuth, ed., London, Academic Press, pp. 135–151.

Hruby, V. J., and Balse, P. M. (2000). Conformational and topographical considerations in designing agonist peptidomimetics from peptide leads. Curr. Med. Chem. 7:945–970.

Hruby, V. J., and Boteju, L. W. (1997). Design of conformationally constrained peptides and mimics. In Molecular Biology and Technology, R. A. Meyers, ed., New York, VCH Publishers Inc., pp. 371–382.

Hruby, V. J., and Smith, C. W. (1987). Structure–function studies of neurohypophyseal hormones. In The Peptides: Analysis, Synthesis, Biology, Vol. 8: Chemistry, Biology and Medicine of Neurohypophyseal Hormones and Their Analogs, C. W. Smith, ed., New York, Academic Press, pp. 77–207.

Hruby, V. J., Sawyer, T. K., Yang, Y. C. S., Bregman, M. D., Hadley, M. E., and Heward C. B. (1980). Synthesis and structure–function studies of melanocyte stimulating hormone (MSH) analogues modified in the 2 and 4(7) position: Comparison of activities on frog skin melanophores and melanoma adenylate cyclase. J. Med. Chem. 23:1432–1437.

Hruby, V. J., Kao, L.-F., Pettitt, B. M., and Karplus, M. (1988). The conformational properties of the delta opioid peptide [D-Pen^2, D-Pen^5]-enkephalin in aqueous solution determined by NMR and energy minimization calculations. J. Am. Chem. Soc. 110:3351–3359.

Hruby, V. J., Al-Obeidi, F., and Kazmierski, W. (1990a). Emerging approaches in the molecular design of receptor selective peptide ligands:

Conformational, topographical and dynamic considerations. Biochem. J. 268:249–262.

Hruby, V. J., Fang, S. Knapp, R., Kazmierski, W. M., Lui, G. K., and Yamamura, H. I. (1990b). Cholecystokinin analogues with high affinity and selectivity for brain membrane receptors. Int. J. Peptide Protein Res. 35:566–573.

Hruby, V. J., Kazmierski, W., Kawasaki, A. M., and Matsunaga, T. (1991a). Synthetic peptide chemistry and the design of peptide based drugs. In *Peptide Pharmaceuticals: Approaches to the Design of Novel Drugs*, D. J. Ward, ed., London, Open University Press, pp. 135–184.

Hruby, V. J., Töth, G., Gehrig, C. A., Kao, L.-F., Knapp, R., Lui, G. K., Yamamura, H. I., Kramer, T. H., Davis, P, and Burks T. F. (1991b). Topographically designed analogues of [D-Pen^2,D-Pen^5]enkephalin. J. Med. Chem. 34:1823–1830.

Hruby, V. J., Li, G., Haskell-Luevano, C., and Shenderovich, M. D. (1997). Design of peptides, proteins and peptidomimetics in chi space. Biopolymers (Peptide Sci.) 43:219–266.

Huang, C. K., and Richards, F. M. (1977). Reaction of a lipid soluble, unsymmetrical, cleavable cross-linking reagent with muscle aldolase and erythrocyte membrane protein. J. Biol. Chem. 252:5514–5521.

Hughes, J., Kosterlitz, H. W., and Leslie, F. M. (1975). Effect of morphine on adrenergic transmission in the mouse vas deferens. Assessment of agonist and antagonist potencies of narcotic analgesics. Br. J. Pharmacol. 53:371–381.

Jaffe, C. L., Lis, H., and Sharon, N. (1979). Identification of peanut agglutinin receptors on human erythrocyte ghosts by affinity crosslinking using a cleavable heterobifunctional reagent. Biochem. Biophys. Res. Commun. 91:402–409.

Jaffe, C. L., Lis, H., and Sharon, N. (1980). New cleavable photoreactive heterobifunctional crosslinking reagent for studying membrane organization. Biochemistry 19:4423–4429.

Ji, T. H. (1977). A novel approach to the identification of surface receptors. The use of photosensitive hetero-bifunctional cross-linking reagent. J. Biol. Chem. 252:1566–1570.

Ji, T. H. (1979). The application of chemical crosslinking for studies on cell membranes and the identification of surface receptors. Biochem. Biophys. Acta 559:39–69.

Ji, T. H., and Ji, I. (1982). Macromolecular photoaffinity labeling with radioactive photoactivable heterobifunctional reagents. Anal. Biochem. 121:286–289.

Jones, D. S. (1972). Polypeptides. Part XIII. Peptides related to the C-terminal tertapeptide sequence of gastrins by complementary reading of the genetic message. J. Chem. Soc. Perkin Trans. I:1407–1415.

Jost, K. (1977). *Handbook of Neurohypophyseal Hormone Analogues*, Vol. 1, Part 2, K. Jost, M. Lebl, and F. Brtnik, eds. Boca Raton, Fla., CRC Press.

Kahn, M. (1993). Peptide secondary structure mimetics: Recent advances and future challenges. SYNLETT 11:821–826.

Karplus, P. A., and Schulz, G. E. (1985). Prediction of chain flexibility in proteins. Naturwissenschaften 72:212–222.

Kaurov, O. A., Martynov, V. F., Milhailov, Y. D., and Auna, Z. P. (1972). Synthesis of the new analogues of oxytocin, modified in position 2 [in Russian]. Zh. Obshch. Khim. 42:1654.

Kawasaki, A. M., Knapp, R. J., Kramer, T. H., Wire, W. S., Vasquez, O. S., Yamamura, H. I., Burks, F., and Hruby, V. J. (1990). Design and synthesis of highly potent and selective cyclic dynorphin A analogues. J. Med. Chem. 33:1874–1879.

Kayser, S. (1983). Thrombosis. In *Applied Therapeutics: The Clinical Use of Drugs*, B. S. Katcher, L. Y. Young, and M. A. Koda-Kimble, eds., San Francisco, Applied Therapeutics Inc., pp. 333–360.

Kazmierski, W. M., and Hruby, V. J. (1988). A new approach to receptor ligand design: synthesis and conformation of a new class of potent and highly selective μ opioid antagonists utilizing tetrahydroisoquinoline carboxylic acid. Tetrahedron 44:697–710.

Kazmierski, W. M., Wire, W. S., Lui, G. K., Knapp, R. J., Shook, J. E., Burks, T. F., Yamamura H. I., and Hruby, V. J. (1988). Design and synthesis of somatostatin analogues with topographical properties that lead to highly potent and specific μ opioid receptor antagonists with greatly reduced binding at somatostatin receptors. J. Med. Chem. 31:2170–2177.

Kessler, H. (1982). Conformation and biological activity of cyclic peptides. Angew. Chem. Int. Ed. Engl. 21:512–523.

Kitagawa, T., and Aikawa, T. (1976). Enzyme coupled immunoassay of insulin using a novel coupling reagent. J. Biochem. 79:233–236.

Kosterlitz, H. W., Lord, J. A. H, Paterson, S. J., and Waterfield, A. A. (1980). Effects of changes in the structure of enkephalins and of narcotic analgesic drugs on their interactions with μ and δ receptors. Br. J. Pharmacol. 68:333–342.

Krstenansky, J. L., and Mao, S. J. T. (1987). Antithrombin properites of C-terminus of hirudin using synthetic unsulfated N^{α}-acetyl-hirudin$_{45-65}$. FEBS Lett. 211(1):10–16.

Kruszynski, M., Lammek, B., Manning, M., Seto, J., Haldar, J., and Sawyer, W. H. (1980). [1-(β-Mercapto-β, β-cyclopentamethylenepropanoic acid), 2(O-methyl)tyrosine]arginine-vasopressin and [1-(β-mercapto-β,β-cyclopentamethylenepropanoic acid)]arginine-vasopressin, two highly potent antagonists of the vasopressor response to arginine-vasopressin. J. Med. Chem. 23:364–368.

Kyte, J., and Doolittle, R. F. (1982). A simple method for displaying hydropatic character of a protein. J. Mol. Biol. 157:105–132.

Lam, K. S., Salmon, S. E., Hersh, E. M. Hruby, V. J., Kazmierski, W. M., and Knapp, R. J. (1991). A new type of synthetic peptide library for identifying ligand-binding activities. Nature 354:82–84.

Lam, K. S., Salmon, S. E., Hersh, E. M., Al-Obeidi, F., and Hruby, V. J. (1992). The selectide process: Rapid generation of large synthetic peptide libraries linked to identification and structure determination of acceptor binding ligands. In *Peptides: Chemistry and Biology*, J. A. Smith and J. E. Rivier, eds., Leiden, Escom, pp. 492–495.

Lebl, M., Barth, T., Servitova, L., Slaninova, J., and Jost, K. (1985). Oxytocin analogues with inhibitory properties, containing in position 2 A hydrophobic amino acid of D-configuration. Collec. Czech. Chem. Commun. 50:132–145.

Leslie, F. M., Chavkin, C., and Cox, B.M (1980). Opioid binding properties of brain and periperal tissues: Evidence for heterogeneity in opioid ligand binding sites. J. Pharm. Exp. Ther. 214:395–402.

Levy, D. (1973). Preparation of photo-affinity probes for the insulin receptor site in adipose and liver cell membranes. Biochem. Biophys. Acta 322:329–336.

Liao, M., Lu, Y., Xiao, Y., Dierich, M. P., and Chen, Y.-H. (2000). Induction of high level of specific antibody response to the neutralizaing epitope ELDKWA on HIV-1 gp41 by peptide-vaccine. Peptides (N.Y.) 21(4):463–468.

Liao, S., Shenderovich, M. D., Zhang, Z., Maletinska, L., Slaninova, J., and Hruby, V. J. (1998) Substitution of the side-chain constrained amino acids β-methyl-2′,6′-dimethyl-4′-methoxytyrosine in position 2 of a bicyclic oxytocin analogue provides unique insights into the bioactive topography of oxytocin antagonists. J. Am. Chem. Soc. 120:7393–7394.

Lin, Y., Trivedi, D., Siegel, M., and Hruby, V. J. (1992). Conformationally constrained glucagon analogues: New evidence for the conformational features important to glucagon-receptor interactions. In *Peptides: Chemistry and Biology.* J. A. Smith and J. E. Rivier, eds., Leiden, Escom, pp. 439–440.

Lord, I. A. H., Waterfield, A. A., Hughes, J., and Kosterlitz, H. W. (1977). Endogenous opioid peptides: multiple agonists and receptors. Nature 267:495–499.

Lowbridge, J., Manning, M., Seto, J., Haldar, J., and Sawyer, W. H. (1979). Synthetic Antagonists of in vivo responses by the rat uterus to oxytocin. J. Med. Chem. 22:565–569.

Ma, S., and Spatola, A. F. (1993). Conformations of $\psi[CH_2NH]$-pseudopeptide-cyclo[Gly-Proψ[CH$_2$NH]Gly-D-Phe-Pro]-TFA and cyclo-[Gly-Proψ]CH$_2$NH]Gly-D-Phe-Pro]. Int. J. Peptide Protein Res. 41:204–206.

Mann, K. G. (1987). The assembly of blood clotting complexes on membranes. Trends Biochem. Sci. 12:229–233.

Mao, S. J. T., Yates, M. T., Owen, T. J., and Krstenansky, J. L. (1988). Interaction of hirudin with thrombin: Identification of a minimal binding domain of hirudin that inhibits clotting activity. Biochemistry 27:8170–8173.

Maraganore, J. M. (1993b). Thrombotic process and emerging drugs for its control. Tex. Heart Inst. J. 20:43–47.

Marshall, G. R. (1993). A hierarchical approach to peptidomimetic design. Tetrahedron 49:3547–3558.

Martin, W. R., Eades, C. G., Thompson, J. A., Huppler, R. E., and Gilbert, P. E. (1976). The effects of morphine and nalorphine-like drugs in the non-dependent and morphine-dependent chronic spinal dog. J. Pharmacol. Exp. Ther. 197:517–532.

Martin, F. J., and Papahadjopoulos, D. (1982). Irreversible coupling of immunoglobulin fragments to preformed vesicles. An improved method for liposome targeting. J. Biol. Chem. 257:286–288.

Matsunaga, T. O., de Lauro Castrucci, A. M., Hadley, M. E., and Hruby, V. J. (1989). Melanin concentrating hormone (MCH): Synthesis and bioactivity studies of MCH fragment analogues. Peptides 10:349–354.

McRobbie, M., Meth-Cohn, O., and Suschitzky, H. (1976a). Competitive cyclizations of singlet and triplet nitrenes. Part I. Cyclization of 1-(2-nitrophenyl)pyrazoles. Tetrahedron Lett. 925–928.

McRobbie, M., Meth-Cohn, O., and Suschitzky, H. (1976b). Competitive cyclizations of singlet and triplet nitrenes. Part II. Cyclization of 2-nitrophenyl-, thiophens-, benzo thiazoles-, and benzimidazoles. Tetrahedron Lett. 929–932.

Mekler, L. B. (1969). On specific selective interactions between amino acid residues of poly peptide chain. Biophys. USSR (Engl. Trans.) 14: 613–617.

Melin, P., Vilhardt, H., and Akerlund, M. (1983). Uteronic oxytocin and vasopressin antagonists with minimal structure modifications. In *Peptides: Structure and Function*, V. J. Hruby and D. H. Rich, eds., Rockford, Ill., Pierce Chemical Co.

Meraldi, J. P., Yamamoto, D., Hruby, V. J., and Brewster, A. I. R. (1975). Conformational studies of neurohypophyseal hormones and inhibitors using high resolution NMR spectroscopy. In *Peptides: Chemistry, Structure and Biology*. R. Walter and J. Meinhofer, eds., Ann Arbor, Mich., Ann Arbor Sci. Publ., pp. 803–814.

Meraldi, J. P., Hruby, V. J., and Brewster, A. I. R. (1977). Relative conformational rigidity in oxytocin and [1-penacillamine]-oxytocin: A proposal for the relationship of conformational flexibility to peptide hormone agonism and antagonism. Proc. Natl. Acad. Sci. USA 74:1373–1377.

Merrifield, R. B. (1963). Solid phase peptide synthesis. I. The synthesis of a tetrapeptide J. Am. Chem. Soc. 85:2149–2152.

Moreland, R. B., Smith, P. K., Fujimoto, E. K., and Dockter, M. E. (1982). Synthesis and characterisation of N-[4-azido phenylthio] phaliimidate: a cleavable photoactivable crosslinking reagent that reacts with sulfhydryl groups. Anal. Biochem. 121:321–326.

Morgan, B. A., Bowers, J. D., Guest, K. P., Handa, B. K., Metcalf, G., and Smith, C. F. C. (1977). Structure–activity relationships of enkephalin analogues. In *Peptides*, M. Goodman and J. Meienhofer, eds., New York, Wiley.

Mosberg, H. I., Hurst, R., Hruby, V. J., Galligan, J. J., Burks, T. F., Gee, K., and Yamamura, H. I. (1983a). Conformationally constrained cyclic enkephalin analogs with pronounced delta opioid receptor agonist selectivity. Life Sciences 32:2565–2569.

Mosberg, H. I., Hurst, R., Hruby, V. J., Gee, K., Yamamura, H. I., Galligan, J. J., and Burks. T. F. (1983b). Bis-penicillamine enkephalins possess highly improved sensitivity toward delta opioid receptors. Proc. Natl. Acad. Sci. USA 80:5871–5874.

Mulchahey, J. J., Neill, J. D., Dion, L. D., Bost, K. L., and Blalock, J. E. (1986). Antibodies to the binding site of the receptor for luteinizing hormone-releasing hormone (LHRH): generation with a synthetic decapeptide encoded by an RNA complementary to LHRH mRNA. Proc. Natl. Acad. Sci. USA 83:9714–9718.

Muller, S. and Briand, J. P. (1998). On the potential of retro-inverso peptides in vaccine design. Res. Immunol. 149:55–57.

Muramoto, K., and Ramachandran, J. (1980). Photoreactive derivative of corticotropin 2. Preparation and characterisation of 2-nitro-4(5)-azido phenylsulfenyl derivatives of corticotropin. Biochemistry 19:3280–3286.

Nassal, M. (1983). 4(1-Azi-2,2,2,trifluoroethyl)benzoic acid, a highly photo-labile carbene generating label readily fixable to biochemical agents. Liebigs Ann. Chem. 1510–1523.

Nikiforovich, G. V., Prakash, O., Gehrig, C. A., and Hruby, V. J. (1993). Solution conformations of the peptide backbone for DPDPE and its β-Me-Phe^4-substituted analogs. Int. J. Peptide Protein Res. 41:347–361.

Ngo, T. T., Yam, C. F., Lenhoff, H. M., and Ivy, I. I. (1981). p-Azidophenylglyoxal. A heterobifunctional photoactivable cross-linking reagent selective for arginyl residues. J. Biol. Chem. 256:11313–11318.

Ogasawara, K. (1999). Synthetic peptide vaccines effective in preventing virus infection. Microbiol. Immunol. 43(10):915–923.

Omichinski, J. G., Olson, A. D., Thorgeirsson, S. S., and Fassina, G. (1989). Computer assisted design of recognition peptides. In Techniques in Protein Chemistry, T. E. Hugli, ed., San Diego, Calif., Academic Press, pp. 430–438.

Palker, T. J., Clark, M. E., Langlois, A. J., Matthews, T. J., Weinhold, K. J., Randall, R. R., Bolognesi, D. P., and Haynes, B. F. (1988). Type-specific neutralization of the human immunodeficiency virus with antibodies to env-encoded synthetic peptides. Proc. Natl. Acad. Sci. USA 85:1932–1936.

Pelton, J. T., Gulya, K., Hruby, V. J., Duckles, S. P., and Yamamura, H. I. (1985). Conformationally restricted analogues of somatostatin with high μ-opiate receptor specificity. Proc. Natl. Acad. Sci. USA 82:236–239.

Pettitt, B. M., Matsunaga, T. O., Al-Obeidi, F., Gehrig, C. A., Hruby, V. J., and Karplus, M. (1991). Dynamic search for bis-penicillamine enkephalin conformations. Biophys. J. 60:1540–1544.

Pfaff, E., Mussgay, M., Bohm, H. O., Schulz, G. E., and Scaller, H. (1982). Antibodies against a preselected peptide recognize and neutralize foot and mouth disease virus. EMBO J. 1:869–874.

Piercy, M. F., Schroeder, A., Einspahr, F. J., Folkers, K., Xu, J.-C., and Horig, J. (1981). Behavioral evidence that substance P may be a spinal cord nociceptor neurotransmitter. In Peptides: Synthesis–Structure–Function, D. H. Rich and E. Gross, eds., Rockford, Ill., Pierce Chemical Co.

Pierschbacher, M. D., and Ruoslahti, E. (1987). Influence of stereochemistry of the sequence Arg-Gly-Asp-Xaa on binding specificity in cell adhesion. J. Biol. Chem. 262:17294–17298.

Qian, X., Kövér, K. E., Shenderovich, M. D., Lou, B-S., Misicka, A., Zalewska, T., Horvath, R., Davis, P., Bilsky, E. J., Porreca, F., Yamamura, H. I. and Hruby, V. J. (1994). Newly discovered stereoche-mical requirements in side chain conformation of δ opioid agonists for recognizing opioid δ receptors. J. Med. Chem. 37:1746–1757.

Qian, X., Russell, K. C., Boteju, L. W., and Hruby, V. J. (1995). Stereoselective total synthesis of topographically constrained designer amino acids: 2′, 6′-dimethyl-β-methyltyrosines. Tetrahedron 51:1033–1054.

Qian, X., Shenderovich, M. D., Kövér, K. E., Davis, P., Horváth, R., Zalewska, T., Yamamura, H. I., Porreca F., and Hruby, V. J. (1996). Probing the stereochemical requirements for receptor recognition of δ opioid agonists through topographic modifications in position 1. J. Am. Chem. Soc. 118:7280–7290.

Rasmussen, U. B., and Hesch, R.-D. (1987). On antisense peptides: The parathyroid hormone as an experimental example and a critical theoretical view. Biochem. Biophys. Res. Commun. 149:930–938.

Rinke, J., Meinke, M., Brimacombe, R., Fink, G., Rommel, W., and Fasold, H. (1980). The use of azidoarylimidoesters in RNA-protein cross-linking studies with Escherichia coli ribosomes. J. Mol. Biol. 137:301–314.

Ripka, W. C., De Lucca, G. V., Bach II, A. C., Pottorf, R. S., and Blaney, J. M. (1993a). Protein β-turn mimetics I. Design, synthesis, and evaluation in model cyclic peptides. Tetrahedron 49:3593–3608.

Ripka, W. C., De Lucca, G. V., Bach II, A. C., Pottorf, R. S., and Blaney, J. M. (1993b). Protein β-turn mimetics II. Design, synthesis, and evaluation in the cyclic peptide gramicidin S. Tetrahedron 49:3609–3628.

Rizo, J., and Gierasch, L. M. (1992). Constrained peptides: Models of bioactive-peptides and protein structures. Annu. Rev. Biochem. 61:387–418.

Rodda, S. J., Geysen, H. M., Manson, I.J, and Schaafs, P. G. (1986). The antibody response to myoglobin-I. Systematic synthesis of myoglobin peptides reveals location and substructure of species dependent continuous antigenic determinants. Mol. Immunol. 23:603–610.

Rodriguez, M., Lignon, M.-F., Galas, M.-C., and Martinez, J. (1990). Highly selective cyclic cholecystokinin analogs for central receptors. In *Peptides: Chemistry, Structure, and Biology.* J. E. Rivier and G. R. Marshall, eds., Leiden, Escom.

Rothbard, J. B., and Taylor, W. R. (1988). A sequence pattern common to T-cell epitopes. EMBO J. 7:93–101.

Rudinger, J., Pliska, V., and Krejci, I. (1972). Oxytocin analogues in the analysis of some phases of hormone action. Rec. Prog. Hormone Res. 28:131–172.

Ruoslahti, E. (1991) Integrins. J. Clin. Invest. 87:1–5.

Rusche, J. R., Javaherian, K., McDanal, C., Petro, J., Lynn, D. L., Grimaila, R., Langlois, A., Gallo, R. C., Arthur, L. O., Fischinger, P. J., Bolognesi, D. P., Putney, S. D., and Matthews, T. J. (1988). Antibodies that inhibit fusion of HIV infected cells bind a 24 amino acid sequence of the viral envelope, gp120. Proc. Natl. Acad. Sci. USA 85:3198–3202.

Salamon, Z., Cowell, S., Varga, E., Yamamura, H. I., Hruby, V. J., and Tollin, G. (2000). Plasmon resonance studies of agonist/antagonist binding to the human δ-opioid receptor: New structural insights into receptor-ligand interactions. Biophys. J. 79:2463–2474.

Sawyer, T. K. (1995). Peptidomimetic design and chemical approaches to peptide metabolism. In *Peptide-Based Drug Design: Controlling Transport and Metabolism*, M. D. Taylor and G. L. Amidon, eds., Washington, D.C., American Chemical Society, pp. 387–422.

Sawyer, T. K. (1977). Peptidomimetic and nonpeptide drug discovery: Impact of structure-based drug design. In *Structural-Based Drug Design:*

Diseases, Targets, Techniques and Developments, P. Veerapandian, ed., New York, Marcel Dekker, pp. 559–634.

Sawyer, T. K., Sanfilippo, P. J. Hruby, V. J., Engel, M. H., Heward, C. B., Burnett, C. B., and Hadley, M. E. (1980). [Nle^4, D-Phe^7]-α-Melanocyte stimulating hormone: A highly potent α-melanotropin with ultralong biological activity. Proc. Natl. Acad. Sci. USA 77:5754–5758.

Sawyer, T. K., Hruby, V. J., Darman, P. S., and Hadley, M. E. (1982). [half-Cys^4, half-Cys^{10}]-α-melanocyte stimulating hormone: A cyclic α-melanotropin exhibiting superagonist biological activity. Proc. Natl. Acad. Sci. USA 79:1751–1755.

Scarborough, R. M., Naughton, M. A., Teng, W., Rose, J. W., Philips, D. R., Nannizzi, L., Arfsten, A., Campbell, A. M., and Charo, I. F. (1993). Design of potent and specific integrin antagonists. J. Biol. Chem. 268(2):1066–1073.

Schiller, P. W., Eggimann, B., DiMaio, J., Lemieux, C., and Nguyen, T. M.-D. (1981). Cyclic enkephalin analogs containing a cystine bridge. Biochem. Biophys. Res. Commun. 101:337–343.

Schiller, P. W., Nguyen, T. M.-D., Maziak, L. A., and Lemieux C. (1985). Novel cyclic opioid peptide analog showing high preference for μ-receptors. Biochem. Biophys. Res. Commun. 127:558–564.

Schneider, J. P. and Kelly, J. W. (1995). Templates that induce α-helical, β-sheet, and loop conformations. Chem. Rev. 95:2169–2187.

Schulz, H., and du Vigneaud, V. (1966). Synthesis of 1-L-penacillamine-oxytocin, 1-D-penacillamine-oxytocin, and 1-deamino-penacillamine-oxytocin, potent inhibitors of oxytocic response to oxytocin. J. Med. Chem. 9:647–650.

Schwartz, I., and Ofengand, J. (1974). Photoaffinity labeling of tRNA binding site in macromolecules. I. Linking of the phenacyl-p-azide of 4-thiouridine in (*Escherichia coli*) valyl-*t*RNA to 16S RNA at the ribosomal P site. Proc. Natl. Acad. Sci. USA 71:3951–3955.

Schwyzer, R. (1977). ATCH: A short introductory review. Ann. N.Y. Acad. Sci. 297:3–26.

Scott, J. K., and Smith, G. P. (1990). Searching for peptide ligands with an epitope library. Science 249:386–390.

Seela, F., and Rosemeyer, H. (1977). 5-Azido-T-bromo-2-nitroacetophenone. A cross-linking reagent with groups of selective reactivity. Hoppe-Seyler's Z. Physiol. Chem. 358:129–131.

Sette, A., Doria, G., and Adauin, L. (1986). A microcomputer program for hydrophilicity and amphipathicity analysis of protein antigens. Mol. Immunol. 23:807–815.

Shai, Y., Flashner, M., and Chaiken, I. M. (1987). Anti-sense peptide recognition of sense peptide: Direct quantitative characterization with the ribonuclease S-peptide system using analytical high-performance affinity chromatography. Biochemistry 26:669–675.

Shai, Y., Brunck, T. K., and Chaiden, I. M. (1989). Antisense peptide recognition of sense peptides: Sequence simplification and evaluation of forces underlying the interaction. Biochemistry 28:8804–8811.

Sharma, S. D., Nikiforovich, G. V., Jiang, J., de Lauro Castrucci, A.-M., Hadley, M. E., and Hruby, V. J. (1993). A new class of positively charged melanotropin analogs: A new concept in peptide design. In *Peptides 1992*, C. H. Schneider and A. N. Eberle, eds., Leiden, Escom, pp. 95–96.

Shenderovich, M. D., Liao, S., Qian, X., and Hruby, V. J. 2000. A three dimensional model of the δ opioid pharmacophore: Comparative molecular modeling of peptide and non-peptide ligands. Biopolymers 53:565–580.

Siemion, I. Z., Szewezuk, Z., Herman, Z. S., and Stachura, Z. (1980). To the problem of the biologically active conformation of enkephalin. Mol. Cell. Biochem. 34:23–29.

Sigrist, H., and Zahler, P. (1980). Heterobifunctional crosslinking of bacteriorhodopsin by azidophenylisothiocanate. FEBS Lett. 113:307–311.

Singh, A., Thornton, E. R., and Westheimer, F. H. (1962). The photolysis of diazoacetyl chymotrypsin. J. Biol. Chem. 237:PC3006–PC3008.

Slomczynska, U., Chalmers, D. K., Cornille, F., Smythe, M. L., Beusen, D. D., Moeller, K. D., and Marshall, G. R. (1996). Electrochemical cyclization of dipeptides to form novel bicyclic, reverse-turn peptidomimetics. 2. Synthesis and conformational analysis of 6,5-bicyclic systems. J. Org. Chem. 61:1198–1204.

Smith, J. A. (1989). Synthetic peptides: Tools for elucidating mechanism of protein antigen processing and presentation. In *Synthetic Peptides: Approaches to Biological Problems*, J. P. Tam and E. T. Kaiser, eds., New York, Alan R. Liss, pp. 31–42.

Smith, J. A., Hunnell J. G. R., and Leach, S. J. (1977). A novel method for delineating antigenic determinants: Peptide synthesis and radioimmunoassay using the same solid support. Immunochemistry 14:565–572.

Smith, L. R., Bost, K. L., and Blalock, J. E. (1987). Generation of idiotypic and anti-idiotypic antibodies by immunnization with peptides encoded by complementary RNA: A possible molecular basis for the network theory. J. Immunol. 138:7–9.

Smith, R. A. G., and Knowles, J. R. (1974). The utility of photoaffinity labels as mapping reagents. Biochem. J. 141:51–56.

Sonder, S. A., and Fenton, J. W. (1984). Proflavin binding within the fibrinopeptide groove adjacent to the catalytic site of human α-thrombin. Biochemistry 23:1818.

Stone, S. R., and Hofsteenge, J. (1986). Kinetics of the inhibition of thrombin by hirudin. Biochemistry 25:4622.

Strand, F. L. (1999). *Neuropeptides: Regulators of Physiological Processes*, Cambridge, Mass., MIT Press.

Struthers, R. S., Rivier, J., and Hagler, A. (1984). Molecular dynamics and minimum energy of GnRH and analogs: A methodology for computeraided drug design. Ann. N.Y. Acad. Sci. 439:81–96.

Stryer, L. (1975). *Biochemistry*, San Francisco, W. H. Freeman & Co.

Takahashi, H., Merli, S., Putney, S. D., Houghten, R., Moss, B., Germain, R. N., and Berzofsky, J. A. (1989). A single amino acid interchange yields reciprocal CTL specificities for HIV gp160. Science 246:118–121.

Thornton, J. M., Edward, M. S., Taylor, W. R., and Barlow, D. J. (1986). Location of "continuous" antigenic determinants in the protruding regions of proteins. EMBO J. 5:409–415.

Toniolo, C. (1990). Conformationally restricted peptides through short-range cyclizations. Int. J. Peptide Protein Res. 35:287–300.

Torchiana, M. L., Cook, P. G., Weise, S. R., Saperstein, R., and Veber, D. F. (1978). Subcutaneous administration of somatostatin analogs as a major factor in the enhancement of drug duration of action. Arch. Int. Pharmacol. Ther. 235:170–176.

Töth, G., Russell, K. C., Landis, G., Kramer, T. H., Fang, L., Knapp, R., Davis, P., Burks, T. F., Yamamura, H. I., and Hruby, V. J. (1992). Ring substituted and other conformationally constrained tyrosine analogues of [D-Pen^2,D-Pen^5]enkephalin with δ opioid receptor selectivity. J. Med. Chem. 35:2384–2391.

Trivedi, D., Lin, Y., Ahn, J-M., Siegel, M., Mollova, N. N., Schram, K. H., and Hruby, V. J. (2000). Design and synthesis of conformationally constrained glucagon analogues. J. Med. Chem. 43:1714–1722.

Trommer, W. E., and Hendrick, M. (1973). The formation of maleimides by a new mild cyclization process. Synthesis 484–485.

Turro, N. J. (1965). Molecular Photochemistry, New York, Benjamin.

Turro, N. J. (1980). Structure and dynamics of important reactive intermediates involved in photobiological systems. Ann. N.Y. Acad. Sci. 346:1–17.

Unanue, E. R., and Allen, P. M. (1987). The basis for the immunoregulatory role of macrophages and other accesory cells. Nature 236:551–557.

van Regenmortel, M. H. V., Briand, J. P., Muller, S., and Plause, S. (1988). Synthetic polypeptides as antigens. In Laboratory Techniques in Biochemistry and Molecular Biology, R. H. Burdon and P. H. von Knippenberg, eds., Amsterdam, Elsevier.

Vanin, E. F., and Ji, T. H. (1981). Synthesis and applications of cleavable photoactivable heterbifunctional reagents. Biochemistry 20:6754–6760.

Vanin, E. F., Burkhard, S. J., and Kaiser, I. I. (1981). p-Azidophenylglyoxal: A heterobifunctional photosensitive reagent. FEBS Lett. 124:89–92.

Vaughan, R. J., and Westheimer, F. H. (1969). A method for marking the hydrophobic binding sites of enzymes. An insertion into the methyl group of an alanine residue of trypsin. J. Am. Chem. Soc. 91:217–218.

Vavrek, R. J., Ferger, M. F., Allen, G. A., Rich, D. H., Blomquist, A. T., and du Vigneaud, V. (1972). Synthesis of three oxytocin analogs related to [1-deaminopenacillamine]oxytocin possessing antioxytocic activity. J. Med. Chem. 15:123–126.

Veber, D. F., Freidinger, R. M., Prelow, D. S., Poleveda, W. J., Holly, F. W., Strachan, R. G., Nutt, R. F., Arison, B. H., Homnick, C., Randall, W. C., Glitzer, M. S., Saperstein, R., and Hirchmann, R. (1981). A potent cyclic hexapeptide of somatostatin. Nature (London) 292:55–57.

Virgilio, A. A., and Ellman, J. A. (1994). Simultaneous solid-phase synthesis of β-turn mimetics incorporating side-chain functionality. J. Am. Chem. Soc. 116:11580–11581.

Virgilio, A. A., Schurer, S. C., and Ellman, J. A. (1997a). Expedient solid-phase synthesis of putative β-turn mimetics incorporating the $i+1$, $i+2$, and $i+3$ sidechains. Tetrahedron Lett. 37:6961–6964.

Virgilio, A. A., Bray, A. A., Zhang, W., Trinh, L., Snyder, M., Morrissey, M. M., and Ellman, J. A. (1997b). Synthesis and evaluation of a library of peptidomimetics based upon the β-turn. Tetrahedron 53:6635–6644.

Walker, B., Wikstrom, P., and Shaw, E. (1985). Evaluation of inhibitor constants and alkylation rates for a series of thrombin affinity labels. Biochem J. 230:245.

Walter, R. (1977). Identification of sites in oxytocin involved in uterine receptor recognition and activation. Fed. Proc. 36:1872–1878.

Wang, W., Borchardt, R. T., and Wang, B. (2000). Orally active peptidomimetic RGD analogs that are glycoprotein IIb/IIIa antagonists. Curr. Med. Chem. 7:437–453.

Ward, D. J., ed. (1990). *Peptide Pharmaceuticals: Approaches to the Design of Novel Drugs*, London, Open University Press.

Whittle, P. J., and Blundell, T. L. (1994). Protein/structure-based drug design. Ann. Rev. Biophys. Biomol. Struct. 23:349–375.

Wiley, R. A., and Rich, D. H. (1993). Peptidomimetics derived from natural products. Med. Res. Rev. 13:327–384.

Wunderlin R., Minakakis P., Tun-Kyi, A., Sharma, S. D., and Schwyzer, R. (1985a). Melanotropin receptors. I. Synthesis and biological activity of N^α-(5-bromovaleryl)-N^α-desacetyl-α-melanotropin. Helv. Chim. Acta 68:1–11.

Wunderlin R., Sharma, S. D., Minakakis P., and Schwyzer, R. (1985b). Melanotropin receptors. II. Syntheisis and biological activity of α-melanotropin/tobacco mosaic virus disulfide conjugates. Helv. Chim. Acta 68:12–22.

Yip, C. C., Yeung, C. W. T., and Moule, M. L. (1980). Photoaffinity labeling of insulin receptor proteins of liver plasma membrane preparations. Biochemistry 19:70–76.

Yoshitake, S., Yamada Y., Ishikawa, E., and Masseyeff, R. (1979). Conjugation of glucose oxidase from Aspergillus niger and rabbit antibodies using N-hydroxysuccinimide ester of N-(4-carboxy cyclohexylmethyl)-maleimide. Eur. J. Biochem. 101:395–399.

Index

377

Printed in the United States
By Bookmasters